Probability and Its Applications

Published in association with the Applied Probability Trust

Editors: S. Asmussen, J. Gani, P. Jagers, T.G. Kurtz

Probability and Its Applications

Leszek Gawarecki · Vidyadhar Mandrekar

Stochastic Differential Equations in Infinite Dimensions

with Applications to Stochastic Partial Differential Equations

 Springer

Leszek Gawarecki
Department of Mathematics
Kettering University
1700 University Ave
Flint, MI 48504
USA
lgawarec@kettering.edu

Vidyadhar Mandrekar
Department of Statistics and Probability
Michigan State University
A436 Wells Hall
East Lansing, MI 48823
USA
mandrekar@stt.msu.edu

Series Editors:

Søren Asmussen
Department of Mathematical Sciences
Aarhus University
Ny Munkegade
8000 Aarhus C
Denmark
asmus@imf.au.dk

Peter Jagers
Mathematical Statistics
Chalmers University of Technology
and University of Gothenburg
412 96 Göteborg
Sweden
jagers@chalmers.se

Joe Gani
Centre for Mathematics and its Applications
Mathematical Sciences Institute
Australian National University
Canberra, ACT 0200
Australia
gani@maths.anu.edu.au

Thomas G. Kurtz
Department of Mathematics
University of Wisconsin - Madison
480 Lincoln Drive
Madison, WI 53706-1388
USA
kurtz@math.wisc.edu

ISSN 1431-7028
ISBN 978-3-642-26634-8   ISBN 978-3-642-16194-0 (eBook)
DOI 10.1007/978-3-642-16194-0
Springer Heidelberg Dordrecht London New York

Mathematics Subject Classification (2010): 35-XX, 60-XX

Cover design: VTEX, Vilnius

Printed on acid-free paper

Springer is part of Springer Science+Business Media (www.springer.com)

We dedicate this book to our wives
Edyta and Veena
for their encouragement and patience during
the preparation of this book

Preface

Stochastic differential equations are playing an increasingly important role in applications to finance, numerical analysis, physics, and biology. In the finite-dimensional case, there are two definitive books: one by Gikhman and Skorokhod [25], which studies the existence and uniqueness problem along with probabilistic properties of solutions, and another by Khasminskii [39], which studies the asymptotic behavior of the solutions using the Lyapunov function method. Our object in this book is to study these topics in the infinite-dimensional case. The two main problems one faces are the invalidity of the Peano theorem in the infinite-dimensional case and the appearance of unbounded operators if one wants to apply finite-dimensional techniques to stochastic partial differential equations (SPDEs).

Motivated by these difficulties, we discuss the theory in the deterministic case from two points of view. The first method (see Pazy [63] and Butzer and Berens [6]) involves semigroups generated by unbounded operators and results in constructing mild solutions. The other sets up the equation in a Gelfand triplet $V \hookrightarrow H \hookrightarrow V^*$ of Hilbert spaces with the space V as the domain of the unbounded operator and V^* its continuous dual. In this case variational solutions are produced. This approach is studied by Agmon [1] and Lions [48], who assume that either the injection $V \hookrightarrow H$ is compact and the unbounded operator is coercive or that the unbounded operator is coercive and monotone.

The systematic study of the first approach to SPDEs was first undertaken by Ichikawa [32, 33] and is explained in the timely monographs of Da Prato and Zabczyk [11, 12]. The approach of J.P. Lions was first used by Viot [75] (see also Metivier [56] and Metivier and Viot [58]). Working under the assumption that the embedding $V \hookrightarrow H$ is compact and that the coefficients of the equation are coercive, the existence of a weak solution was proven. These results were generalized by Pardoux [62], who assumed coercivity and monotonicity, and used the crucial deterministic result of Lions [48] to produce the strong solution. Later, Krylov, and Rozovskii [42] established the above-mentioned result of Lions in the stochastic case and also produced strong solutions. The initial presentation was given by Rozovskii in [66]. However, rather rigorous and complete exposition in a slightly general form is provided by Prévôt and Röckner [64].

In addition to presenting these results on SPDEs, we discuss the work of Leha and Ritter [46, 47] on SDEs in $\mathbb{R}^\infty$ with applications to interacting particle systems and the related work of Albeverio et al. [2, 3], and also of Gawarecki and Mandrekar [20, 21] on the equations in the field of Glauber dynamics for quantum lattice systems. In both cases the authors study infinite systems of SDEs.

We do not present here the approach used in Kalliapur and Xiong [37], as it requires introducing additional terminology for nuclear spaces. For this type of problem (referred to as "type 2" equations by K. Itô in [35]), we refer the reader to [22, 23], and [24], as well as to [37].

A third approach, which involves solutions being Hida distribution is presented by Holden et al. in the monograph [31].

The book is divided into two parts. We begin Part I with a discussion of the semigroup and variational methods for solving PDEs. We simultaneously develop stochastic calculus with respect to a Q-Wiener process and a cylindrical Wiener process, relying on the classic approach presented in [49]. These foundations allow us to develop the theory of semilinear partial differential equations. We address the case of Lipschitz coefficients first and produce unique mild solutions as in [11]; however, we then extend our research to the case where the equation coefficients depend on the entire "past" of the solution, invoking the techniques of Gikhman and Skorokhod [25]. We also prove Markov and Feller properties for mild solutions, their dependence on the initial condition, and the Kolmogorov backward equation for the related transition semigroup. Here we have adapted the work of B. Øksendal [61], S. Cerrai [8], and Da Prato and Zabczyk [11].

To go beyond the Lipschitz case, we have adapted the method of approximating continuous functions by Lipschitz functions $f : [0, T] \times \mathbb{R}^n \to \mathbb{R}^n$ from Gikhman and Skorokhod [25] to the case of continuous functions $f : [0, T] \times H \to H$ [22]. This technique enabled us to study the existence of weak solutions for SDEs with continuous coefficients, with the solution identified in a larger Hilbert space, where the original Hilbert space is compactly embedded. This arrangement is used, as we have already mentioned, due to the invalidity of the Peano theorem. In addition, we study martingale solutions to semilinear SDEs in the case of a compact semigroup and for coefficients depending on the entire past of the solution.

The variational method is addressed in Chap. 4, where we study both the weak and strong solutions. The problem of the existence of weak variational solutions is not well addressed in the existent literature, and our original results are obtained with the help of the ideas presented in Kallianpur et al. [36]. We have followed the approach of Prévôt and Röckner in our presentation of the problem of the existence and uniqueness of strong solutions.

We conclude Part I with an interesting problem of an infinite system of SDEs that does not arise from a stochastic partial differential equation and serves as a model of an interacting particle system and in Glauber dynamics for quantum lattice systems.

In Part II of the book, we present the asymptotic behaviors of solutions to infinite-dimensional stochastic differential equations. The study of this topic was undertaken for specific cases by Ichikawa [32, 33] and Da Prato and Zabczyk [12] in the case of mild solutions. A general Lyapunov function approach for strong solutions in a

Gelfand triplet setting for exponential stability was originated in the work of Khasminskii and Mandrekar [40] (see also [55]). A generalization of this approach for mild and strong solutions involving exponential boundedness was put forward by R. Liu and Mandrekar [52, 53]. This work allows readers to study the existence of invariant measure [52] and weak recurrence of the solutions to compact sets [51]. Some of these results were presented by K. Liu in a slightly more general form in [50].

Although we have studied the existence and uniqueness of non-Markovian solutions, we do not investigate the ergodic properties of these processes, as the techniques in this field are still in development [28].

East Lansing   *Leszek Gawarecki*
October, 2010   *Vidyadhar Mandrekar*

Acknowledgements

During the time we were working on this book, we were helped by discussions with various scholars. We thank Professors A.V. Skorokhod and R. Khasminskii for providing insights in the problems studied in Parts I and II, respectively. We are indebted to Professor B. Øksendal for giving us ideas on the organization of the book. One of us had the privilege of visiting Professors G. Da Prato and Jürgen Pothoff. The discussions with them helped in understanding several problems. The latter provided an opportunity to present a preliminary version of the first part to his students. Clearly, all the participants' comments improved the presentation. The content of Chap. 5 bears the influence of Professor Sergio Albeverio, from whom we got an insight into applications to physics. Professor Kallianpur provided preliminary knowledge of the field and encouragement by extending an invitation to a conference on SPDEs in Charlotte. We will be amiss if we do not thank the participants of the seminar on the subject given at Michigan State University. Our special thanks go to Professors Peter Bates, Sheldon Newhouse, and Shlomo Levental, whose comments and questions led to cleaning up some confusion in the book.

Finally, we thank two referees for their insightful comments which led to a significant improvement of our presentation. We are grateful to Dr. M. Reizakis of Springer Verlag for her timely administrative support and encouragement.

Contents

Acronyms

$(H, \|\cdot\|_H)$	Real separable Hilbert space
$(X, \|\cdot\|_X)$	Banach space
$(X^*, \|\cdot\|_{X^*})$	Dual of a Banach space
T^*	Adjoint of a linear operator T
$B(H)$	Collection of Borel-measurable real-valued functions on H
$B_b(H)$	Collection of bounded Borel-measurable real-valued functions on H
$C^m(H)$	Collection of real-valued m-times continuously Fréchet differentiable functions on H
$C_0^m(H)$	Subspace of $C^m(H)$ consisting of functions with compact support
$C_b^m(H)$	Subspace of $C^m(H)$ consisting of functions whose derivatives of order $k = 0, \ldots, m$ are bounded
$C(H) = C^0(H)$	Collection of real-valued continuous functions on H
$C([0, T], H)$	Banach space of H-valued continuous functions on $[0, T]$
$C^m([0, T], H)$	H-valued m-times continuously differentiable functions on $[0, T]$
$C^\infty([0, T], H)$	H-valued infinitely many times continuously differentiable functions on $[0, T]$
$L^p([0, T], H)$	Banach space of H-valued Borel measurable functions on $[0, T]$, Bochner integrable in the pth power
$L^\infty([0, T], H)$	Banach space of H-valued essentially bounded functions on $[0, T]$
$\mathcal{M}_T^2(H)$	Hilbert space of H-valued continuous square-integrable martingales
$W^{m,p}(\mathcal{O})$	Sobolev space of functions $f \in L^p(\mathcal{O})$, with $D^\alpha f \in L^p(\mathcal{O})$, $|\alpha| \leq m$
$W_0^{m,p}(\mathcal{O})$	Completion of $C_0^m(\mathcal{O})$ in $W^{m,p}(\mathcal{O})$
$\mathcal{B}(H)$	Borel σ-field on $(H, \|\cdot\|_H)$
$\mathcal{E}(\mathcal{L}(K, H))$	$\mathcal{L}(K, H)$-valued elementary processes

$\mathscr{L}(X, Y)$ Bounded linear operators from X to Y
$\mathscr{L}_1(H)$ Trace-class operators on H
$\mathscr{L}_2(H_1, H_2)$ Hilbert–Schmidt operators from H_1 to H_2
A_λ Yosida approximation of an operator A
$\mathscr{D}(A)$ Domain of an operator A
$\mathscr{R}(A)$ Range of an operator A
$R(\lambda, A)$ Resolvent of an operator A
$\rho(A)$ Resolvent set of an operator A
$\mathrm{tr}(A)$ Trace of an operator A
$\tau(A)$ Trace norm of an operator A
$S \star \Phi(t)$ Stochastic convolution
Q_t Covariance of the stochastic convolution $\int_0^t S(t-s)\,dW_s$
W_t Q-Wiener process
$\tilde{W}_t$ Cylindrical Wiener process
$\int_0^t \Phi(s)\,dW_s$ Stochastic integral with respect to a Q-Wiener process
$\int_0^t \Phi(s)\,d\tilde{W}_s$ Stochastic integral with respect to a cylindrical Wiener process
$\langle M \rangle_t$ Increasing process of a martingale M_t
$\langle\langle M \rangle\rangle_t$ Quadratic variation process of a martingale M_t

Part I
Stochastic Differential Equations in Infinite Dimensions

Chapter 1
Partial Differential Equations as Equations in Infinite Dimensions

The purpose of this chapter is to explain how infinite-dimensional equations arise from finite-dimensional partial differential equations and to consider their classical and mild solutions. The mild solutions are studied using the semigroup methods as in [63], and strong solutions are studied using variational methods as in [48] (see also [71]).

1.1 The Heat Equation as an Abstract Cauchy Problem

Let us consider a PDE and explain how it is related to a semigroup of linear operators.

Example 1.1 Consider the one-dimensional heat equation

$$\begin{cases} u_t(t,x) = u_{xx}(t,x), & t > 0, \\ u(0,x) = \varphi(x), & x \in \mathbb{R}, \end{cases} \tag{1.1}$$

with the initial temperature distribution φ being modeled by a bounded, uniformly continuous function on $\mathbb{R}$.

A function $u(t,x)$ is said to be a solution of (1.1) if it satisfies (1.1) for $t > 0$, u, u_t, u_x, u_{xx} are bounded, uniformly continuous functions in $\mathbb{R}$ for every $t > 0$ and $\lim_{t \to 0^+} u(t,x) = \varphi(x)$ uniformly in x.

The uniqueness of the solution and the major premise of the Hughen's principle lead to the following implications. If the temperature distribution $u(t,x)$ at $t > 0$ is uniquely determined by the initial condition $\varphi(x)$, then $u(t,x)$ can also be obtained by first calculating $u(s,x)$ for some intermediate time $s < t$ and then by using $u(s,x)$ as the initial condition. Thus, there exist transformations $G(t)$ on φ defined by $(G(t)\varphi)(x) = u^{\varphi}(t,x)$ satisfying the semigroup property

$$G(t)\varphi = G(t-s)\big(G(s)\varphi\big)$$

L. Gawarecki, V. Mandrekar, *Stochastic Differential Equations in Infinite Dimensions*, Probability and Its Applications, DOI 10.1007/978-3-642-16194-0_1, © Springer-Verlag Berlin Heidelberg 2011 and the strong continuity property

$$\lim_{t \to 0^+} = \| G(t)\varphi - \varphi \| = 0.$$

Above, $\|\varphi\| = \sup_{x \in \mathbb{R}} |\varphi(x)|$.

The transformations are linear due to the linear dependence of the problem on the initial condition. The relationship between the semigroup $G(t)$ and the differential operator $\partial^2/\partial x^2$ in (1.1) can be explained by calculating

$$\lim_{h \to 0^+} \left\| \frac{u^\varphi(\cdot, t+h) - u^\varphi(\cdot, t)}{h} \right\| = \lim_{h \to 0^+} \left\| G(t)\left(\frac{G(h) - I}{h}\right)\varphi \right\|$$

$$= \lim_{h \to 0^+} \left\| \left(\frac{G(h) - I}{h}\right) G(t)\varphi \right\| = \lim_{h \to 0^+} \left\| \left(\frac{G(h) - I}{h}\right) u^\varphi(\cdot, t) \right\|,$$

where I denotes the identity. Let us define the linear operator

$$A\varphi = \lim_{t \to 0^+} \frac{G(t)\varphi - \varphi}{t} \quad \text{(in norm)}$$

and denote by $\mathscr{D}(A)$, called the domain of A, the collection of all those functions φ for which the limit exists. We arrive at an abstract formulation of the heat equation in the form of an *abstract Cauchy problem* in the Banach space $(X, \|\cdot\|)$ of bounded uniformly continuous functions on $\mathbb{R}$,

$$\begin{cases} \dfrac{du(t)}{dt} = Au(t), & t > 0, \\ u(0) = \varphi \in X, \end{cases} \tag{1.2}$$

where the differentiation is understood in the Banach space X.

For a solution $u(t)$, it is required that for $t > 0$, $u(t) \in \mathscr{D}(A)$, u is continuously differentiable and $\lim_{t \to 0^+} \|u(t) - \varphi\| = 0$. In the case of (1.1), there is an explicit form of the semigroup $G(t)$, given by the Gaussian semigroup

$$\big(G(t)\varphi\big)(x) = \begin{cases} \dfrac{1}{(4\pi t)^{1/2}} \displaystyle\int_{\mathbb{R}} \exp\{-|x - y|^2/4t\}\varphi(y)\,dy, & t > 0, \\ \varphi(x), & t = 0. \end{cases} \tag{1.3}$$

The solution to (1.1) is known to be (see [7] Chap. 3 for detailed presentation)

$$u(t, x) = \big(G(t)\varphi\big)(x).$$

The operator $A = d^2/dx^2$, and

$$\mathscr{D}(A) = \left\{ f \in X : f, \frac{df}{dx} \text{ are continuously differentiable, and } \frac{d^2 f}{dx^2} \in X \right\}.$$

Exercise 1.1 Show that the operators $G(t)$ defined in (1.3) have the semigroup property.

We now review the fundamentals of the theory of semigroups of linear operators with the goal of studying the existence of classical and mild solutions to an abstract Cauchy problem.

1.2 Elements of Semigroup Theory

In this section we review the fundamentals of semigroup theory and refer the reader to [6, 63], or [78] for proofs.

Let $(X, \| \cdot \|_X)$ and $(Y, \| \cdot \|_Y)$ be Banach spaces. Denote by $\mathscr{L}(X, Y)$ the family of bounded linear operators from X to Y. $\mathscr{L}(X, Y)$ becomes a Banach space when equipped with the norm

$$\|T\|_{\mathscr{L}(X,Y)} = \sup_{x \in X, \|x\|_X = 1} \|Tx\|_Y, \quad T \in \mathscr{L}(X, Y).$$

For brevity, $\mathscr{L}(X)$ will denote the Banach space of bounded linear operators on X. The identity operator on X is denoted by I.

Let X^* denote the dual space of all bounded linear functionals x^* on X. X^* is again a Banach space under the supremum norm

$$\|x^*\|_{X^*} = \sup_{x \in X, \|x\|_X = 1} |\langle x, x^* \rangle|,$$

where $\langle \cdot, \cdot \rangle$ denotes the duality on $X \times X^*$.

For $T \in \mathscr{L}(X, Y)$, the adjoint operator $T^* \in \mathscr{L}(Y^*, X^*)$ is defined by

$$\langle x, T^* y^* \rangle = \langle Tx, y^* \rangle, \quad x \in X, \ y^* \in Y^*.$$

Let H be a real Hilbert space. A linear operator $T \in \mathscr{L}(H)$ is called *symmetric* if for all $h, g \in H$,

$$\langle Th, g \rangle_H = \langle h, Tg \rangle_H.$$

A symmetric operator T is called *nonnegative definite* if for every $h \in H$,

$$\langle Th, h \rangle_H \geq 0.$$

Definition 1.1 A family $S(t) \in \mathscr{L}(X)$, $t \geq 0$, of bounded linear operators on a Banach space X is called a strongly continuous semigroup (or a C_0-semigroup, for short) if

(S1) $S(0) = I$,
(S2) (Semigroup property) $S(t + s) = S(t)S(s)$ for every $t, s \geq 0$,
(S3) (Strong continuity property) $\lim_{t \to 0^+} S(t)x = x$ for every $x \in X$.

Let $S(t)$ be a C_0-semigroup on a Banach space X. Then, there exist constants $\alpha \geq 0$ and $M \geq 1$ such that

$$\|S(t)\|_{\mathscr{L}(X)} \leq M e^{\alpha t}, \quad t \geq 0. \tag{1.4}$$

If $M = 1$, then $S(t)$ is called a *pseudo-contraction semigroup*. If $\alpha = 0$, then $S(t)$ is called *uniformly bounded*, and if $\alpha = 0$ and $M = 1$ (i.e., $\|S(t)\|_{\mathscr{L}(X)} \leq 1$), then $S(t)$ is called a *semigroup of contractions*. If for every $x \in X$, the mapping $t \to S(t)x$ is differentiable for $t > 0$, then $S(t)$ is called a *differentiable* semigroup. A semigroup of linear operators $\{S(t),\ t \geq 0\}$ is called *compact* if the operators $S(t)$, $t > 0$, are compact.

For any C_0-semigroup $S(t)$ and arbitrary $x \in X$, the mapping

$$\mathbb{R}_+ \ni t \to S(t)x \in X$$

is continuous.

Definition 1.2 Let $S(t)$ be a C_0-semigroup on a Banach space X. The linear operator A with domain

$$\mathscr{D}(A) = \left\{ x \in X\ :\ \lim_{t \to 0^+} \frac{S(t)x - x}{t}\ \text{exists} \right\} \tag{1.5}$$

defined by

$$Ax = \lim_{t \to 0^+} \frac{S(t)x - x}{t} \tag{1.6}$$

is called the *infinitesimal generator* of the semigroup $S(t)$.

A semigroup $S(t)$ is called *uniformly continuous* if

$$\lim_{t \to 0^+} \big\| S(t) - I \big\|_{\mathscr{L}(X)} = 0.$$

Theorem 1.1 *A linear operator A is the infinitesimal generator of a uniformly continuous semigroup $S(t)$ on a Banach space X if and only if $A \in \mathscr{L}(X)$. We have*

$$S(t) = e^{tA} = \sum_{n=0}^{\infty} \frac{(tA)^n}{n!},$$

the series converging in norm for every $t \geq 0$.

We will however be mostly interested in the case where $A \notin \mathscr{L}(X)$, as in (1.2). The following theorem provides useful facts about semigroups.

Theorem 1.2 *Let A be an infinitesimal generator of a C_0-semigroup $S(t)$ on a Banach space X. Then*

(1) *For $x \in X$,*

$$\lim_{h \to 0} \frac{1}{h} \int_t^{t+h} S(t)x\, ds = S(t)x. \tag{1.7}$$

(2) *For $x \in \mathscr{D}(A)$, $S(t)x \in \mathscr{D}(A)$ and*

$$\frac{d}{dt}S(t)x = AS(t)x = S(t)Ax. \tag{1.8}$$

(3) *For $x \in X$, $\int_0^t S(s)x\,ds \in \mathscr{D}(A)$, and*

$$A\left(\int_0^t S(s)x\,ds\right) = S(t)x - x. \tag{1.9}$$

(4) *If $S(t)$ is differentiable then for $n = 1, 2, \ldots$, $S(t) : X \to \mathscr{D}(A^n)$ and*

$$S^{(n)}(t) = A^n S(t) \in \mathscr{L}(X).$$

(5) *If $S(t)$ is compact then $S(t)$ is continuous in the operator topology for $t > 0$, i.e.,*

$$\lim_{s \to t, \, s,t>0} \|S(s) - S(t)\|_{\mathscr{L}(H)} = 0. \tag{1.10}$$

(6) *For $x \in \mathscr{D}(A)$,*

$$S(t)x - S(s)x = \int_s^t S(u)Ax\,du = \int_s^t AS(u)x\,du. \tag{1.11}$$

(7) *$\mathscr{D}(A)$ is dense in X, and A is a closed linear operator.*
(8) *The intersection $\bigcap_{n=1}^{\infty} \mathscr{D}(A^n)$ is dense in X.*
(9) *Let X be a reflexive Banach space. Then the adjoint semigroup $S(t)^*$ of $S(t)$ is a C_0-semigroup whose infinitesimal generator is A^*, the adjoint of A.*

If $X = H$, a real separable Hilbert space, then for $h \in H$, define the graph norm

$$\|h\|_{\mathscr{D}(A)} = \left(\|h\|_H^2 + \|Ah\|_H^2\right)^{1/2}. \tag{1.12}$$

Then $(\mathscr{D}(A), \|\cdot\|_{\mathscr{D}(A)})$ is a real separable Hilbert space.

Exercise 1.2 Let A be a closed linear operator on a real separable Hilbert space. Prove that $(\mathscr{D}(A), \|\cdot\|_{\mathscr{D}(A)})$ is a real separable Hilbert space.

Let $\mathscr{B}(H)$ denote the Borel σ-field on H. Then $\mathscr{D}(A) \in \mathscr{B}(H)$, and

$$A : \left(\mathscr{D}(A), \mathscr{B}(H)|_{\mathscr{D}(A)}\right) \to \left(H, \mathscr{B}(H)\right).$$

Consequently, the restricted Borel σ-field $\mathscr{B}(H)|_{\mathscr{D}(A)}$ coincides with the Borel σ-field on the Hilbert space $(\mathscr{D}(A), \|\cdot\|_{\mathscr{D}(A)})$, and measurability of $\mathscr{D}(A)$-valued functions can be understood with respect to either Borel σ-field.

Theorem 1.3 *Let $f : [0, T] \to \mathscr{D}(A)$ be measurable, and let $\int_0^t \|f(s)\|_{\mathscr{D}(A)} < \infty$. Then*

$$\int_0^t f(s)\,ds \in \mathscr{D}(A) \quad and \quad \int_0^t Af(s)\,ds = A \int_0^t f(s)\,ds. \qquad (1.13)$$

Exercise 1.3 Prove Theorem 1.3.

Conditions under which an operator A can be an infinitesimal generator of a C_0-semigroup involve the resolvent of A.

Definition 1.3 The *resolvent set* $\rho(A)$ of a closed linear operator A on a Banach space X is the set of all complex numbers λ for which $\lambda I - A$ has a bounded inverse, i.e., the operator $(\lambda I - A)^{-1} \in \mathscr{L}(X)$. The family of bounded linear operators

$$R(\lambda, A) = (\lambda I - A)^{-1}, \quad \lambda \in \rho(A), \qquad (1.14)$$

is called the *resolvent* of A.

We note that $R(\lambda, A)$ is a one-to-one transformation of X onto $\mathscr{D}(A)$, i.e.,

$$\begin{aligned}(\lambda I - A)R(\lambda, A)x &= x, & x \in X, \\ R(\lambda, A)(\lambda I - A)x &= x, & x \in \mathscr{D}(A).\end{aligned} \qquad (1.15)$$

In particular,

$$AR(\lambda, A)x = R(\lambda, A)Ax, \quad x \in \mathscr{D}(A). \qquad (1.16)$$

In addition, we have the following commutativity property:

$$R(\lambda_1, A)R(\lambda_2, A) = R(\lambda_2, A)R(\lambda_1, A), \quad \lambda_1, \lambda_2 \in \rho(A). \qquad (1.17)$$

The following statement is true in greater generality; however, we will use it only in the real domain.

Proposition 1.1 *Let $S(t)$ be a C_0-semigroup with infinitesimal generator A on a Banach space X. If $\alpha_0 = \lim_{t \to \infty} t^{-1} \ln \|S(t)\|_{\mathscr{L}(X)}$, then any real number $\lambda > \alpha_0$ belongs to the resolvent set $\rho(A)$, and*

$$R(\lambda, A)x = \int_0^\infty e^{-\lambda t} S(t)x\,dt, \quad x \in X. \qquad (1.18)$$

Furthermore, for each $x \in X$,

$$\lim_{\lambda \to \infty} \|\lambda R(\lambda, A)x - x\|_X = 0. \qquad (1.19)$$

Theorem 1.4 (Hille–Yosida) *Let $A : \mathscr{D}(A) \subset X \to X$ be a linear operator on a Banach space X. Necessary and sufficient conditions for A to generate a C_0-semigroup $S(t)$ are*

(1) *A is closed and $\overline{\mathscr{D}(A)} = X$.*
(2) *There exist real numbers M and α such that for every $\lambda > \alpha$, $\lambda \in \rho(A)$ (the resolvent set) and*

$$\left\|\left(R(\lambda, A)\right)^r\right\|_{\mathscr{L}(X)} \le M(\lambda - \alpha)^{-r}, \quad r = 1, 2, \dots. \tag{1.20}$$

In this case, $\|S(t)\|_{\mathscr{L}(X)} \le Me^{\alpha t}, t \ge 0$.

We will now introduce an important approximation of an operator A and of the C_0-semigroup it generates.

For $\lambda \in \rho(A)$, consider the family of operators

$$R_\lambda = \lambda R(\lambda, A). \tag{1.21}$$

Since the range $\mathscr{R}(R(\lambda, A)) \subset \mathscr{D}(A)$, we can define the *Yosida approximation* of A by

$$A_\lambda x = A R_\lambda x, \quad x \in X. \tag{1.22}$$

Note that by (1.16)

$$A_\lambda x = R_\lambda A x, \quad x \in \mathscr{D}(A).$$

Since $\lambda(\lambda I - A)R(\lambda, A) = \lambda I$, we have $\lambda^2 R(\lambda, A) - \lambda I = \lambda A R(\lambda, A)$, so that

$$A_\lambda x = \lambda^2 R(\lambda, A) - \lambda I,$$

proving that $A_\lambda \in \mathscr{L}(X)$. Denote by $S_\lambda(t)$ the (uniformly continuous) semigroup generated by A_λ,

$$S_\lambda(t)x = e^{t A_\lambda} x, \quad x \in X. \tag{1.23}$$

Using the commutativity of the resolvent (1.17), we have

$$A_{\lambda_1} A_{\lambda_2} = A_{\lambda_2} A_{\lambda_1} \tag{1.24}$$

and, by the definition of $S_\lambda(t)$ (1.23),

$$A_\lambda S_\lambda(t) = S_\lambda(t) A_\lambda. \tag{1.25}$$

Proposition 1.2 (Yosida Approximation) *Let A be an infinitesimal generator of a C_0-semigroup $S(t)$ on a Banach space X. Then*

$$\lim_{\lambda \to \infty} R_\lambda x = x, \quad x \in X, \tag{1.26}$$

$$\lim_{\lambda \to \infty} A_\lambda x = A x, \quad x \in \mathscr{D}(A), \tag{1.27}$$

and

$$\lim_{\lambda \to \infty} S_\lambda(t)x = S(t)x, \quad x \in X. \tag{1.28}$$

The convergence in (1.28) *is uniform on compact subsets of* $\mathbb{R}_+$. *The following esti-mate holds*:

$$\left\| S_\lambda(t) \right\|_{\mathscr{L}(X)} \leq M \exp\{t\lambda\alpha/(\lambda - \alpha)\} \tag{1.29}$$

with the constants M, α *determined by the Hille–Yosida theorem*.

1.3 Commonly Used Function Spaces

We define commonly used function spaces. Let $(H, \|\cdot\|_H)$ be a real separable Hilbert space, and $\mathscr{B}(H)$ be its Borel σ-field. The collection of measurable (respectively, bounded measurable) real-valued functions on H will be denoted by $B(H)$ (respectively, $B_b(H)$).

For a positive constant T, let $C([0, T], H)$ be the Banach space of H-valued continuous functions on $[0, T]$, with the norm $\|f\|_{C([0,T],H)} = \sup_{0 \leq t \leq T} \|f(t)\|_H$. For a positive integer m, $C^m([0, T], H)$ and $C^\infty([0, T], H)$ denote the spaces of H-valued, respectively m-times and infinitely many times continuously differentiable functions on $[0, T]$.

By $C^m(H)$ we denote the collection of real-valued m-times continuously Fréchet differentiable functions on H, and let $C_0^m(H)$ and $C_b^m(H)$ be its subspaces consisting respectively of those functions which have compact support and of those whose derivatives of order $k = 0, \ldots, m$ are bounded. It is typical to suppress the superscript m if $m = 0$ and write, for example, $C(H)$ for $C^0(H)$. If the Hilbert space H is replaced with an interval $[0, T[$ (resp. $]0, T]$), we consider the right derivatives at $t = 0$ (resp. left derivatives at $t = T$).

The Banach space $L^p([0, T], H)$ is the space of H-valued Borel-measurable functions on $[0, T]$ that are Bochner integrable in the pth power, with the norm

$$\|f\|_{L^p([0,T],H)} = \left(\int_0^T \|f(t)\|_H^p \, dt \right)^{1/p}.$$

The space of square-integrable functions $L^2([0, T], H)$ equipped with the scalar product

$$\langle f, g \rangle_{L^2(H)} = \int_0^T \langle f(t), g(t) \rangle_H \, dt$$

is a Hilbert space. We refer the reader to [14] for details of Bochner integration in a Hilbert space.

The Banach space of H-valued essentially bounded functions on $[0, T]$ is denoted by $L^\infty([0, T], H)$ and consists of functions f for which the norm

$$\|f\|_{L^\infty([0,T],H)} = \text{ess sup}\{|f(t)|, \ t \in [0, T]\}$$

is finite.

We now define Sobolev spaces. For nonnegative integers $\alpha_1, \ldots, \alpha_d$, consider the multiindex $\alpha = (\alpha_1, \ldots, \alpha_d)$ of order $|\alpha| = \sum_{i=1}^d \alpha_i$ and for $x = (x_1, x_2, \ldots, x_d) \in \mathbb{R}^d$, denote

$$D^\alpha = \frac{\partial^{\alpha_1}}{\partial x_1^{\alpha_1}} \frac{\partial^{\alpha_2}}{\partial x_2^{\alpha_2}} \cdots \frac{\partial^{\alpha_d}}{\partial x_d^{\alpha_d}}.$$

Let $\mathcal{O}$ be an open subset of $\mathbb{R}^d$. For a function $f \in C^m(\mathcal{O})$ with positive integer m, define the norm

$$\|f\|_{m,p} = \left(\int_{\mathcal{O}} \sum_{|\alpha| \le m} |D^\alpha f(x)|^p \, dx \right)^{1/p} < \infty,$$

where the differentiation is in the sense of distributions. The Sobolev space $W^{m,p}(\mathcal{O})$ with $1 \le p < \infty$ is the completion in the norm $\| \cdot \|_{m,p}$ of the subspace of $C^m(\mathcal{O})$ consisting of all functions f such that $\|f\|_{m,p} < \infty$. Similarly, the space $W_0^{m,p}(\mathcal{O})$ is the completion in the norm $\| \cdot \|_{m,p}$ of $C_0^m(\mathcal{O})$. The Sobolev spaces $W^{m,p}(\mathcal{O})$ defined above consist of functions $f \in L^p(\mathcal{O})$ whose distributional derivatives $D^\alpha f$ of order up to m are also in $L^p(\mathcal{O})$.

The spaces $(W^{m,p}(\mathcal{O}), \| \cdot \|_{m,p})$ and $(W_0^{m,p}(\mathcal{O}), \| \cdot \|_{m,p})$ are Banach spaces, and $W^{m,2}(\mathcal{O})$ and $W_0^{m,2}(\mathcal{O})$ are Hilbert spaces with the scalar product

$$\langle f, g \rangle_{W^{m,2}(\mathcal{O})} = \int_{\mathcal{O}} \sum_{|\alpha| \le m} D^\alpha f(x) D^\alpha g(x) \, dx.$$

Note that the Sobolev space $W^{m,p}(\mathcal{O})$ is a subset of $L^p(\mathcal{O})$. If $\mathcal{O} = \mathbb{R}^d$, then it is known ([30], Chap. 10, Proposition 1.5) that

$$W_0^{1,2}(\mathbb{R}^d) = W^{1,2}(\mathbb{R}^d). \tag{1.30}$$

1.4 The Abstract Cauchy Problem

Let A be a linear operator on a real separable Hilbert space H, and let us consider the abstract Cauchy problem given by

$$\begin{cases} \dfrac{du(t)}{dt} = Au(t), & 0 < t < T, \\ u(0) = x, & x \in H. \end{cases} \tag{1.31}$$

Definition 1.4 A function $u : [0, T[\to H$ is a (classical) solution of the problem (1.31) on $[0, T[$ if u is continuous on $[0, T[$, continuously differentiable and $u(t) \in \mathscr{D}(A)$ for $t \in]0, T[$, and (1.31) is satisfied on $[0, T[$.

Exercise 1.4 Argue why if $x \notin \overline{\mathscr{D}(A)}$, then (1.31) cannot have a solution.

If A is an infinitesimal generator of a C_0-semigroup $\{S_t, t \geq 0\}$, then for any $x \in \mathscr{D}(A)$, the function $u^x(t) = S(t)x, t \geq 0$, is a solution of (1.31) ([63], Chap. 4, Theorem 1.3). On the other hand, if $x \notin \mathscr{D}(A)$, then the continuity at zero may not be a problem (see Exercise 1.4), but $u^x(t)$ does not have to be differentiable, unless the C_0-semigroup has additional properties, for example, when it is a differentiable semigroup (refer to (4) in Theorem 1.2).

In this case, $u^x(t) = S(t)x$ is not a solution in the usual sense, but it can be viewed as a "generalized solution," which will be called a "mild solution." In fact, the concept of mild solution can be introduced to study the following nonhomogeneous initial-value problem:

$$\begin{cases} \dfrac{du(t)}{dt} = Au(t) + f(t), & 0 < t < T, \\ u(0) = x, & x \in H, \end{cases} \tag{1.32}$$

where $f : [0, T[\to H$.

We assume that A is an infinitesimal generator of a C_0-semigroup so that the homogeneous equation (1.31) has a unique solution for all $x \in \mathscr{D}(A)$. The definition of a classical solution, Definition 1.4, extends to the case of the nonhomogeneous initial-value problem by requiring that in this case, the solution satisfies (1.32).

We now define the concept of a mild solution.

Definition 1.5 Let A be an infinitesimal generator of a C_0-semigroup $S(t)$ on H, $x \in H$, and $f \in L^1([0, T], H)$ be the space of Bochner-integrable functions on $[0, T]$ with values in H. The function $u \in C([0, T], H)$ given by

$$u(t) = S(t)x + \int_0^t S(t - s)f(s)\,ds, \quad 0 \leq t \leq T,$$

is the *mild solution* of the initial-value problem (1.32) on $[0, T]$.

Exercise 1.5 Prove that the function $u(t)$ in Definition 1.5 is continuous.

Note that for $x \in H$ and $f \equiv 0$, the mild solution is $S(t)x$, which is not in general a classical solution.

When $x \in \mathscr{D}(A)$, the continuity of f is insufficient to assure the existence of a classical solution. To see this, following [63], consider $f(t) = S(t)x$ for $x \in H$ such that $S(t)x \notin \mathscr{D}(A)$. Then (1.32) may not have a classical solution even if $u(0) = 0 \in \mathscr{D}(A)$, as the mild solution

$$u(t) = \int_0^t S(t - s)S(s)x\,ds = tS(t)x$$

is not, in general, differentiable.

One has the following theorem ([63], Chap. 4, Theorem 2.4).

Theorem 1.5 *Let A be an infinitesimal generator of a C_0-semigroup $\{S(t), t > 0\}$, let $f \in L^1([0, T], H)$ be continuous on $]0, T]$, and let*

$$v(t) = \int_0^t S(t - s) f(s) \, ds, \quad 0 \le t \le T.$$

The mild solution u to the initial–value problem (1.32) is a (classical) solution on $[0, T[$ for every $x \in \mathscr{D}(A)$ if

(1) $v(t)$ *is continuously differentiable on $]0, T[$.*
(2) $v(t) \in \mathscr{D}(A)$ *for $0 < t < T$, and $Av(t)$ is continuous on $]0, T[$.*

If (1.32) has a solution u on $[0, T[$ for some $x \in \mathscr{D}(A)$, then v satisfies (1) and (2).

Exercise 1.6 Show that if f is continuously differentiable on $[0, T]$, then

$$v(t) = \int_0^t S(t - s) f(s) \, ds = \int_0^t S(s) f(t - s) \, ds$$

is continuously differentiable for $t > 0$, and its derivative is given by

$$v'(t) = S(t) f(0) + \int_0^t S(t) f'(t - s) \, ds = S(t) f(0) + \int_0^t S(t - s) f'(s) \, ds.$$

Conclude that in this case the initial-value problem (1.32) has a solution for every $x \in \mathscr{D}(A)$.

We conclude with examples of the heat equation in $\mathbb{R}^d$ and in a bounded domain $\mathcal{O} \subset \mathbb{R}^d$.

Example 1.2 Consider the heat equation in $\mathbb{R}^d$

$$\begin{cases} u_t(t, x) = \Delta u(t, x), & 0 < t < T, \\ u(0, x) = \varphi(x), \end{cases} \tag{1.33}$$

$x \in \mathbb{R}^d$. The Gaussian family of operators

$$(G_d(t)\varphi)(x) = \begin{cases} \dfrac{1}{(4\pi t)^{d/2}} \displaystyle\int_{\mathbb{R}} \exp\{-\|x - y\|_{\mathbb{R}^d}^2 / 4t\} \varphi(y) \, dy, & t > 0, \\ \varphi(x), & t = 0, \end{cases} \tag{1.34}$$

defines a C_0-semigroup of contractions on $H = L^2(\mathbb{R}^d)$ with the infinitesimal generator

$$\Delta\varphi = \sum_{i=1}^d \frac{\partial^2}{\partial x_i^2} \varphi,$$

whose domain is $\mathscr{D}(\Delta) = W^{2,2}(\mathbb{R}^d)$. Note the difference between a bounded region and $\mathbb{R}^d$. Here the domain of the infinitesimal generator has this simple form due to (1.30). Consider the related abstract Cauchy problem

$$\begin{cases} \dfrac{du}{dt} = \Delta u, & 0 < t < T, \\ u(0) = \varphi \in L^2(\mathbb{R}^d). \end{cases} \tag{1.35}$$

It is known that $(G_d(t)\varphi)(x)$ is a classical solution of problem (1.35) for any $\varphi \in H = L^2(\mathbb{R}^d)$ in the sense of Definition 1.4, since the semigroup $G_d(t)$ on H is differentiable ([63], Chap. 7, Theorem 2.7 and Remark 2.9).

Example 1.3 Let $\mathcal{O} \subset \mathbb{R}^d$ be a bounded domain with smooth boundary $\partial\mathcal{O}$ (i.e., for each $x \in \partial\mathcal{O}$ and some ball B centered at x, $\partial\mathcal{O} \cap B$ has the form $x_i = f(x_1, \ldots, x_{i-1}, x_{i+1}, \ldots, x_d)$ for some $1 \le i \le d$ with f being $k \ge 1$ times differentiable). Then the heat equation has an abstract Cauchy form

$$\begin{cases} \dfrac{du}{dt} = \Delta u, & t > 0, \\ u(0) = \varphi \in W^{2,2}(\mathcal{O}) \cap W_0^{1,2}(\mathcal{O}). \end{cases} \tag{1.36}$$

The Laplace operator Δ is an infinitesimal generator of a C_0-semigroup of contractions on $H = L^2(\mathcal{O})$, and the initial-value problem (1.36) has a unique solution $u(t,x) \in C([0,\infty[, W^{2,2}(\mathcal{O}) \cap W_0^{1,2}(\mathcal{O}))$. This is a consequence of a more general result for strongly elliptic operators ([63], Chap. 7, Theorem 2.5 and Corollary 2.6).

To study nonlinear equations, one also needs to look at the Peano theorem in an infinite-dimensional Hilbert space, that is, to study the existence of a solution of the equation

$$\frac{du}{dt}(t) = G\big(u(t)\big), \quad u(0) = x \in H,$$

where G is a continuous function on H. We note that due to the failure of the Arzela–Ascoli theorem in $C([0,T], H)$, the proof in the finite-dimensional case fails (see the proofs in [15] and [29]). In fact the Peano theorem in a Hilbert space is not true [26]. However, if we look at semilinear equations

$$\begin{cases} \dfrac{du(t)}{dt} = Au(t) + G(u(t)), & t > 0, \\ u(0) = x, & x \in H, \end{cases}$$

we can salvage the theorem if $\{S(t), t > 0\}$ is a semigroup of compact operators. We present this theorem in the general case of SPDEs in Chap. 2.

1.5 The Variational Method

We shall now consider problem (1.32) with $f(t) \equiv 0$ and with further condition
of coercivity on A. In this case, we follow the variational method due to Lions
[48]. However, we use the approach of Tanabe [71]. We present the basic technique
of Lions, without presenting the proofs. However, we present one crucial lemma
of Lions [48] on which Pardoux's [62] approach is based. This lemma has been
appropriately generalized for the stochastic case by Krylov and Rozovskii [42]. We
present the method where the work of Krylov and Rozovskii is essential.

Let us start with the variational set up. We have three Hilbert spaces

$$V \hookrightarrow H \hookrightarrow V^*$$

with V a dense subspace of H, V^* the continuous dual of V, and the embeddings
$\hookrightarrow$ being continuous. Also, with $\langle \cdot, \cdot \rangle$ denoting the duality on $V \times V^*$, we assume
that $\langle v, v' \rangle = \langle v, v' \rangle_H$ if $v \in V$ and $v' \in H$.

Let us consider now an analogue of problem (1.32). Let A be a linear operator,
$A : V \to V^*$ such that

$$\|Av\|_{V^*} \le M\|v\|_V$$

and

$$2\langle v, Av \rangle \le \lambda \|v\|_H^2 - \alpha \|v\|_V^2, \quad v \in V,$$

for some real number λ and $M, \alpha > 0$. The following theorem is due to Lions [48].

Theorem 1.6 (Lions) *Let $x \in H$ and $f \in L^2([0, T]), V^*)$. Then there exists a
unique function $u \in L^2([0, T], V)$ with $du(t)/dt \in L^2([0, T], V^*)$ and satisfying*

$$\begin{cases} \dfrac{du(t)}{dt} = Au(t) + f(t), & t > 0, \\ u(0) = x. \end{cases} \tag{1.37}$$

For the proof, see [48], p. 150.

The crucial lemma needed for proving that the solution $u \in C([0, T], H)$ and to
work out an analogue of Itô's formula for the function $\|u(t)\|_H^2$ is the following,

Lemma 1.1 (Lions) *If $u \in L^2([0, T], V)$ and $du(t)/dt \in L^2([0, T], V^*)$, then $u \in
C([0, T], H)$. Furthermore,*

$$\frac{d}{dt}\|u(t)\|_H^2 = 2\left\langle u(t), \frac{du(t)}{dt} \right\rangle. \tag{1.38}$$

Proof Let $0 < a < T$; we extend u to $(-a, T + a)$ by putting $u(t) = u(-t)$ for $a <
t < 0$ and $u(t) = u(2T - t)$ for $T < t < T + a$. Observe that $u \in L^2((-a, T + a), V)$
and $u' \in L^2((-a, T + a), V^*)$. Define

$$w(t) = \theta(t)u(t),$$

where $\theta(t)$ is continuously differentiable function, $\theta \equiv 1$ in $[0, T]$ and $\theta \equiv 0$ in the neighborhood of $-a$ and $T + a$. By using a mollifier j_n, we define

$$w_n(t) = \int_{-a}^{T+a} j_n(t - s)w(s)\,ds.$$

Then, as $n \to \infty$, we have that $w_n \to w$ in $L^2((-a, T + a), V)$ and $dw_n/dt \to dw/dt$ in $L^2((-a, T + a), V^*)$. Note that $dw_n/dt \in V$, so that we have

$$\left\| w_n(t) - w_m(t) \right\|_H^2 = \int_{-a}^{t} \frac{d}{ds} \left\| w_n(s) - w_m(s) \right\|_H^2 ds$$

$$= \int_{-a}^{t} 2\left\langle w_n(s) - w_m(s), \frac{dw_n}{ds}(s) - \frac{dw_m}{ds}(s) \right\rangle ds$$

$$\leq \int_{-a}^{T+a} \left\| \frac{dw_n}{ds}(s) - \frac{dw_m}{ds}(s) \right\|_{V^*}^2 ds + \int_{-a}^{T+a} \left\| w_n(s) - w_m(s) \right\|_V^2 ds.$$

Thus, in addition to the L^2 convergence in V, $\{w_n\}$ is a Cauchy sequence in $C([-a, T + a], H)$. Hence, modifying the values on a null set, we get that $w(t) \in H$ and $w_n \to w$ uniformly on $[-a, T + a]$. This gives $u(t) \in C([0, T], H)$, by the choice of θ. Let $\varphi \in C_0^\infty([0, T])$ and consider the integral

$$\int_0^T \langle w_n(t), v(t) \rangle \frac{d\varphi}{dt}(t)\,dt$$

$$= \int_0^T \left\langle v(t), \frac{d}{dt}(\varphi(t)w_n(t)) \right\rangle dt - \int_0^T \left\langle v(t), \varphi(t)\frac{dw_n}{dt}(t) \right\rangle dt$$

$$= -\int_0^T \left\langle \varphi(t)w_n(t), \frac{dv}{dt}(t) \right\rangle dt - \int_0^T \left\langle v(t), \varphi(t)\frac{dw_n}{dt}(t) \right\rangle_H dt.$$

Taking the limit, we obtain

$$\int_0^T \langle u(t), v(t) \rangle_H \frac{d\varphi}{dt}(t)\,dt = -\int_0^T \left\{ \left\langle u(t), \frac{dv}{dt}(t) \right\rangle + \left\langle v(t), \frac{du}{dt}(t) \right\rangle \right\} \varphi(t)\,dt.$$

For $v(t) = u(t)$, this shows that the statement (1.38) is valid in terms of distributions. Since the RHS in (1.38) is an integrable function of t, we conclude that $\|u(t)\|_H^2$ is absolutely continuous. □

Chapter 2
Stochastic Calculus

2.1 Hilbert-Space-Valued Process, Martingales, and Cylindrical Wiener Process

2.1.1 Cylindrical and Hilbert-Space-Valued Gaussian Random Variables

We first introduce cylindrical Gaussian random variables and Hilbert-space-valued Gaussian random variables and then define cylindrical Wiener process and Hilbert-space-valued Wiener process in a natural way. Let $(\Omega, \mathscr{F}, P)$ be a probability space, and K be a real separable Hilbert space with the norm and scalar product denoted by $\| \cdot \|_K$ and $\langle \cdot, \cdot \rangle_K$. We will always assume that $(\Omega, \mathscr{F}, P)$ is *complete*, i.e., that $\mathscr{F}$ contains all subsets A of Ω with P-outer measure zero,

$$P^*(A) = \inf\{P(F) : A \subset F \in \mathscr{F}\} = 0.$$

Definition 2.1 We say that $\tilde{X}$ is a *cylindrical standard Gaussian random variable* on K if $\tilde{X} : K \to L^2(\Omega, \mathscr{F}, P)$ satisfies the following conditions:

(1) The mapping $\tilde{X}$ is linear.
(2) For an arbitrary $k \in K$, $\tilde{X}(k)$ is a Gaussian random variable with mean zero and variance $\|k\|_K^2$.
(3) If $k, k' \in K$ are orthogonal, i.e., $\langle k, k' \rangle_K = 0$, then the random variables $\tilde{X}(k)$ and $\tilde{X}(k')$ are independent.

Note that if $\{f_j\}_{j=1}^{\infty}$ is an orthonormal basis (ONB) in K, then $\{\tilde{X}(f_j)\}_{j=1}^{\infty}$ is a sequence of independent Gaussian random variables with mean zero and variance one. By linearity of the mapping $\tilde{X} : K \to L^2(\Omega, \mathscr{F}, P)$, we can represent $\tilde{X}$ as

$$\tilde{X}(k) = \sum_{j=1}^{\infty} \langle k, f_j \rangle_K \tilde{X}(f_j),$$

L. Gawarecki, V. Mandrekar, *Stochastic Differential Equations in Infinite Dimensions*,   17
Probability and Its Applications,
DOI 10.1007/978-3-642-16194-0_2, © Springer-Verlag Berlin Heidelberg 2011

with the series convergent P-a.s. by Kolmogorov's three-series theorem ([5], Theorem 22.3).

However, it is not true that there exists a K-valued random variable X such that

$$\tilde{X}(k)(\omega) = \langle X(\omega), k \rangle_K.$$

This can be easily seen since we can express

$$\|X(\omega)\|_K^2 = \sum_{j=1}^{\infty} \langle X(\omega), f_j \rangle_K^2,$$

with the series being P-a.s. divergent by the strong law of large numbers.

In order to produce a K-valued Gaussian random variable, we proceed as follows. Denote by $\mathscr{L}_1(K)$ the space of trace-class operators on K,

$$\mathscr{L}_1(K) = \left\{ L \in \mathscr{L}(K) : \tau(L) := \operatorname{tr}\left((LL^*)^{1/2}\right) < \infty \right\}, \tag{2.1}$$

where the trace of the operator $[L] = (LL^*)^{1/2}$ is defined by

$$\operatorname{tr}([L]) = \sum_{j=1}^{\infty} \langle [L] f_j, f_j \rangle_K$$

for an ONB $\{f_j\}_{j=1}^{\infty} \subset K$. It is well known [68] that $\operatorname{tr}([L])$ is independent of the choice of the ONB and that $\mathscr{L}_1(K)$ equipped with the trace norm τ is a Banach space. Let $Q : K \to K$ be a symmetric nonnegative definite trace-class operator.

Assume that $X : K \to L^2(\Omega, \mathscr{F}, P)$ satisfies the following conditions:

(1) The mapping X is linear.
(2) For an arbitrary $k \in K$, $X(k)$ is a Gaussian random variable with mean zero.
(3) For arbitrary $k, k' \in K$, $E(X(k)X(k')) = \langle Qk, k' \rangle_K$.

Let $\{f_j\}_{j=1}^{\infty}$ be an ONB in K diagonalizing Q, and let the eigenvalues corresponding to the eigenvectors f_j be denoted by λ_j, so that $Q f_j = \lambda_j f_j$. We define

$$X(\omega) = \sum_{j=1}^{\infty} X(f_j)(\omega) f_j.$$

Since $\sum_{j=1}^{\infty} \lambda_j < \infty$, the series converges in $L^2((\Omega, \mathscr{F}, P), H)$ and hence P-a.s. In this case, P-a.s.,

$$\langle X(\omega), k \rangle_K = X(k)(\omega),$$

so that $X : \Omega \to K$ is $\mathscr{F}/\mathscr{B}(K)$-measurable, where $\mathscr{B}(K)$ denotes the Borel σ-field on K.

Definition 2.2 We call $X : \Omega \to K$ defined above a *K-valued Gaussian random variable with covariance Q*.

Definition 2.3 Let K be a separable Hilbert space. The measure $P \circ X^{-1}$ induced by a K-valued Gaussian random variable X with covariance Q on the measurable Hilbert space $(K, \mathscr{B}(K))$ is called a *Gaussian measure with covariance Q on K*.

Exercise 2.1 Let K be a separable Hilbert space, $Q : K \to K$ be a symmetric non-negative definite trace-class operator, and X be a cylindrical Gaussian random variable on K. Show that $Y = X \circ Q^{1/2} = \sum_{j=1}^{\infty} X(Q^{1/2}(f_j))f_j$ is a K-valued Gaussian random variable with covariance Q.

2.1.2 Cylindrical and Q-Wiener Processes

Let $(\Omega, \mathscr{F}, \{\mathscr{F}_t\}_{t \geq 0}, P)$ be a *filtered probability space*, and, as above, K be a real separable Hilbert space. We will always assume that the filtration $\mathscr{F}_t$ satisfies the *usual conditions*

(1) $\mathscr{F}_0$ contains all $A \in \mathscr{F}$ such that $P(A) = 0$,
(2) $\mathscr{F}_t = \bigcap_{s>t} \mathscr{F}_s$.

Definition 2.4 A K-valued stochastic process $\{X_t\}_{t \geq 0}$ defined on a probability space $(\Omega, \mathscr{F}, P)$ is called *Gaussian* if for any positive integer n and $t_1, \ldots, t_n \geq 0$, $(X_{t_1}, \ldots, X_{t_n})$ is a K^n-valued Gaussian random variable.

A standard cylindrical Wiener process can now be introduced using the concept of a cylindrical random variable.

Definition 2.5 We call a family $\{\tilde{W}_t\}_{t \geq 0}$ defined on a filtered probability space $(\Omega, \mathscr{F}, \{\mathscr{F}_t\}_{t \geq 0}, P)$ a *cylindrical Wiener process* in a Hilbert space K if:

(1) For an arbitrary $t \geq 0$, the mapping $\tilde{W}_t : K \to L^2(\Omega, \mathscr{F}, P)$ is linear;
(2) For an arbitrary $k \in K$, $\tilde{W}_t(k)$ is an $\mathscr{F}_t$-Brownian motion;
(3) For arbitrary $k, k' \in K$ and $t \geq 0$, $E(\tilde{W}_t(k)\tilde{W}_t(k')) = t\langle k, k'\rangle_K$.

For every $t > 0$, $\tilde{W}_t/\sqrt{t}$ is a standard cylindrical Gaussian random variable, so that for any $k \in K$, $\tilde{W}_t(k)$ can be represented as a P-a.s. convergent series

$$\tilde{W}_t(k) = \sum_{j=1}^{\infty} \langle k, f_j\rangle_K \tilde{W}_t(f_j), \tag{2.2}$$

where $\{f_j\}_{j=1}^{\infty}$ is an ONB in K.

Exercise 2.2 Show that $E(\tilde{W}_t(k)\tilde{W}_s(k')) = (t \wedge s)\langle k, k'\rangle_K$ and conclude that $\tilde{W}_t(f_j)$, $j = 1, 2, \ldots$, are independent Brownian motions.

For the same reason why a cylindrical Gaussian random variable cannot be realized as a K-valued random variable, there is no K-valued process W_t such that

$$\tilde{W}_t(k)(\omega) = \langle W_t(\omega), k \rangle_K.$$

However, if Q is a nonnegative definite symmetric trace-class operator on K, then a K-valued Q-Wiener process can be defined.

Definition 2.6 Let Q be a nonnegative definite symmetric trace-class operator on a separable Hilbert space K, $\{f_j\}_{j=1}^{\infty}$ be an ONB in K diagonalizing Q, and let the corresponding eigenvalues be $\{\lambda_j\}_{j=1}^{\infty}$. Let $\{w_j(t)\}_{t \geq 0}$, $j = 1, 2, \ldots$, be a sequence of independent Brownian motions defined on $(\Omega, \mathscr{F}, \{\mathscr{F}_t\}_t, P)$. The process

$$W_t = \sum_{j=1}^{\infty} \lambda_j^{1/2} w_j(t) f_j \tag{2.3}$$

is called a *Q-Wiener process* in K.

We can assume that the Brownian motions $w_j(t)$ are continuous. Then, the series (2.3) converges in $L^2(\Omega, C([0, T], K))$ for every interval $[0, T]$, see Exercise 2.3. Therefore, the K-valued Q-Wiener process can be assumed to be continuous. We denote

$$W_t(k) = \sum_{j=1}^{\infty} \lambda_j^{1/2} w_j(t) \langle f_j, k \rangle_K$$

for any $k \in K$, with the series converging in $L^2(\Omega, C([0, T], \mathbb{R}))$ on every interval $[0, T]$.

Exercise 2.3 Use Doob's inequality, Theorem 2.2, for the submartingale

$$\left\| \sum_{j=m}^{n} \lambda_j^{1/2} w_j(t) f_j \right\|_K$$

to prove that the partial sums of the series (2.3) defining the Q-Wiener process are a Cauchy sequence in $L^2(\Omega, C([0, T], K))$.

Remark 2.1 A stronger convergence result can be obtained for the series (2.3). Since

$$P\left(\sup_{0 \leq t \leq T} \left\| \sum_{j=m}^{n} \lambda_j^{1/2} w_j(t) f_j \right\|_K > \varepsilon \right) \leq \frac{1}{\varepsilon^2} E \left\| \sum_{j=m}^{n} \lambda_j^{1/2} w_j(T) f_j \right\|_K^2$$

$$= \frac{T}{\varepsilon^2} \sum_{j=m}^{n} \lambda_j \to 0$$

with $m \leq n$, $m, n \to \infty$, the series (2.3) converges uniformly on $[0, T]$ in probability P, and hence, by the Lévy–Itô–Nisio theorem ([45], Theorem 2.4), it also converges P-a.s. uniformly on $[0, T]$.

Basic properties of a Q-Wiener process are summarized in the next theorem.

Theorem 2.1 *A K-valued Q-Wiener process $\{W_t\}_{t \geq 0}$ has the following properties:*

(1) $W_0 = 0$.
(2) W_t *has continuous trajectories in K.*
(3) W_t *has independent increments.*
(4) W_t *is a Gaussian process with the covariance operator Q, i.e., for any $k, k' \in K$ and $s, t \geq 0$,*

$$E\left(W_t(k) W_s(k')\right) = (t \wedge s)\langle Qk, k'\rangle_K.$$

(5) *For an arbitrary $k \in K$, the law $\mathscr{L}((W_t - W_s)(k)) \sim N(0, (t - s)\langle Qk, k\rangle_K)$.*

Exercise 2.4 Consider a cylindrical Wiener process $\tilde{W}_t(k) = \sum_{j=1}^{\infty} \langle k, f_j\rangle_K \tilde{W}_t(f_j)$ and a Q-Wiener process $W_t = \sum_{j=1}^{\infty} \lambda_j^{1/2} w_j(t) f_j$, as defined in (2.2) and (2.3), respectively. Show that

(a) $W_t^1 = \tilde{W}_t \circ Q^{1/2} = \sum_{j=1}^{\infty} \lambda_j^{1/2} \tilde{W}_t(f_j) f_j$ defines a Q-Wiener process;
(b) $\tilde{W}_t^1(k) = \sum_{j=1}^{\infty} \langle k, f_j\rangle_K w_j(t)$ defines a cylindrical Wiener process.

2.1.3 Martingales in a Hilbert Space

Definition 2.7 Let H be a separable Hilbert space considered as a measurable space with its Borel σ-field $\mathscr{B}(H)$. We fix $T > 0$ and let $(\Omega, \mathscr{F}, \{\mathscr{F}_t\}_{t \leq T}, P)$ be a filtered probability space and $\{M_t\}_{t \leq T}$ be an H-valued process adapted to the filtration $\{\mathscr{F}_t\}_{t \leq T}$. Assume that M_t is integrable, $E\|M_t\|_H < \infty$. Then M_t is called a *martingale* if for any $0 \leq s \leq t$,

$$E(M_t | \mathscr{F}_s) = M_s, \quad P\text{-a.s.}$$

We note that because H is separable, the measurability of M_t with respect to the σ-fields $\mathscr{F}_t$ and $\mathscr{B}(H)$ is equivalent to the measurability of $\langle M_t, h\rangle_H$ with respect to $\mathscr{F}_t$ and $\mathscr{B}(\mathbb{R})$ for all $h \in H$, which implies the measurability of $\|M_t\|_H$. If $E\|M_t\|_H^p < \infty$, we will also write $M_t \in L^p(\Omega, \mathscr{F}_t, P)$. The condition for $M_t \in L^1(\Omega, \mathscr{F}_t, P)$ to be a martingale is equivalent to

$$E\left(\langle M_t, h\rangle_H | \mathscr{F}_s\right) = \langle M_s, h\rangle_H, \quad \forall h \in H, \quad P\text{-a.s.}$$

if $0 \leq s \leq t$.

If $M_t \in L^p(\Omega, \mathscr{F}, P)$ is an H-valued martingale than for $p \geq 1$, the process $\|M_t\|_H^p$ is a real-valued submartingale. We have therefore the following theorem.

Theorem 2.2 (Doob's Maximal Inequalities) *If $M_t \in L^p(\Omega, \mathscr{F}, P)$ is an H-valued martingale, then*

(1) $P(\sup_{0 \leq t \leq T} \|M_t\|_H > \lambda) \leq \frac{1}{\lambda^p} E \|M_T\|_H^p,\, p \geq 1,\, \lambda > 0;$
(2) $E(\sup_{0 \leq t \leq T} \|M_t\|_H^p) \leq \left(\frac{p}{p-1}\right)^p E \|M_T\|_H^p,\, p > 1.$

We will now introduce the Hilbert space of square-integrable H-valued martingales. Note that by Doob's inequality, Theorem 2.2, we have

$$E\left(\sup_{t \leq T} \|M_t\|_H^2\right) \leq 4E \|M_T\|_H^2.$$

Definition 2.8 A martingale $\{M_t\}_{0 \leq t \leq T}$ is called *square integrable* if

$$E \|M_T\|_H^2 < \infty.$$

The class of continuous square-integrable martingales will be denoted by $\mathscr{M}_T^2(H)$.

Since $M_t \in \mathscr{M}_T^2(H)$ is determined by the relation $M_t = E(M_T | \mathscr{F}_t)$, the space $\mathscr{M}_T^2(H)$ is a Hilbert space with scalar product

$$\langle M, N \rangle_{\mathscr{M}_T^2(H)} = E(\langle M_T, N_T \rangle_H).$$

In the case of real-valued martingales $M_t, N_t \in \mathscr{M}_T^2(\mathbb{R})$, there exist unique *quadratic variation* and *cross quadratic variation* processes, denoted by $\langle M \rangle_t$ and $\langle M, N \rangle_t$, respectively, such that $M_t^2 - \langle M \rangle_t$ and $M_t N_t - \langle M, N \rangle_t$ are continuous martingales. For Hilbert-space-valued martingales, we have the following definition.

Definition 2.9 Let $M_t \in \mathscr{M}_T^2(H)$. We denote by $\langle M \rangle_t$ the unique adapted continuous *increasing process* starting from 0 such that $\|M_t\|_H^2 - \langle M \rangle_t$ is a continuous martingale. We define a *quadratic variation process* $\langle\!\langle M \rangle\!\rangle_t$ of M_t as an adapted continuous process starting from 0, with values in the space of nonnegative definite trace-class operators on H, such that for all $h, g \in H$,

$$\langle M_t, h \rangle_H \langle M_t, g \rangle_H - \langle \langle\!\langle M \rangle\!\rangle_t(h), g \rangle_H$$

is a martingale.

Lemma 2.1 *The quadratic variation process of a martingale $M_t \in \mathscr{M}_T^2(H)$ exists and is unique. Moreover,*

$$\langle M \rangle_t = \operatorname{tr}\langle\!\langle M \rangle\!\rangle_t. \tag{2.4}$$

Proof The lemma follows by applying the classical one-dimensional results. We can assume without loss of generality that $M_0 = 0$. Denote

$$M_t^i = \langle M_t, e_i \rangle_H,$$

where $\{e_i\}_{i=1}^{\infty}$ is an ONB in H. Note that the quadratic variation process has to satisfy

$$\left\langle \langle\!\langle M \rangle\!\rangle_t(e_i), e_j \right\rangle_H = \langle M_i, M_j \rangle_t.$$

Consequently, we define the quadratic variation process by

$$\left\langle \langle\!\langle M \rangle\!\rangle_t(h), g \right\rangle_H = \sum_{i,j=1}^{\infty} \langle M_i, M_j \rangle_t \langle e_i, h \rangle_H \langle e_j, g \rangle_H. \tag{2.5}$$

The sum in (2.5) converges P-a.s. and defines a nonnegative definite trace-class operator on H, since

$$E \operatorname{tr} \langle\!\langle M \rangle\!\rangle_t = E \sum_{i=1}^{\infty} \langle M_i \rangle_t = E \sum_{i=1}^{\infty} \langle M_t, e_i \rangle_H^2 = E \| M_t \|_H^2 < \infty.$$

Now equality (2.4) follows from (2.5).   □

Exercise 2.5 Show that an H-valued Q-Wiener process $\{W_t\}_{t \leq T}$ is a continuous square-integrable martingale with $\langle W \rangle_t = t(\operatorname{tr} Q)$ and $\langle\!\langle W \rangle\!\rangle_t = tQ$.

Exercise 2.6 Let $0 < t_1, \ldots, t_n < t \leq T$ be a partition of the interval $[0, t]$, $t \leq T$, and $\max\{t_{j+1} - t_j, 1 \leq j \leq n-1\} \to 0$. Denote $\Delta W_j = W_{t_{j+1}} - W_{t_j}$. Show that

$$\sum_{j=1}^{n} \| \Delta W_j \|_K^2 \to t(\operatorname{tr} Q), \quad P\text{-a.s.}$$

2.2 Stochastic Integral with Respect to a Wiener Process

We will introduce the concept of Itô's stochastic integral with respect to a Q-Wiener process and with respect to a cylindrical Wiener process simultaneously.

Let K and H be separable Hilbert spaces, and Q be either a symmetric non-negative definite trace-class operator on K or $Q = I_K$, the identity operator on K. In case Q is trace-class, we will always assume that its all eigenvalues $\lambda_j > 0$, $j = 1, 2, \ldots$; otherwise we can start with the Hilbert space $\ker(Q)^{\perp}$ instead of K. The associated eigenvectors forming an ONB in K will be denoted by f_k.

Then the space $K_Q = Q^{1/2}K$ equipped with the scalar product

$$\langle u, v \rangle_{K_Q} = \sum_{j=1}^{\infty} \frac{1}{\lambda_j} \langle u, f_j \rangle_K \langle v, f_j \rangle_K$$

is a separable Hilbert space with an ONB $\{\lambda_j^{1/2} f_j\}_{j=1}^{\infty}$.

If H_1, H_2 are two real separable Hilbert spaces with $\{e_i\}_{i=1}^{\infty}$ an ONB in H_1, then the space of Hilbert–Schmidt operators from H_1 to H_2 is defined as

$$\mathscr{L}_2(H_1, H_2) = \left\{ L \in \mathscr{L}(H_1, H_2) : \sum_{i=1}^{\infty} \|L e_i\|_{H_2}^2 < \infty \right\}. \tag{2.6}$$

It is well known (see [68]) that $\mathscr{L}_2(H_1, H_2)$ equipped with the norm

$$\|L\|_{\mathscr{L}_2(H_1, H_2)} = \left(\sum_{i=1}^{\infty} \|L e_i\|_{H_2}^2 \right)^{1/2}$$

is a Hilbert space. Since the Hilbert spaces H_1 and H_2 are separable, the space $\mathscr{L}_2(H_1, H_2)$ is also separable, as Hilbert–Schmidt operators are limits of sequences of finite-dimensional linear operators.

Consider $\mathscr{L}_2(K_Q, H)$, the space of Hilbert–Schmidt operators from K_Q to H. If $\{e_j\}_{j=1}^{\infty}$ is an ONB in H, then the Hilbert–Schmidt norm of an operator $L \in \mathscr{L}_2(K_Q, H)$ is given by

$$\|L\|_{\mathscr{L}_2(K_Q, H)}^2 = \sum_{j,i=1}^{\infty} \langle L(\lambda_j^{1/2} f_j), e_i \rangle_H^2 = \sum_{j,i=1}^{\infty} \langle L Q^{1/2} f_j, e_i \rangle_H^2$$

$$= \|L Q^{1/2}\|_{\mathscr{L}_2(K, H)}^2 = \mathrm{tr}\big((L Q^{1/2})(L Q^{1/2})^*\big). \tag{2.7}$$

The scalar product between two operators $L, M \in \mathscr{L}_2(K_Q, H)$ is defined by

$$\langle L, M \rangle_{\mathscr{L}_2(K_Q, H)} = \mathrm{tr}\big((L Q^{1/2})(M Q^{1/2})^*\big). \tag{2.8}$$

Since the Hilbert spaces K_Q and H are separable, the space $\mathscr{L}_2(K_Q, H)$ is also separable.

Let $L \in \mathscr{L}(K, H)$. If $k \in K_Q$, then

$$k = \sum_{j=1}^{\infty} \langle k, \lambda_j^{1/2} f_j \rangle_{K_Q} \lambda_j^{1/2} f_j,$$

and L, considered as an operator from K_Q to H, defined as

$$Lk = \sum_{j=1}^{\infty} \langle k, \lambda_j^{1/2} f_j \rangle_{K_Q} \lambda_j^{1/2} L f_j,$$

has a finite Hilbert–Schmidt norm, since

$$\|L\|_{\mathscr{L}_2(K_Q, H)}^2 = \sum_{j=1}^{\infty} \|L(\lambda_j^{1/2} f_j)\|_H^2 = \sum_{j=1}^{\infty} \lambda_j \|L f_j\|_H^2 \leq \|L\|_{\mathscr{L}(K, H)}^2 \, \mathrm{tr}(Q).$$

Thus, $\mathscr{L}(K, H) \subset \mathscr{L}_2(K_Q, H)$. If $L, M \in \mathscr{L}(K, H)$, formulas (2.7) and (2.8) reduce to

$$\|L\|^2_{\mathscr{L}_2(K_Q, H)} = \mathrm{tr}\left(LQL^*\right) \qquad (2.9)$$

and

$$\langle L, M \rangle_{\mathscr{L}_2(K_Q, H)} = \mathrm{tr}\left(LQM^*\right), \qquad (2.10)$$

allowing for separation of $Q^{1/2}$ and L^*. This is usually exploited in calculations where $L \in \mathscr{L}_2(K_Q, H)$ is approximated with a sequence $L_n \in \mathscr{L}(K, H)$.

The space $\mathscr{L}_2(K_Q, H)$ consists of linear operators $L : K \to H$, not necessarily bounded, with domain $\mathscr{D}(L) \supset Q^{1/2}K$, and such that $\mathrm{tr}((LQ^{1/2})(LQ^{1/2})^*)$ is finite. If $Q = I_K$, then $K_Q = K$. We note that the space $\mathscr{L}_2(K_Q, H)$ contains genuinely unbounded linear operators from K to H.

Exercise 2.7 Give an example of an unbounded linear operator from K to H, which is an element of $\mathscr{L}_2(K_Q, H)$.

2.2.1 Elementary Processes

Let $\mathscr{E}(\mathscr{L}(K, H))$ denote the class of $\mathscr{L}(K, H)$-valued elementary processes adapted to the filtration $\{\mathscr{F}_t\}_{t \leq T}$ that are of the form

$$\Phi(t, \omega) = \phi(\omega) 1_{\{0\}}(t) + \sum_{j=0}^{n-1} \phi_j(\omega) 1_{(t_j, t_{j+1}]}(t), \qquad (2.11)$$

where $0 = t_0 \leq t_1 \leq \cdots \leq t_n = T$, and ϕ, ϕ_j, $j = 0, 1, \ldots, n - 1$, are respectively $\mathscr{F}_0$-measurable and $\mathscr{F}_{t_j}$-measurable $\mathscr{L}_2(K_Q, H)$-valued random variables such that $\phi(\omega), \phi_j(\omega) \in \mathscr{L}(K, H)$, $j = 0, 1, \ldots, n - 1$ (recall that $\mathscr{L}(K, H) \subset \mathscr{L}_2(K_Q, H)$).

Note that if $Q = I_K$, then the random variables ϕ_j are, in fact, $\mathscr{L}_2(K, H)$-valued.

We shall say that an elementary process $\Phi \in \mathscr{E}(\mathscr{L}(K, H))$ is bounded if it is bounded in $\mathscr{L}_2(K_Q, H)$.

We have defined elementary processes to be left-continuous as opposed to being right-continuous. There is no difference if the Itô stochastic integral is constructed with respect to a Wiener process. Our choice, however, is consistent with the construction of a stochastic integral with respect to square-integrable martingales.

2.2.2 Stochastic Itô Integral for Elementary Processes

For an elementary process $\Phi \in \mathscr{E}(\mathscr{L}(K, H))$, we define the Itô stochastic integral with respect to a Q-Wiener process W_t by

$$\int_0^t \Phi(s)\, dW_s = \sum_{j=0}^{n-1} \phi_j (W_{t_{j+1} \wedge t} - W_{t_j \wedge t})$$

for $t \in [0, T]$. The term ϕW_0 is neglected since $P(W_0 = 0) = 1$. This stochastic integral is an H-valued stochastic process.

We define the Itô cylindrical stochastic integral of an elementary process $\Phi \in \mathscr{E}(\mathscr{L}(K, H))$ with respect to a cylindrical Wiener process $\tilde{W}$ by

$$\left(\int_0^t \Phi(s)\, d\tilde{W}_s \right)(h) = \sum_{j=0}^{n-1} \left(\tilde{W}_{t_{j+1} \wedge t}\left(\phi_j^*(h)\right) - \tilde{W}_{t_j \wedge t}\left(\phi_j^*(h)\right) \right) \qquad (2.12)$$

for $t \in [0, T]$ and $h \in H$. The following proposition states Itô's isometry, which is essential in furthering the construction of the stochastic integral.

Proposition 2.1 *For a bounded elementary process $\Phi \in \mathscr{E}(\mathscr{L}(K, H))$,*

$$E \left\| \int_0^t \Phi(s)\, dW_s \right\|_H^2 = E \int_0^t \|\Phi(s)\|_{\mathscr{L}_2(K_Q, H)}^2\, ds < \infty \qquad (2.13)$$

for $t \in [0, T]$.

Proof The proof resembles calculations in the real case. Without loss of generality we can assume that $t = T$, then

$$E \left\| \int_0^T \Phi(s)\, dW_s \right\|_H^2 = E \left\| \sum_{j=1}^{n-1} \phi_j (W_{t_{j+1}} - W_{t_j}) \right\|_H^2$$

$$= E \left(\sum_{j=1}^{n-1} \|\phi_j (W_{t_{j+1}} - W_{t_j})\|_H^2 + \sum_{i \neq j=1}^{n-1} \langle \phi_j (W_{t_{j+1}} - W_{t_j}), \phi_i (W_{t_{i+1}} - W_{t_i}) \rangle_H \right).$$

Consider first the single term $E \|\phi_j (W_{t_{j+1}} - W_{t_j})\|_H^2$. We use the fact that the random variable ϕ_j and consequently for a vector $e_m \in H$, the random variable $\phi_j^* e_m$ is $\mathscr{F}_{t_j}$-measurable, while the increment $W_{t_{j+1}} - W_{t_j}$ is independent of this σ-field. With $\{f_l\}_{l=1}^\infty$ and $\{e_m\}_{m=1}^\infty$, orthonormal bases in Hilbert spaces K and H, we have

$$E \|\phi_j (W_{t_{j+1}} - W_{t_j})\|_H^2 = E \sum_{m=1}^\infty \langle \phi_j (W_{t_{j+1}} - W_{t_j}), e_m \rangle_H^2$$

$$= \sum_{m=1}^\infty E \left(E \left(\langle \phi_j (W_{t_{j+1}} - W_{t_j}), e_m \rangle_H^2 \, \big| \, \mathscr{F}_{t_j} \right) \right)$$

$$= \sum_{m=1}^{\infty} E\left(E\left(\langle W_{t_{j+1}} - W_{t_j}, \phi_j^* e_m \rangle_K^2 \,\middle|\, \mathscr{F}_{t_j}\right)\right)$$

$$= \sum_{m=1}^{\infty} E\left(E\left(\left(\sum_{l=1}^{\infty} \langle W_{t_{j+1}} - W_{t_j}, f_l \rangle_K \langle \phi_j^* e_m, f_l \rangle_K\right)^2 \,\middle|\, \mathscr{F}_{t_j}\right)\right)$$

$$= \sum_{m=1}^{\infty} E\left(E\left(\left(\sum_{l=1}^{\infty} \langle W_{t_{j+1}} - W_{t_j}, f_l \rangle_K^2 \langle \phi_j^* e_m, f_l \rangle_K^2\right) \,\middle|\, \mathscr{F}_{t_j}\right)\right)$$

$$+ \sum_{m=1}^{\infty} E\left(E\left(\left(\sum_{l \neq l'=1}^{\infty} \langle W_{t_{j+1}} - W_{t_j}, f_l \rangle_K \langle \phi_j^* e_m, f_l \rangle_K \right.\right.\right.$$

$$\times \langle W_{t_{j+1}} - W_{t_j}, f_{l'} \rangle_K \langle \phi_j^* e_m, f_{l'} \rangle_K\bigg)\,\bigg|\, \mathscr{F}_{t_j}\bigg)\bigg)$$

$$= \sum_{m=1}^{\infty} (t_{j+1} - t_j) \sum_{l=1}^{\infty} \lambda_l \langle \phi_j^* e_m, f_l \rangle_K^2$$

$$= (t_{j+1} - t_j) \sum_{m,l=1}^{\infty} \langle \phi_j(\lambda_l^{1/2} f_l), e_m \rangle_H^2 = (t_{j+1} - t_j) \|\phi_j\|_{\mathscr{L}_2(K_Q, H)}^2.$$

Similarly, for the single term $E\langle \phi_j (W_{t_{j+1}} - W_{t_j}), \phi_i(W_{t_{i+1}} - W_{t_i}) \rangle_H$, we obtain that

$$E\langle \phi_j (W_{t_{j+1}} - W_{t_j}), \phi_i(W_{t_{i+1}} - W_{t_i}) \rangle_H$$

$$= E \sum_{m=1}^{\infty} E\left(\sum_{l,l'=1}^{\infty} \langle W_{t_{j+1}} - W_{t_j}, f_l \rangle_K \langle \phi_j^* e_m, f_l \rangle_K \right.$$

$$\times \langle W_{t_{i+1}} - W_{t_i}, f_{l'} \rangle_K \langle \phi_i^* e_m, f_{l'} \rangle_K \,\bigg|\, \mathscr{F}_{t_j}\bigg) = 0$$

if $i < j$. Now we can easily reach the conclusion. $\qquad\square$

We have the following counterpart of (2.13) for the Itô cylindrical stochastic integral of a bounded elementary process $\Phi \in \mathscr{E}(\mathscr{L}(K, H))$:

$$E\left(\left(\int_0^t \Phi(s)\, d\tilde{W}_s\right)(h)\right)^2 = \int_0^t E\|\Phi^*(s)(h)\|_K^2\, ds < \infty. \qquad (2.14)$$

Exercise 2.8 Prove (2.14).

The idea now is to extend the definition of the Itô stochastic integral and cylindrical stochastic integral to a larger class of stochastic processes utilizing the fact that the mappings $\Phi \to \int_0^T \Phi(s)\,dW_s$ and $\Phi^*(\cdot)(h) \to (\int_0^T \Phi(s)\,d\tilde{W}_s)(h)$ are isometries due to. (2.13) and (2.14).

When the stochastic integral with respect to martingales is constructed, then the natural choice of the class of integrands is the set of predictable processes (see, for example, [57]), and sometimes this restriction is applied when the martingale is a Wiener process. We will however carry out a construction which will allow the class of integrands to be simply adapted and not necessarily predictable processes.

Let $\Lambda_2(K_Q, H)$ be a class of $\mathscr{L}_2(K_Q, H)$-valued processes measurable as mappings from $([0, T] \times \Omega, \mathscr{B}([0, T]) \otimes \mathscr{F})$ to $(\mathscr{L}_2(K_Q, H), \mathscr{B}(\mathscr{L}_2(K_Q, H)))$, adapted to the filtration $\{\mathscr{F}_t\}_{t \leq T}$ (thus $\mathscr{F}$ can be replaced with $\mathscr{F}_T$), and satisfying the condition

$$E \int_0^T \|\Phi(t)\|_{\mathscr{L}_2(K_Q, H)}^2 \, dt < \infty. \tag{2.15}$$

Obviously, elementary processes satisfying the above condition (2.15) are elements of Λ_2.

We note that $\Lambda_2(K_Q, H)$ equipped with the norm

$$\|\Phi\|_{\Lambda_2(K_Q, H)} = \left(E \int_0^T \|\Phi(t)\|_{\mathscr{L}_2(K_Q, H)}^2 \, dt \right)^{1/2} \tag{2.16}$$

is a Hilbert space. The following proposition shows that the class of bounded elementary processes is dense in $\Lambda_2(K_Q, H)$. It is valid for Q, a trace-class operator, and for $Q = I_K$.

Proposition 2.2 *If $\Phi \in \Lambda_2(K_Q, H)$, then there exists a sequence of bounded elementary processes $\Phi_n \in \mathscr{E}(\mathscr{L}(K, H))$ approximating Φ in $\Lambda_2(K_Q, H)$, i.e.,*

$$\|\Phi_n - \Phi\|_{\Lambda_2(K_Q, H)}^2 = E \int_0^T \|\Phi_n(t) - \Phi(t)\|_{\mathscr{L}_2(K_Q, H)}^2 \, dt \to 0$$

as $n \to \infty$.

Proof We follow the idea in [61].

(a) We can assume that $\|\Phi(t, \omega)\|_{\mathscr{L}_2(K_Q, H)} < M$ for all t, ω. Otherwise, we define

$$\Phi_n(t, \omega) = \begin{cases} n \dfrac{\Phi(t, \omega)}{\|\Phi(t, \omega)\|_{\mathscr{L}_2(K_Q, H)}} & \text{if } \|\Phi(t, \omega)\|_{\mathscr{L}_2(K_Q, H)} > n, \\ \Phi(t, \omega) & \text{otherwise.} \end{cases}$$

Then $\|\Phi_n - \Phi\|_{\Lambda_2(K_Q, H)} \to 0$ by the Lebesgue dominated convergence theorem (DCT).

(b) We can assume that $\Phi(t, \omega) \in \mathcal{L}(K, H)$ since every operator $L \in \mathcal{L}_2(K_Q, H)$ can be approximated in $\mathcal{L}_2(K_Q, H)$ by the operators $L_n \in \mathcal{L}(K, H)$ defined by

$$L_n k = \sum_{j=1}^{n} L(\lambda_j^{1/2} f_j)\langle \lambda_j^{1/2} f_j, k \rangle_{K_Q}.$$

Indeed, for $k \in K$, we have

$$\|L_n k\|_H^2 = \sum_{i=1}^{\infty} \left\langle \sum_{j=1}^{n} L(\lambda_j^{1/2} f_j)\langle \lambda_j^{1/2} f_j, k \rangle_{K_Q}, e_i \right\rangle_H^2$$

$$\leq \sum_{i,j=1}^{\infty} \langle L(\lambda_j^{1/2} f_j), e_i \rangle_H^2 \sum_{j=1}^{n} \frac{1}{\lambda_j} \langle f_j, k \rangle_K^2$$

$$\leq C_n \|L\|_{\mathcal{L}_2(K_Q, H)}^2 \|k\|_K^2,$$

so that $L_n \in \mathcal{L}(K, H)$, and

$$\left\| \Phi(t) - \Phi_n(t) \right\|_{\mathcal{L}_2(K_Q, H)}^2 = \sum_{j=n+1}^{\infty} \left\| \Phi(t)(\lambda_j^{1/2} f_j) \right\|_H^2 \to 0.$$

Then

$$E \int_0^T \left\| \Phi(t) - \Phi_n(t) \right\|_{\mathcal{L}_2(K_Q, H)}^2 dt \to 0$$

as $n \to \infty$ by the Lebesgue DCT, so that $\Phi_n \to \Phi$ in $\Lambda_2(K_Q, H)$.

(c) Now assume, in addition to (a) and (b), that for each ω, the function $\Phi(\cdot, \omega) : [0, T] \to \mathcal{L}_2(K_Q, H)$ is continuous. For a partition $0 = t_0 \leq t_1 \leq \cdots \leq t_n = T$, define the elementary process

$$\Phi_n(t, \omega) = \Phi(0, \omega) 1_{\{0\}}(t) + \sum_{j=1}^{n-1} \Phi(t_j, \omega) 1_{(t_j, t_{j+1}]}(t).$$

With the size of the partition $\max\{|t_{j+1} - t_j| : j = 0, \ldots, n\} \to 0$ as $n \to \infty$, we have that $\Phi_n(t, \omega) \to \Phi(t, \omega)$ and

$$\int_0^T \left\| \Phi(t, \omega) - \Phi_n(t, \omega) \right\|_{\mathcal{L}_2(K_Q, H)}^2 dt \to 0$$

due to the continuity of $\Phi(\cdot, \omega)$. Consequently, due to the Lebesgue DCT,

$$E \int_0^T \left\| \Phi(t, \omega) - \Phi_n(t, \omega) \right\|_{\mathcal{L}_2(K_Q, H)}^2 dt \to 0.$$

(d) If $\Phi(t, \omega) \in \mathscr{L}(K, H)$ is bounded for every (t, ω) but not necessarily continuous, then we first extend Φ to the entire $\mathbb{R}$ by assigning $\Phi(t, \omega) = 0$ for $t < 0$ and
$t > T$. Next, we define bounded continuous approximations of Φ by

$$\Phi_n(t, \omega) = \int_0^t \psi_n(s - t)\Phi(s, \omega)\, ds, \quad 0 \le t \le T.$$

Here $\psi_n(t) = n\psi(nt)$, and $\psi(t)$ is any nonnegative bounded continuous function on
$\mathbb{R}$ with support in $[-1, 0]$ and such that $\int_{\mathbb{R}} \psi(t)\, dt = 1$. We will use the technique
of an approximate identity (refer to [73], Chap. 9).

The functions $\Phi_n(t, \omega)$ are clearly bounded, and we now justify their continuity.
With $t + h \le T$, we have

$$\left\| \Phi_n(t + h, \omega) - \Phi_n(t, \omega) \right\|_{\mathscr{L}_2(K_Q, H)}$$

$$= \left\| \int_0^{t+h} \psi_n\big(s - (t + h)\big)\Phi(s, \omega)\, ds - \int_0^t \psi_n(s - t)\Phi(s, \omega)\, ds \right\|_{\mathscr{L}_2(K_Q, H)}$$

$$\le \left\| \int_0^t \big(\psi_n(s - (t + h)) - \psi_n(s - t)\big)\Phi(s, \omega)\, ds \right\|_{\mathscr{L}_2(K_Q, H)}$$

$$+ \left\| \int_0^T \psi_n\big(s - (t + h)\big)\Phi(s, \omega)1_{[t, t+h]}(s)\, ds \right\|_{\mathscr{L}_2(K_Q, H)}.$$

The second integral converges to zero with $h \to 0$ by the Lebesgue DCT. The first
integral is dominated by

$$\int_{\mathbb{R}} \big|\psi_n(s - (t + h)) - \psi_n(s - t)\big| \big\|\Phi(s, \omega)\big\|_{\mathscr{L}_2(K_Q, H)} 1_{[0, t]}(s)\, ds$$

$$= \int_{\mathbb{R}} \big|\psi_n(u + h) - \psi_n(u)\big| \big\|\Phi(u + t, \omega)\big\|_{\mathscr{L}_2(K_Q, H)} 1_{[-t, 0]}(u)\, du$$

$$= \left(\int_{\mathbb{R}} \big|\psi_n(u + h) - \psi_n(u)\big|^2 du \right)^{1/2}$$

$$\times \left(\int_{\mathbb{R}} \big\|\Phi(u + t, \omega))\big\|_{\mathscr{L}_2(K_Q, H)}^2 1_{[-t, 0]}(u)\, du \right)^{1/2},$$

so that it converges to zero due to the continuity of the shift operator in $L^2(\mathbb{R})$ (see
Theorem 8.19 in [73]). Left continuity in T follows by a similar argument.

Since the process $\Phi(t, \omega)$ is adapted to the filtration $\mathscr{F}_t$, we deduce from the
definition of $\Phi_n(t, \omega)$ that it is also $\mathscr{F}_t$-adapted.

We will now show that

$$\int_0^T \big\|\Phi_n(t, \omega) - \Phi(t, \omega)\big\|_{\mathscr{L}_2(K_Q, H)}^2\, dt \to 0$$

as $n \to \infty$. Consider

$$\left\| \Phi_n(t,\omega) - \Phi(t,\omega) \right\|_{\mathscr{L}_2(K_Q,H)} \leq \left\| \int_0^t \psi_n(s-t)\big(\Phi(s,\omega) - \Phi(t,\omega)\big)\,ds \right\|_{\mathscr{L}_2(K_Q,H)}$$

$$+ \left\| \int_0^t \big(\psi_n(s-t) - 1\big)\Phi(t,\omega)\,ds \right\|_{\mathscr{L}_2(K_Q,H)}.$$

For a fixed ω, denote $w_n(t) = \| \int_0^t (\psi_n(s-t) - 1)\Phi(t,\omega)\,ds \|_{\mathscr{L}_2(K_Q,H)}$. Then $w_n(t)$ converges to zero for every t as $n \to \infty$ and is bounded by $C\|\Phi(t,\omega)\|_{\mathscr{L}_2(K_Q,H)}$ for some constant C. From now on, the constant C can change its value from line to line. The first integral is dominated by

$$\int_{\mathbb{R}} \psi_n(s-t)\|\Phi(s,\omega) - \Phi(t,\omega)\|_{\mathscr{L}_2(K_Q,H)} 1_{[0,t]}(s)\,ds$$

$$= \int_{\mathbb{R}} \psi_n(u)\|\Phi(u+t,\omega) - \Phi(t,\omega)\|_{\mathscr{L}_2(K_Q,H)} 1_{[-t,0]}(u)\,du$$

$$\leq \int_{\mathbb{R}} \|\Phi(u+t,\omega) - \Phi(t,\omega)\|_{\mathscr{L}_2(K_Q,H)} \psi_n^{1/2}(u)\psi_n^{1/2}(u) 1_{[-t,0]}(u)\,du$$

$$\leq \left(\int_{\mathbb{R}} \|\Phi(u+t,\omega) - \Phi(t,\omega)\|_{\mathscr{L}_2(K_Q,H)}^2 \psi_n(u) 1_{[-t,0]}(u)\,du \right)^{1/2}$$

by the Schwarz–Bunyakovsky inequality and the property that $\psi_n(t)$ integrates to one.

We arrive at

$$\int_0^T \|\Phi_n(t,\omega) - \Phi(t,\omega)\|_{\mathscr{L}_2(K_Q,H)}^2\,dt$$

$$\leq 2 \int_0^T \int_{\mathbb{R}} \|\Phi(u+t,\omega) - \Phi(t,\omega)\|_{\mathscr{L}_2(K_Q,H)}^2 \psi_n(u) 1_{[-t,0]}(u)\,du\,dt$$

$$+ 2 \int_0^T w_n^2(t,\omega)\,dt.$$

The second integral converges to zero as $n \to \infty$ by the Lebesgue DCT and is dominated by $C \int_0^T \|\Phi(t,\omega)\|_{\mathscr{L}_2(K_Q,H)}^2\,dt$. Now we change the order of integration in the first integral:

$$\int_0^T \int_{\mathbb{R}} \|\Phi(u+t,\omega) - \Phi(t,\omega)\|_{\mathscr{L}_2(K_Q,H)}^2 \psi_n(u) 1_{[-t,0]}(u)\,du\,dt$$

$$= \int_{\mathbb{R}} \psi_n(u) \int_0^T 1_{[-t,0]}(u)\|\Phi(u+t,\omega) - \Phi(t,\omega)\|_{\mathscr{L}_2(K_Q,H)}^2\,dt\,du$$

$$= \int_{\mathbb{R}} \psi(v) \int_0^T 1_{[-t,0]}\left(\frac{v}{n}\right)\left\|\Phi\left(\frac{v}{n}+t,\omega\right) - \Phi(t,\omega)\right\|_{\mathscr{L}_2(K_Q,H)}^2\,dt\,dv.$$

We note that, again by the continuity of the shift operator in $L^2([0, T], \mathscr{L}_2(K_Q, H))$,

$$\int_0^T 1_{[-t,0]}\left(\frac{v}{n}\right)\left\|\Phi\left(\frac{v}{n}+t, \omega\right) - \Phi(t, \omega)\right\|_{\mathscr{L}_2(K_Q,H)}^2 dt$$

$$\leq \int_0^T \left\|\Phi\left(\frac{v}{n}+t, \omega\right) - \Phi(t, \omega)\right\|_{\mathscr{L}_2(K_Q,H)}^2 dt \to 0$$

as $n \to \infty$, so that the function

$$\psi(v) \int_0^T 1_{[-t,0]}\left(\frac{v}{n}\right)\left\|\Phi\left(\frac{v}{n}+t, \omega\right) - \Phi(t, \omega)\right\|_{\mathscr{L}_2(K_Q,H)}^2 dt$$

converges to zero and is bounded by $C\|\Phi(\cdot, \omega)\|_{L^2([0,T],\mathscr{L}_2(K_Q,H))}^2$. This proves that

$$r_n(\omega) = \int_0^T \left\|\Phi_n(t, \omega) - \Phi(t, \omega)\right\|_{\mathscr{L}_2(K_Q,H)}^2 dt \to 0$$

as $n \to \infty$ and $r_n(\omega) \leq C\|\Phi(t, \omega)\|_{L^2([0,T],\mathscr{L}_2)(K_Q,H)}^2$. Therefore,

$$E \int_0^T \left\|\Phi_n(t, \omega) - \Phi(t, \omega)\right\|_{\mathscr{L}_2(K_Q,H)}^2 dt = Er_n(\omega) \to 0$$

as $n \to \infty$ by the Lebesgue DCT.   □

We will need the following lemma when using the Yosida approximation of an unbounded operator.

Lemma 2.2 (a) *Let* $T, T_n \in \mathscr{L}(H)$ *be such that for every* $h \in H$, $T_n h \to Th$, *and let* $L \in \mathscr{L}_2(K_Q, H)$. *Then*

$$\|T_n L - TL\|_{\mathscr{L}_2(K_Q,H)} \to 0. \tag{2.17}$$

(b) *Let* A *be the generator of a* C_0-*semigroup* $S(t)$ *on a real separable Hilbert space* H, *and* $A_n = AR_n$ *be the Yosida approximations of* A *as defined in* (1.22). *Then, for* $\Phi(t) \in \Lambda_2(K_Q, H)$ *such that* $E \int_0^T \|\Phi(t)\|_{\mathscr{L}_2(K_Q,H)}^{2p} dt < \infty$, $p \geq 1$,

$$\lim_{n\to\infty} \sup_{t\leq T} E \int_0^t \left\|\left(e^{(t-s)A_n} - S(t-s)\right)\Phi(s)\right\|_{\mathscr{L}_2(K_Q,H)}^{2p} ds \to 0. \tag{2.18}$$

(c) *Under the assumptions in part* (b),

$$\lim_{n\to\infty} E \int_0^t \sup_{0\leq s\leq T} \left\|\left(e^{sA_n} - S(s)\right)\Phi(s)\right\|_{\mathscr{L}_2(K_Q,H)}^{2p} ds \to 0.$$

Proof (a) Note that by the Banach–Steinhaus theorem,

$$\max\left\{\|T\|_{\mathscr{L}(H)}, \sup_n \|T_n\|_{\mathscr{L}(H)}\right\} < C$$

for some constant C. Hence, for an ONB $\{f_j\}_{j=1}^{\infty} \subset K$, we have

$$\|T_n L - TL\|_{\mathscr{L}_2(K_Q, H)}^2 = \sum_{j=1}^{\infty} \|(T_n - T)LQ^{1/2} f_j\|_H^2$$

$$\leq \sum_{j=1}^{\infty} \|T_n - T\|_{\mathscr{L}(H)}^2 \|LQ^{1/2} f_j\|_H^2$$

$$\leq \sum_{j=1}^{\infty} 4C^2 \|LQ^{1/2} f_j\|_H^2 = 4C^2 \|L\|_{\mathscr{L}_2(K_Q, H)}^2.$$

The terms $\|(T_n - T)LQ^{1/2} f_j\|_H^2$ converge to zero as $n \to \infty$ and are bounded by $4C^2 \|LQ^{1/2} f_j\|_H^2$, so that (a) follows by the Lebesgue DCT with respect to the counting measure δ_j.

(b) We will use two facts about the semigroup $S_n(t) = e^{tA_n}$. By Proposition 1.2, we have

$$\sup_n \|S_n(t)\|_{\mathscr{L}(H)} \leq M e^{2\alpha t}, \quad n > 2\alpha,$$

and $\lim_{n\to\infty} S_n(t)x = S(t)x$, $x \in H$, uniformly on finite intervals.

Now, for $n > 2\alpha$,

$$\sup_{0 \leq t \leq T} E \int_0^t \|(S_n(t-s) - S(t-s))\Phi(s)\|_{\mathscr{L}_2(K_Q, H)}^{2p} \, ds$$

$$= \sup_{0 \leq t \leq T} E \int_0^t \left(\sum_{j=1}^{\infty} \|(S_n(t-s) - S(t-s))\Phi(s)Q^{1/2} f_j\|_H^2\right)^p \, ds$$

$$= \sup_{0 \leq t \leq T} E \int_0^T \left(\sum_{j=1}^{\infty} \|(S_n(t-s) - S(t-s))\Phi(s)Q^{1/2} f_j\|_H^2 1_{[0,t]}(s)\right)^p \, ds$$

$$\leq E \int_0^T \left[\sum_{j=1}^{\infty} \left(\sup_{0 \leq t \leq T} \{\|(S_n(t-s) - S(t-s))\Phi(s)Q^{1/2} f_j\|_H^2 1_{[0,t]}(s)\}\right)\right]^p \, ds$$

$$\leq E \int_0^T \left[\sum_{j=1}^{\infty} \left(\sup_{n > 2\alpha} \sup_{0 \leq t \leq T} \{\|(S_n(t-s) - S(t-s))\Phi(s)Q^{1/2} f_j\|_H^2 1_{[0,t]}(s)\}\right)\right]^p \, ds$$

$$\leq E \int_0^T \left[\sum_{j=1}^{\infty} \left(4M^2 e^{4\alpha T} \|\Phi(s)Q^{1/2}f_j\|_H^2\right) \right]^p ds$$

$$= \left(4M^2 e^{4\alpha T}\right)^p E \int_0^T \|\Phi(s)\|_{\mathscr{L}_2(K_Q,H)}^{2p} ds < \infty.$$

The term

$$\sup_{0 \leq t \leq T} \left\{ \left\| \left(S_n(t-s) - S(t-s)\right)\Phi(s)Q^{1/2}f_j \right\|_H^2 1_{[0,t]}(s) \right\} \to 0$$

as $n \to \infty$ and is bounded by $4M^2 e^{4\alpha T} \|\Phi(s)Q^{1/2}f_j\|_H^2$; hence, (b) follows by the Lebesgue DCT relative to the counting measure δ_j and then relative to $dP \otimes dt$.

(c) The proof of part (c) follows the arguments in the proof of part (b). □

2.2.3 Stochastic Itô Integral with Respect to a Q-Wiener Process

We are ready to extend the definition of the Itô stochastic integral with respect to a Q-Wiener process to adapted stochastic processes $\Phi(s)$ satisfying the condition

$$E \int_0^T \|\Phi(s)\|_{\mathscr{L}_2(K_Q,H)}^2 ds < \infty,$$

which will be further relaxed to the condition

$$P\left(\int_0^T \|\Phi(s)\|_{\mathscr{L}_2(K_Q,H)}^2 ds < \infty \right) = 1.$$

Definition 2.10 The *stochastic integral* of a process $\Phi \in \Lambda_2(K_Q, H)$ with respect to a K-valued Q-Wiener process W_t is the unique isometric linear extension of the mapping

$$\Phi(\cdot) \to \int_0^T \Phi(s)\,dW_s$$

from the class of bounded elementary processes to $L^2(\Omega, H)$, to a mapping from $\Lambda_2(K_Q, H)$ to $L^2(\Omega, H)$, such that the image of $\Phi(t) = \phi 1_{\{0\}}(t) + \sum_{j=0}^{n-1} \phi_j 1_{(t_j, t_{j+1}]}(t)$ is $\sum_{j=0}^{n-1} \phi_j (W_{t_{j+1}} - W_{t_j})$. We define the stochastic integral process $\int_0^t \Phi(s)\,dW_s$, $0 \leq t \leq T$, for $\Phi \in \Lambda_2(K_Q, H)$ by

$$\int_0^t \Phi(s)\,dW_s = \int_0^T \Phi(s)1_{[0,t]}(s)\,dW_s.$$

Theorem 2.3 *The stochastic integral* $\Phi \to \int_0^t \Phi(s)\,dW_s$ *with respect to a* K-valued Q-Wiener process W_t is an isometry between $\Lambda_2(K_Q, H)$ and the space of contin-

uous square-integrable martingales $\mathscr{M}_T^2(H)$,

$$E\left\|\int_0^t \Phi(s)\,dW_s\right\|_H^2 = E\int_0^t \|\Phi(s)\|_{\mathscr{L}_2(K_Q,H)}^2\,ds < \infty \qquad (2.19)$$

for $t \in [0,T]$.

The quadratic variation process of the stochastic integral process $\int_0^t \Phi(s)\,dW_s$ and the increasing process related to $\|\int_0^t \Phi(s)\,dW_s\|_H^2$ are given by

$$\left\langle\!\left\langle \int_0^\cdot \Phi(s)\,dW_s \right\rangle\!\right\rangle_t = \int_0^t \left(\Phi(s)Q^{1/2}\right)\left(\Phi(s)Q^{1/2}\right)^*\,ds$$

and

$$\left\langle \int_0^\cdot \Phi(s)\,dW_s \right\rangle_t = \int_0^t \mathrm{tr}\left(\left(\Phi(s)Q^{1/2}\right)\left(\Phi(s)Q^{1/2}\right)^*\right)\,ds$$

$$= \int_0^t \|\Phi(s)\|_{\mathscr{L}_2(K_Q,H)}^2\,ds.$$

Proof We note that the stochastic integral process for a bounded elementary process in $\mathscr{E}(\mathscr{L}(K,H))$ is a continuous square-integrable martingale. Let the sequence of bounded elementary processes $\{\Phi_n\}_{n=1}^\infty \subset \mathscr{E}(\mathscr{L}(K,H))$ approximate $\Phi \in \Lambda_2(K_Q,H)$. We can assume that $\Phi_1 = 0$ and

$$\|\Phi_{n+1} - \Phi_n\|_{\Lambda_2(K_Q,H)} < \frac{1}{2^n}. \qquad (2.20)$$

Then by Doob's inequality, Theorem 2.2, we have

$$\sum_{n=1}^\infty P\left(\sup_{t\le T}\left\|\int_0^t \Phi_{n+1}(s)\,dW_s - \int_0^t \Phi_n(s)\,dW_s\right\|_H > \frac{1}{n^2}\right)$$

$$\le \sum_{n=1}^\infty n^4 E\left\|\int_0^T \left(\Phi_{n+1}(s) - \Phi_n(s)\right)dW_s\right\|_H^2$$

$$= \sum_{n=1}^\infty n^4 E\int_0^T \|\Phi_{n+1}(s) - \Phi_n(s)\|_{\mathscr{L}_2(K_Q,H)}^2\,ds \le \sum_{n=1}^\infty \frac{n^4}{2^n}.$$

By the Borel–Cantelli lemma,

$$\sup_{t\le T}\left\|\int_0^t \Phi_{n+1}(s)\,dW_s - \int_0^t \Phi_n(s)\,dW_s\right\|_H \le \frac{1}{n^2}, \quad n > N(\omega),$$

for some $N(\omega)$, P-a.s. Consequently, the series

$$\sum_{n=1}^\infty \left(\int_0^t \Phi_{n+1}(s)\,dW_s - \int_0^t \Phi_n(s)\,dW_s\right)$$

converges to $\int_0^t \Phi(s)\,dW_s$ in $L^2(\Omega, H)$ for every $t \leq T$ and converges P-a.s. as a series of H-valued continuous functions to a continuous version of $\int_0^t \Phi(s)\,dW_s$.

Thus, the mapping $\Phi \to \int_0^t \Phi(s)\,dW_s$ is an isometry from the subset of bounded elements in $\mathscr{E}(\mathscr{L}(K, H))$ into the space of continuous square-integrable martingales $\mathscr{M}_T^2(H)$, and it extends to $\Lambda_2(K_Q, H)$ with images in $\mathscr{M}_T^2(H)$ by the completeness argument. We only need to verify the formula for the quadratic variation process. Note (see Exercise 2.9) that using representation (2.3), we have for $h \in H$,

$$\left\langle \int_0^t \Phi(s)\,dW_s, h \right\rangle_H = \sum_{j=1}^{\infty} \int_0^t \langle \lambda_j^{1/2} \Phi(s) f_j, h \rangle_H \, dw_j(t),$$

with the series convergent in $L^2(\Omega, \mathbb{R})$. If $h, g \in H$, then

$$\langle (\Phi(s)Q^{1/2})(\Phi(s)Q^{1/2})^* h, g \rangle_H = \sum_{j=1}^{\infty} \langle h, \Phi(s)Q^{1/2} f_j \rangle_H \langle g, \Phi(s)Q^{1/2} f_j \rangle_H$$

$$= \sum_{j=1}^{\infty} \lambda_j \langle h, \Phi(s) f_j \rangle_H \langle g, \Phi(s) f_j \rangle_H.$$

Now, for $0 \leq u \leq t$,

$$E\left(\left\langle \int_0^t \Phi(s)\,dW_s, h \right\rangle_H \left\langle \int_0^t \Phi(s)\,dW_s, g \right\rangle_H \right.$$

$$\left. - \left\langle \left(\int_0^t (\Phi(s)Q^{1/2})(\Phi(s)Q^{1/2})^* \, ds \right)(h), g \right\rangle_H \Big| \mathscr{F}_u \right)$$

$$= E\left(\sum_{i=1}^{\infty} \int_0^t \lambda_i^{1/2} \langle \Phi(s) f_i, h \rangle_H \, dw_i(s) \sum_{j=1}^{\infty} \int_0^t \lambda_j^{1/2} \langle \Phi(s) f_j, g \rangle_H \, dw_j(s) \right.$$

$$\left. - \sum_{j=1}^{\infty} \int_0^t \lambda_j \langle h, \Phi(s) f_j \rangle_H \langle g, \Phi(s) f_j \rangle_H \, ds \Big| \mathscr{F}_u \right)$$

$$= E\left(\sum_{j=1}^{\infty} \left(\lambda_j \int_0^t \langle \Phi(s) f_j, h \rangle_H \, dw_j(s) \int_0^t \langle \Phi(s) f_j, g \rangle_H \, dw_j(s) \right) \right.$$

$$\left. - \sum_{j=1}^{\infty} \int_0^t \lambda_j \langle h, \Phi(s) f_j \rangle_H \langle g, \Phi(s) f_j \rangle_H \, ds \Big| \mathscr{F}_u \right)$$

$$+ E\left(\sum_{i \neq j=1}^{\infty} \left(\lambda_i^{1/2} \lambda_j^{1/2} \int_0^t \langle \Phi(s) f_i, h \rangle_H \, dw_i(s) \right.\right.$$

$$\times \int_0^t \langle \Phi(s) f_j, g \rangle_H \, dw_j(s) \Big) \Big| \mathscr{F}_u \Big)$$

$$= \sum_{j=1}^{\infty} \left(\lambda_j \int_0^u \langle \Phi(s) f_j, h \rangle_H \, dw_j(s) \int_0^u \langle \Phi(s) f_j, g \rangle_H \, dw_j(s) \right)$$

$$- \sum_{j=1}^{\infty} \int_0^u \lambda_j \langle h, \Phi(s) f_j \rangle_H \langle g, \Phi(s) f_j \rangle_H \, ds$$

$$+ \sum_{i \neq j=1}^{\infty} \left(\lambda_i^{1/2} \lambda_j^{1/2} \int_0^u \langle \Phi(s) f_i, h \rangle_H \, dw_i(s) \int_0^u \langle \Phi(s) f_j, g \rangle_H \, dw_j(s) \right)$$

$$= \left\langle \int_0^u \Phi(s) \, dW_s, h \right\rangle_H \left\langle \int_0^u \Phi(s) \, dW_s, h \right\rangle_H$$

$$- \left\langle \left(\int_0^u (\Phi(s) Q^{1/2})(\Phi(s) Q^{1/2})^* \, ds \right)(h), g \right\rangle_H.$$

The formula for the increasing process follows from Lemma 2.1. □

Exercise 2.9 Prove that for $\Phi \in \Lambda_2(K_Q, H)$ and $h \in H$,

$$\left\langle \int_0^t \Phi(s) \, dW_s, h \right\rangle_H = \sum_{j=1}^{\infty} \int_0^t \langle \lambda_j^{1/2} \Phi(s) f_j, h \rangle_H \, dw_j(t)$$

with the series convergent in $L^2(\Omega, \mathbb{R})$.

The following corollary follows from the proof of Theorem 2.3.

Corollary 2.1 *For the sequence of bounded elementary processes $\Phi_n \in \mathscr{E}(\mathscr{L}(K, H))$ approximating Φ in $\Lambda_2(K_Q, H)$ and satisfying condition (2.20), the corresponding stochastic integrals converge uniformly with probability one,*

$$P \left(\sup_{t \leq T} \left\| \int_0^t \Phi_n(s) \, dW_s - \int_0^t \Phi(s) \, dW_s \right\|_H \to 0 \right) = 1.$$

Exercise 2.10 Prove the following two properties of the stochastic integral process $\int_0^t \Phi(s) \, dW_s$ for $\Phi \in \Lambda_2(K_Q, H)$:

$$P \left(\sup_{0 \leq t \leq T} \left\| \int_0^t \Phi(s, \omega) \, dW_s \right\|_H > \lambda \right) \leq \frac{1}{\lambda^2} \int_0^T E \| \Phi(s, \omega) \|_{\Lambda_2(K_Q, H)}^2 \, ds, \quad (2.21)$$

$$E \sup_{0 \leq t \leq T} \left\| \int_0^t \Phi(s, \omega) \, dW_s \right\|_H^2 \leq 4 \int_0^T E \| \Phi(s) \|_{\Lambda_2(K_Q, H)}^2 \, ds. \quad (2.22)$$

Conclude that if Φ is approximated by a sequence $\{\Phi_n\}_{n=1}^\infty$ in $\Lambda_2(K_Q, H)$, then for every $t \leq T$, $\int_0^t \Phi_n(s)\,dW_s \to \int_0^t \Phi(s)\,dW_s$ in $L^2(\Omega, H)$.

Remark 2.2 For $\Phi \in \Lambda_2(K_Q, H)$ such that $\Phi(s) \in \mathcal{L}(K, H)$, the quadratic variation process of the stochastic integral process $\int_0^t \Phi(s)\,dW_s$ and the increasing process related to $\| \int_0^t \Phi(s)\,dW_s \|_H^2$ simplify to

$$\left\langle\!\!\left\langle \int_0^{\cdot} \Phi(s)\,dW_s \right\rangle\!\!\right\rangle_t = \int_0^t \Phi(s)Q\Phi(s)^*\,ds$$

and

$$\left\langle \int_0^{\cdot} \Phi(s)\,dW_s \right\rangle_t = \int_0^t \mathrm{tr}\big(\Phi(s)Q\Phi(s)^*\big)\,ds.$$

The final step in constructing the Itô stochastic integral is to extend it to the class of integrands satisfying a less restrictive assumption on their second moments. This extension is necessary if one wants to study Itô's formula even for functions as simple as $x \to x^2$. We use the approach presented in [49] for real-valued processes.

In this chapter, we will only need the concept of a real-valued progressively measurable process, but in Chap. 4 we will have to refer to H-valued progressively measurable processes. Therefore we include a more general definition here.

Definition 2.11 An H-valued stochastic process $X(t), t \geq 0$, defined on a filtered probability space $(\Omega, \mathcal{F}, \{\mathcal{F}_t\}_{t\geq 0}, P)$ is called *progressively measurable* if for every $t \geq 0$, the mapping

$$X(\cdot, \cdot) : \big([0, t], \mathcal{B}([0, t])\big) \times (\Omega, \mathcal{F}_t) \to \big(H, \mathcal{B}(H)\big)$$

is measurable with respect to the indicated σ-fields.

It is well known (e.g., see Proposition 1.13 in [38]) that an adapted right-continuous (or left-continuous) process is progressively measurable.

Exercise 2.11 Show that adapted right-continuous (or left-continuous) processes are progressively measurable.

Let $\mathcal{P}(K_Q, H)$ denote the class of $\mathcal{L}_2(K_Q, H)$-valued stochastic processes adapted to the filtration $\{\mathcal{F}_t\}_{t\leq T}$, measurable as mappings from $([0, T] \times \Omega, \mathcal{B}([0, T]) \otimes \mathcal{F}_T)$ to $(\mathcal{L}_2(K_Q, H), \mathcal{B}(\mathcal{L}_2(K_Q, H)))$, and satisfying the condition

$$P\left\{ \int_0^T \|\Phi(t)\|_{\mathcal{L}_2(K_Q,H)}^2\,dt < \infty \right\} = 1. \tag{2.23}$$

Obviously, $\Lambda_2(K_Q, H) \subset \mathcal{P}(K_Q, H)$. We will show that processes from $\mathcal{P}(K_Q, H)$ can be approximated in a suitable way by processes from $\Lambda_2(K_Q, H)$ and, in fact, by bounded elementary processes from $\mathcal{E}(\mathcal{L}(K, H))$. This procedure will allow us to derive basic properties of the extended stochastic integral.

Lemma 2.3 *Let $\Phi \in \mathcal{P}(K_Q, H)$. Then there exists a sequence of bounded processes $\Phi_n \in \mathcal{E}(\mathcal{L}(K, H)) \subset \Lambda_2(K_Q, H)$ such that*

$$\int_0^T \left\| \Phi(t, \omega) - \Phi_n(t, \omega) \right\|_{\mathcal{L}_2(K_Q, H)}^2 dt \to 0 \quad as\ n \to \infty \qquad (2.24)$$

in probability and P-a.s.

Proof For $\Phi \in \mathcal{P}(K_Q, H)$, define

$$\tau_n(\omega) = \begin{cases} \inf\left\{ t \leq T : \int_0^t \left\| \Phi(s, \omega) \right\|_{\mathcal{L}_2(K_Q, H)}^2 ds \geq n \right\}, \\[2mm] T \quad \text{if } \int_0^T \left\| \Phi(s, \omega) \right\|_{\mathcal{L}_2(K_Q, H)}^2 ds < n. \end{cases} \qquad (2.25)$$

The real-valued process $\int_0^t \left\| \Phi(s, \omega) \right\|_{\mathcal{L}_2(K_Q, H)}^2 ds$ is adapted to the filtration $\mathcal{F}_t$ and continuous, and hence it is progressively measurable. Therefore, τ_n is an $\mathcal{F}_t$-stopping time, and we can define the $\mathcal{F}_t$-adapted process

$$\Phi_n(t, \omega) = \Phi(t, \omega) 1_{\{t \leq \tau_n(\omega)\}}. \qquad (2.26)$$

By the definition of $\tau_n(\omega)$ we have

$$E \int_0^T \left\| \Phi_n(t, \omega) \right\|_{\mathcal{L}_2(K_Q, H)}^2 dt \leq n,$$

so that $\Phi_n \in \Lambda_2(K_Q, H)$. Moreover, in view of (2.26),

$$P\left(\int_0^T \left\| \Phi(t, \omega) - \Phi_n(t, \omega) \right\|_{\mathcal{L}_2(K_Q, H)}^2 dt > 0 \right)$$

$$\leq P\left(\int_0^T \left\| \Phi(t, \omega) \right\|_{\mathcal{L}_2(K_Q, H)}^2 dt > n \right),$$

so that $\Phi_n \to \Phi$ in probability in the sense of the convergence in (2.24).

By Proposition 2.2, for every n, there exists a sequence of bounded elementary processes $\{\Phi_{n,k}\}_{k=1}^\infty \subset \mathcal{E}(\mathcal{L}(K, H))$ such that

$$E \int_0^T \left\| \Phi_n(t, \omega) - \Phi_{n,k}(t, \omega) \right\|_{\mathcal{L}_2(K_Q, H)}^2 dt \to 0 \quad as\ n \to \infty.$$

Then

$$P\left(\int_0^T \left\| \Phi(t, \omega) - \Phi_{n,k}(t, \omega) \right\|_{\mathcal{L}_2(K_Q, H)}^2 dt > \varepsilon \right)$$

$$\leq P\left(2 \int_0^T \left\| \Phi(t, \omega) - \Phi_n(t, \omega) \right\|_{\mathcal{L}_2(K_Q, H)}^2 dt > 0 \right)$$

$$+ P\left(2 \int_0^T \left\|\Phi_n(t, \omega) - \Phi_{n,k}(t, \omega)\right\|_{\mathscr{L}_2(K_Q, H)}^2 dt > \varepsilon\right)$$

$$\leq P\left(\int_0^T \left\|\Phi(t, \omega) - \Phi_n(t, \omega)\right\|_{\mathscr{L}_2(K_Q, H)}^2 dt > 0\right)$$

$$+ P\left(\int_0^T \left\|\Phi_n(t, \omega) - \Phi_{n,k}(t, \omega)\right\|_{\mathscr{L}_2(K_Q, H)}^2 dt > \frac{\varepsilon}{2}\right)$$

$$\leq P\left(\int_0^T \left\|\Phi(t, \omega)\right\|_{\mathscr{L}_2(K_Q, H)}^2 dt > n\right)$$

$$+ \frac{2}{\varepsilon} E \int_0^T \left\|\Phi_n(t, \omega) - \Phi_{n,k}(t, \omega)\right\|_{\mathscr{L}_2(K_Q, H)}^2 dt,$$

which proves the convergence in probability in (2.24) and P-a.s. convergence for a subsequence. $\square$

We can define a class of H-valued elementary processes $\mathscr{E}(H)$ adapted to the filtration $\{\mathscr{F}_t\}_{t \leq T}$ as all processes of the form

$$\Psi(t, \omega) = \psi(\omega) 1_{\{0\}}(t) + \sum_{j=0}^{n-1} \psi_j(\omega) 1_{(t_j, t_{j+1}]}(t), \tag{2.27}$$

where $0 = t_0 \leq t_1 \leq \cdots \leq t_n = T$, ψ is $\mathscr{F}_0$-measurable, and ψ_j $(j = 0, 1, \ldots, n-1)$, are $\mathscr{F}_{t_j}$-measurable H-valued random variables. Applying the same proof as in Lemma 2.3, we can prove the following statement.

Lemma 2.4 *Let $\Psi(t)$, $t \leq T$, be an H-valued, $\mathscr{F}_t$-adapted stochastic process satisfying the condition*

$$P\left(\int_0^T \left\|\Psi(t)\right\|_H dt < \infty\right) = 1.$$

Then there exists a sequence of bounded elementary processes $\Psi_n \in \mathscr{E}(H)$ such that

$$\int_0^T \left\|\Psi(t, \omega) - \Psi_n(t, \omega)\right\|_H dt \to 0 \quad as\ n \to \infty \tag{2.28}$$

in probability and almost surely.

We will need the following useful estimate.

Lemma 2.5 *Let $\Phi \in \Lambda_2(K_Q, H)$. Then for arbitrary $\delta > 0$ and $n > 0$,*

$$P\left(\sup_{t \leq T} \left\|\int_0^t \Phi(s)\, dW_s\right\|_H > \delta\right) \leq \frac{n}{\delta^2} + P\left(\int_0^T \left\|\Phi(s)\right\|_{\Lambda_2(K_Q, H)}^2 ds > n\right). \tag{2.29}$$

Proof Let τ_n be the stopping time defined in (2.25). Then

$$P\left(\sup_{t\leq T}\left\|\int_0^t \Phi(s)\,dW_s\right\|_H > \delta\right)$$

$$= P\left(\sup_{t\leq T}\left\|\int_0^t \Phi(s)\,dW_s\right\|_H > \delta \text{ and } \int_0^T \|\Phi(s)\|^2_{\Lambda_2(K_Q,H)}\,ds > n\right)$$

$$+ P\left(\sup_{t\leq T}\left\|\int_0^t \Phi(s)\,dW_s\right\|_H > \delta \text{ and } \int_0^T \|\Phi(s)\|^2_{\Lambda_2(K_Q,H)}\,ds \leq n\right).$$

The first probability on the right is bounded by $P(\int_0^T \|\Phi(s)\|^2_{\Lambda_2(K_Q,H)}\,ds > n)$, while the second probability does not exceed

$$P\left(\sup_{t\leq T}\left\|\int_0^t 1_{[0,\tau_n]}(s)\Phi(s)\,dW_s\right\|_H > \delta\right) \leq \frac{1}{\delta^2}E\int_0^T \|1_{[0,\tau_n]}(s)\Phi(s)\|^2_{\Lambda_2(K_Q,H)}\,ds$$

$$\leq \frac{n}{\delta^2}$$

by Doob's maximal inequality. $\qquad\square$

We are ready to conclude the construction of the stochastic integral now.

Lemma 2.6 *Let Φ_n be a sequence in $\Lambda_2(K_Q, H)$ approximating a process $\Phi \in \mathscr{P}(K_Q, H)$ in the sense of (2.24), i.e.,*

$$P\left(\int_0^T \|\Phi_n(t,\omega) - \Phi(t,\omega)\|^2_{\mathscr{L}_2(K_Q,H)}\,dt > 0\right) \to 0.$$

Then, there exists an H-valued $\mathscr{F}_T$-measurable random variable, denoted by $\int_0^T \Phi(t)\,dW_t$, such that

$$\int_0^T \Phi_n(t)\,dW_t \to \int_0^T \Phi(t)\,dW_t$$

in probability. The random variable $\int_0^T \Phi(t)\,dW_t$ does not depend (up to stochastic equivalence) on the choice of the approximating sequence.

Proof For every $\varepsilon > 0$, we have

$$\lim_{m,n\to\infty} P\left(\int_0^T \|\Phi_n(t,\omega) - \Phi_m(t,\omega)\|^2_{\mathscr{L}_2(K_Q,H)} > 0\right) = 0.$$

If $\delta > 0$, then by (2.29)

$$\limsup_{m,n\to\infty} P\left(\left\|\int_0^T \Phi_n(t)\,dW_t - \int_0^T \Phi_m(t)\,dW_t\right\|_H > \delta\right)$$

$$\leq \frac{\varepsilon}{\delta^2} + \lim_{m,n\to\infty} P\left(\int_0^T \|\Phi_n(t,\omega) - \Phi_m(t,\omega)\|_{\Lambda_2(K_Q,H)}^2 \, dt > \varepsilon\right) = \frac{\varepsilon}{\delta^2},$$

and since ε was arbitrary, the convergence

$$\lim_{m,n\to\infty} P\left(\left\|\int_0^T \Phi_n(t)\,dW_t - \int_0^T \Phi_m(t)\,dW_t\right\|_H > \delta\right) = 0$$

follows. The limit in probability $\int_0^T \Phi(t)\,dW_t = \lim_{n\to\infty} \int_0^T \Phi_n(t)\,dW_t$ does not depend on the choice of the approximating sequence. If Ψ_n is another approximating sequence, then the two sequences can be merged into one. The resulting limit would then have to coincide with the limits of its all subsequences. $\square$

Definition 2.12 The H-valued random variable $\int_0^T \Phi(t)\,dW_t$ defined in Lemma 2.6 is called the *stochastic integral* of a process in $\mathscr{P}(K_Q, H)$ with respect to a Q-Wiener process. For $0 \leq t \leq T$, we define an H-valued stochastic integral process $\int_0^t \Phi(s)\,dW_s$ by

$$\int_0^t \Phi(s)\,dW_s = \int_0^T \Phi(s)1_{[0,t]}(s)\,dW_s.$$

Exercise 2.12 Prove (2.29) for $\Phi \in \mathscr{P}(K_Q, H)$.

Remark 2.3 We note that for $\Phi \in \mathscr{P}(K_Q, H)$, the stochastic integral process $\int_0^t \Phi(s)\,dW_s$ also has a continuous version.

Indeed, let $\Omega_n = \{\omega : n-1 \leq \int_0^T \|\Phi(s)\|_{\Lambda_2(K_Q,H)}^2 < n\}$; then $P(\Omega - \bigcup_{n=1}^\infty \Omega_n) = 0$, and if Φ_n are defined as in (2.26), then on Ω_n,

$$\Phi_n(t) = \Phi_{n+1}(t) = \cdots = \Phi(t), \quad t \leq T.$$

Therefore, on Ω_n, $\int_0^t \Phi(s)\,dW_s = \int_0^t \Phi_n\,dW_s$ has a continuous version, and hence it has a continuous version on Ω.

The stochastic integral process for $\Phi \in \mathscr{P}(K_Q, H)$ may not be a martingale, but it is a local martingale. We will now discuss this property.

Definition 2.13 A stochastic process $\{M_t\}_{t\leq T}$, adapted to a filtration $\mathscr{F}_t$, with values in a separable Hilbert space H is called a *local martingale* if there exists a sequence of increasing stopping times τ_n, with $P(\lim_{n\to\infty} \tau_n = T) = 1$, such that for every n, $M_{t\wedge\tau_n}$ is a uniformly integrable martingale.

Exercise 2.13 Show an example of $\Phi \in \mathscr{P}(K_Q, H)$ such that $\int_0^t \Phi(s)\,dW_s$ is not a martingale.

We begin with the following lemma.

Lemma 2.7 *Let $\Phi \in \mathscr{P}(K_Q, H)$, and τ be a stopping time relative to $\{\mathscr{F}_t\}_{0 \leq t \leq T}$ such that $P(\tau \leq T) = 1$. Define*

$$\int_0^\tau \Phi(t)\,dW_t = \int_0^u \Phi(t)\,dW_t \quad \text{on the set } \{\omega : \tau(\omega) = u\}.$$

Then

$$\int_0^\tau \Phi(t)\,dW_t = \int_0^T \Phi(t)1_{\{t \leq \tau\}}\,dW_t. \tag{2.30}$$

Proof For an arbitrary process $\Phi \in \mathscr{P}(K_Q, H)$, let Φ_n be a sequence of bounded elementary processes approximating Φ as in Lemma 2.3. Since

$$\int_0^T \left\| \Phi_n(t)1_{\{t \leq \tau\}} - \Phi(t)1_{\{t \leq \tau\}} \right\|_{\Lambda_2(K_Q, H)}^2 \, dt \leq \int_0^T \left\| \Phi_n(t) - \Phi(t) \right\|_{\Lambda_2(K_Q, H)}^2 \, dt,$$

we conclude that

$$\int_0^T \Phi_n(t)1_{\{t \leq \tau\}}\,dW_t \to \int_0^T \Phi(t)1_{\{t \leq \tau\}}\,dW_t$$

in probability.

For bounded elementary processes $\Phi \in \mathscr{E}(\mathscr{L}(K, H))$, equality (2.30) can be verified by inspection, so that $\int_0^\tau \Phi_n(t)\,dW_t = \int_0^T \Phi_n(s)1_{\{s \leq \tau\}}\,dW_s$.

On the set $\{\omega : \tau(\omega) = u\}$, we have $\int_0^\tau \Phi_n(t)\,dW_t = \int_0^u \Phi_n(t)\,dW_t$. Also, for every $u \leq T$,

$$\int_0^u \Phi_n(t)\,dW_t \to \int_0^u \Phi(t)\,dW_t$$

in probability. Thus, for every $u \leq T$,

$$1_{\{\tau = u\}} \int_0^T \Phi_n(t)1_{\{t \leq \tau\}}\,dW_t \to 1_{\{\tau = u\}} \int_0^u \Phi(t)\,dW_t$$

in probability. This implies that for every $u \leq T$,

$$1_{\{\tau = u\}} \int_0^u \Phi(t)\,dW_t = 1_{\{\tau = u\}} \int_0^T \Phi(t)1_{\{t \leq \tau\}}\,dW_t \quad P\text{-a.s.}$$

Since the stochastic integral process $\int_0^u \Phi(t)\,dW_t$, $u \leq T$ is P-a.s. continuous, we get that the above equality holds P-a.s. for all $u \leq T$, and (2.30) follows. □

Now observe that for $\Phi \in \mathscr{P}(K_Q, H)$, the stochastic integral process $\int_0^t \Phi(s)\,dW_s$ is a local martingale, with the localizing sequence of stopping times τ_n defined in (2.25). We have $P(\lim_{n \to \infty} \tau_n = T) = 1$, and, by (2.30), $\int_0^{t \wedge \tau_n} \Phi(s)\,dW_s$ is a martingale with

$$E\left(\left\|\int_0^{t\wedge\tau_n}\Phi(s)\,dW_s\right\|_H^2\right)=E\int_0^T\|\Phi(s)\|_{\mathscr{L}_2(K_Q,H)}^2\mathbf{1}_{\{s\le\tau_n\}}\,ds\le n.$$

This proves that for every n, the process $\int_0^{t\wedge\tau_n}\Phi(s)\,dW_s$ is a square-integrable and hence a uniformly integrable martingale.

2.2.4 Stochastic Itô Integral with Respect to Cylindrical Wiener Process

We now proceed with the definition of the stochastic integral with respect to a cylindrical Wiener process. We will restrict ourselves to the case where the integrand $\Phi(s)$ is a process taking values in $\mathscr{L}_2(K,H)$, following the work in [18]. A more general approach can be found in [57] and [19].

We recall that if $\Phi(s)$ is an elementary process, $\Phi(s)\in\mathscr{E}(\mathscr{L}(K,H))$, then $\Phi(s)\in\mathscr{L}_2(K,H)$, since $Q=I_K$. Assume that $\Phi(s)$ is bounded in the norm of $\mathscr{L}_2(K,H)$. Using (2.14), with $\{e_i\}_{i=1}^\infty$ an ONB in H, we calculate,

$$E\sum_{i=1}^\infty\left(\left(\int_0^t\Phi(s)\,d\tilde{W}_s\right)(e_i)\right)^2=\sum_{i=1}^\infty\int_0^t E\|\Phi^*(s)e_i\|_K^2\,ds$$

$$=E\int_0^t\sum_{i=1}^\infty\|\Phi^*(s)e_i\|_K^2\,ds=E\int_0^t\|\Phi^*(s)\|_{\mathscr{L}_2(H,K)}^2\,ds$$

$$=E\int_0^t\|\Phi(s)\|_{\mathscr{L}_2(K,H)}^2\,ds.$$

Then we define the stochastic integral $\int_0^t\Phi(s)\,d\tilde{W}_s$ of a bounded elementary process $\Phi(s)$ as follows:

$$\int_0^t\Phi(s)\,d\tilde{W}_s=\sum_{i=1}^\infty\left(\left(\int_0^t\Phi(s)\,d\tilde{W}_s\right)(e_i)\right)e_i.\qquad(2.31)$$

By the above calculations, $\int_0^t\Phi(s)\,d\tilde{W}_s\in L^2(\Omega,H)$ and is adapted to the filtration $\mathscr{F}_t$. The equality

$$\left\|\int_0^T\Phi(s)\,d\tilde{W}_s\right\|_{L^2(\Omega,H)}=\|\Phi\|_{\Lambda_2(K,H)}\qquad(2.32)$$

establishes the isometry property of the stochastic integral transformation.

Definition 2.14 The *stochastic integral* of a process $\Phi \in \Lambda_2(K, H)$ with respect to a standard cylindrical Wiener process $\tilde{W}_t$ in a Hilbert space K is the unique isometric linear extension of the mapping

$$\Phi(\cdot) \to \int_0^T \Phi(s) \, d\tilde{W}_s$$

from the class of bounded elementary processes to $L^2(\Omega, H)$ to a mapping from $\Lambda_2(K, H)$ to $L^2(\Omega, H)$, such that the image of $\Phi(t) = \phi 1_{\{0\}}(t) + \sum_{j=0}^{n-1} \phi_j 1_{(t_j, t_{j+1}]}(t)$ is

$$\sum_{i=1}^{\infty} \sum_{j=0}^{n} \left(\tilde{W}_{t_{j+1} \wedge t} \left(\phi_j^*(e_i) \right) - \tilde{W}_{t_j \wedge t} \left(\phi_j^*(e_i) \right) \right) e_i.$$

We define the stochastic integral process $\int_0^t \Phi(s) \, d\tilde{W}_s$, $0 \le t \le T$, for $\Phi \in \Lambda_2(K, H)$ by

$$\int_0^t \Phi(s) \, d\tilde{W}_s = \int_0^T \Phi(s) 1_{[0,t]}(s) \, d\tilde{W}_s.$$

The next theorem can be proved in a similar manner as Theorem 2.3.

Theorem 2.4 *The stochastic integral $\Phi \to \int_0^{\cdot} \Phi(s) \, d\tilde{W}_s$ with respect to a cylindrical Wiener process $\tilde{W}_t$ in K is an isometry between $\Lambda_2(K, H)$ and the space of continuous square-integrable martingales $\mathcal{M}_T^2(H)$. The quadratic variation process of the stochastic integral process $\int_0^t \Phi(s) \, d\tilde{W}_s$ is given by*

$$\left\langle\!\!\left\langle \int_0^{\cdot} \Phi(s) \, d\tilde{W}_s \right\rangle\!\!\right\rangle_t = \int_0^t \Phi(s) \Phi^*(s) \, ds$$

and

$$\left\langle\!\left\langle \int_0^{\cdot} \Phi(s) \, d\tilde{W}_s \right\rangle\!\right\rangle_t = \int_0^t \operatorname{tr} \Phi(s) \Phi^*(s) \, ds = \int_0^t \|\Phi(s)\|_{\mathcal{L}_2(K,H)}^2 \, ds.$$

Similarly as in the case of the stochastic integral with respect to a Q-Wiener process, we now conclude the construction of the integral with respect to a cylindrical Wiener process.

Remark 2.4 Since for $\Phi \in \Lambda_2(K, H)$, the process $\int_0^t \Phi(s) \, d\tilde{W}_s \in \mathcal{M}_T^2(H)$, the conclusion of Lemma 2.5 holds in the cylindrical case.

Define $\mathscr{P}(K, H) = \mathscr{P}(K_Q, H)$ with $Q = I_K$. We can construct the cylindrical stochastic integral $\int_0^T \Phi(t) \, d\tilde{W}_t$ for processes $\Phi \in \mathscr{P}(K, H)$ by repeating the arguments in Lemma 2.6.

Exercise 2.14 Verify the statements in Remark 2.4.

We arrive at the following definition.

Definition 2.15 Let Φ_n be a sequence in $\Lambda_2(K, H)$ approximating a process $\Phi \in \mathscr{P}(K, H)$ in the sense of (2.24), i.e.,

$$P\left(\int_0^T \|\Phi_n(t, \omega) - \Phi(t, \omega)\|^2_{\mathscr{L}_2(K,H)} \, dt > 0\right) \to 0.$$

Denote the limit in probability of the sequence $\int_0^T \Phi_n(t) \, d\tilde{W}_t$ by $\int_0^T \Phi(t) \, d\tilde{W}_t$ and call the limit the stochastic integral of Φ with respect to the cylindrical Wiener process $\tilde{W}_t$. The integral is an H-valued $\mathscr{F}_T$-measurable random variable, and it does not depend (up to stochastic equivalence) on the choice of the approximating sequence.

For $0 \le t \le T$, we define an H-valued stochastic integral process $\int_0^t \Phi(s) \, d\tilde{W}_s$ by

$$\int_0^t \Phi(s) \, d\tilde{W}_s = \int_0^T \Phi(s) 1_{[0,t]}(s) \, d\tilde{W}_s.$$

Exercise 2.15 (a) Prove (2.30) in the cylindrical case, i.e., show that for a process $\Phi \in \mathscr{P}(K, H)$ and a stopping time τ relative to $\{\mathscr{F}_t\}_{0 \le t \le T}$ such that $P(\tau \le T) = 1$,

$$\int_0^\tau \Phi(t) \, d\tilde{W}_t = \int_0^T \Phi(t) 1_{\{t \le \tau\}} \, d\tilde{W}_t, \tag{2.33}$$

where

$$\int_0^\tau \Phi(t) \, d\tilde{W}_t = \int_0^u \Phi(t) \, d\tilde{W}_t \quad \text{on the set } \{\omega : \tau(\omega) = u\}.$$

(b) Show that the stochastic integral $\int_0^t \Phi(s) \, d\tilde{W}_s$ $0 \le t \le T$, is a local martingale and that it has a continuous version.

The following representation of the stochastic integral with respect to a Q-Wiener process and with respect to a cylindrical Wiener process if $Q = I_K$ can be also used as a definition.

Lemma 2.8 *Let W_t be a Q-Wiener process in a separable Hilbert space K, $\Phi \in \Lambda_2(K_Q, H)$, and $\{f_j\}_{j=1}^\infty$ be an ONB in K consisting of eigenvectors of Q. Then*

$$\int_0^t \Phi(s) \, dW_s = \sum_{j=1}^\infty \int_0^t \left(\Phi(s)\lambda_j^{1/2} f_j\right) d\langle W_s, \lambda_j^{1/2} f_j\rangle_{K_Q}$$

$$= \sum_{j=1}^\infty \int_0^t \left(\Phi(s) f_j\right) d\langle W_s, f_j\rangle_K. \tag{2.34}$$

For a cylindrical Wiener process $\tilde{W}_t$, if $\Phi \in \Lambda_2(K, H)$ and $\{f_j\}_{j=1}^{\infty}$ is an ONB in K, then

$$\int_0^t \Phi(s) \, d\tilde{W}_s = \sum_{j=1}^{\infty} \int_0^t \left(\Phi(s) f_j\right) d\tilde{W}_s(f_j). \tag{2.35}$$

Proof We will prove (2.35), the cylindrical case only, since the proof for a Q-Wiener process is nearly identical.

We first note that

$$E \left\| \sum_{j=1}^{\infty} \int_0^t \left(\Phi(s) f_j\right) d\tilde{W}_s(f_j) \right\|_H^2 = E \sum_{i=1}^{\infty} \left\langle \sum_{j=1}^{\infty} \int_0^t \left(\Phi(s) f_j\right) d\tilde{W}_s(f_j), e_i \right\rangle_H^2$$

$$= \sum_{i=1}^{\infty} \sum_{j=1}^{\infty} E \int_0^t \left\langle \Phi(s) f_j, e_i \right\rangle_H^2$$

$$= E \int_0^t \left\| \Phi(s) \right\|_{\mathscr{L}_2(K,H)}^2 < \infty.$$

Thus, $\sum_{j=1}^{\infty} \int_0^t (\Phi(s) f_j) \, d\tilde{W}_s(f_j) \in H$, P-a.s. For a bounded elementary process $\Phi(s) = 1_{\{0\}} \phi + \sum_{k=1}^{n-1} \phi_k 1_{(t_k, t_{k+1}]}(s) \in \mathscr{E}(\mathscr{L}(K, H))$ and any $h \in H$, we have that a.s.

$$\left\langle \int_0^t \Phi(s) \, d\tilde{W}_s, h \right\rangle_H = \sum_{k=1}^{n-1} \left(\tilde{W}_{t_{k+1}}\left(\phi_k^*(h)\right) - \tilde{W}_{t_k}\left(\phi_k^*(h)\right)\right)$$

$$= \sum_{k=1}^{n-1} \sum_{j=1}^{\infty} \left(\tilde{W}_{t_{k+1}}(f_j) - \tilde{W}_{t_k}(f_j)\right)\left\langle f_j, \phi_k^*(h) \right\rangle_K$$

$$= \sum_{j=1}^{\infty} \sum_{k=1}^{n-1} \left\langle \phi_k(f_j)\left(\tilde{W}_{t_{k+1}}(f_j) - \tilde{W}_{t_k}(f_j)\right), h \right\rangle_H$$

$$= \left\langle \sum_{j=1}^{\infty} \sum_{k=1}^{n-1} \phi_k(f_j)\left(\tilde{W}_{t_{k+1}}(f_j) - \tilde{W}_{t_k}(f_j)\right), h \right\rangle_H$$

$$= \left\langle \sum_{j=1}^{\infty} \int_0^t \left(\Phi(s) f_j\right) d\tilde{W}_s(f_j), h \right\rangle_H,$$

so that (2.35) holds in this case.

Let $P_{m+1}^{\perp}$ denote the orthogonal projection onto $\mathrm{span}\{f_{m+1}, f_{m+2}, \ldots\}$. Now for a general $\Phi \in \Lambda_2(K, H)$, we have for an approximating sequence $\Phi_n(s) \in \mathcal{E}(\mathcal{L}(K, H))$, using (2.35) for elementary processes,

$$E \left\| \int_0^t \Phi(s)\, d\tilde{W}_s - \sum_{j=1}^m \int_0^t \left(\Phi(s) f_j \right) d\tilde{W}_s(f_j) \right\|_H^2$$

$$= \lim_{n \to \infty} E \left\| \left(\int_0^t \Phi(s)\, d\tilde{W}_s - \int_0^t \Phi_n(s)\, d\tilde{W}_s \right) + \int_0^t \Phi_n(s)\, d\tilde{W}_s \right.$$

$$\left. - \sum_{j=1}^m \int_0^t \left(\Phi(s) f_j \right) d\tilde{W}_s(f_j) \right\|_H^2$$

$$= \lim_{n \to \infty} E \left\| \sum_{j=m+1}^{\infty} \int_0^t \left(\Phi_n(s) f_j \right) d\tilde{W}_s(f_j) + \sum_{j=1}^m \int_0^t \left(\Phi_n(s) f_j \right) d\tilde{W}_s(f_j) \right.$$

$$\left. - \sum_{j=1}^m \int_0^t \left(\Phi(s) f_j \right) d\tilde{W}_s(f_j) \right\|_H^2$$

$$= \lim_{n \to \infty} E \left\| \sum_{j=m+1}^{\infty} \int_0^t \left(\Phi_n(s) f_j \right) d\tilde{W}_s(f_j) \right\|_H^2$$

$$= \lim_{n \to \infty} E \left\| \sum_{j=1}^{\infty} \int_0^t \left((\Phi_n(s) P_{m+1}^{\perp}) f_j \right) d\tilde{W}_s(f_j) \right\|_H^2$$

$$= \lim_{n \to \infty} E \left\| \int_0^t \left(\Phi_n(s) P_{m+1}^{\perp} \right) d\tilde{W}_s \right\|_H^2$$

$$= E \left\| \int_0^t \left(\Phi(s) P_{m+1}^{\perp} \right) d\tilde{W}_s \right\|_H^2$$

$$= E \int_0^t \left\| \Phi(s) P_{m+1}^{\perp} \right\|_{\mathcal{L}_2(K,H)}^2 ds$$

$$= E \int_0^t \sum_{j=m+1}^{\infty} \left\| \Phi(s) f_j \right\|_H^2 \to 0, \quad \text{as } m \to \infty,$$

where we have used the fact that $\Phi_n P_{m+1}^{\perp} \to \Phi P_{m+1}^{\perp}$ in $\Lambda_2(K, H)$ as $n \to \infty$. This concludes the proof. $\square$

2.2.5 The Martingale Representation Theorem

We recall (Definition 2.6) that an H-valued Q-Wiener process is defined by

$$W_t = \sum_{j=1}^{\infty} \lambda_j^{1/2} w_j(t) f_j,$$

where $\{f_j\}_{j=1}^{\infty}$ is an ONB in a separable Hilbert space K, $\{w_j(t)\}_{j=1}^{\infty}$ are independent Brownian motions, and $\{\lambda_j\}_{j=1}^{\infty}$ are summable and assumed to be strictly positive without loss of generality. Let us denote

$$\mathscr{F}_t^j = \sigma\{w_j(s) : s \le t\}, \qquad \mathscr{G}_t = \sigma\left\{\bigcup_{j=1}^{\infty} \mathscr{F}_t^j\right\},$$

and

$$\mathscr{F}_t^W = \sigma\{W_s(k) : k \in K, \text{ and } s \le t\}.$$

Then clearly $\mathscr{F}_T^W = \mathscr{G}_T$ and

$$L^2(\Omega, \mathscr{F}_T^W, P) = L^2(\Omega, \mathscr{G}_T, P) = \bigoplus_{j=1}^{\infty} L^2(\Omega, \mathscr{F}_T^j, P).$$

Exercise 2.16 Prove that if $\mathscr{F}_i$, $i = 1, 2, \ldots$, are independent σ-fields, then

$$L^2\left(\Omega, \sigma\left\{\bigcup_{i=1}^{\infty} \mathscr{F}_i\right\}, P\right) = \bigoplus_{i=1}^{\infty} L^2(\Omega, \mathscr{F}_i, P).$$

In view of the fact that the linear span

$$\text{span}\left\{e^{\int_0^T h(t)\,dw_j(t) - \frac{1}{2}\int_0^T h^2(t)\,dt} : h \in L^2([0, T], \mathbb{R})\right\}$$

is dense in $L^2(\Omega, \mathscr{F}_T^j, P)$ ([61], Lemma 4.3.2), we deduce that the linear span

$$\text{span}\left\{e^{\int_0^T h(t)\,dw_j(t) - \frac{1}{2}\int_0^T h^2(t)\,dt} : h \in L^2([0, T], \mathbb{R}), \; j = 1, 2, \ldots\right\}$$

is dense in $L^2(\Omega, \mathscr{F}_T^W, P)$.

Now following the proof of Theorem 4.3.4 in [61], we conclude that every real-valued $\mathscr{F}_t^W$-martingale m_t in $L^2(\Omega, \mathscr{F}_T^W, P)$ has a unique representation

$$m_t(\omega) = E m_0 + \sum_{j=1}^{\infty} \int_0^t \lambda_j^{1/2} \phi_j(s, \omega)\,dw_j(s), \tag{2.36}$$

where $\sum_{j=1}^{\infty} \lambda_j E \int_0^T \phi_j^2(s, \omega)\,ds < \infty$.

Let H be a separable Hilbert space, and M_t be an $\mathscr{F}_t^W$-martingale, such that $E\|M_t\|_H^2 < \infty$. Choose an ONB $\{e_j\}_{j=1}^{\infty} \subset H$. Then there exist unique processes $\phi_j^i(t, \omega)$ as in the representation (2.36) such that

$$\langle M_t, e_i \rangle_H = E \langle M_0, e_i \rangle_H + \sum_{j=1}^{\infty} \int_0^t \lambda_j^{1/2} \phi_j^i(s, \omega) \, dw_j(s).$$

Since $E \sum_{i=1}^{\infty} \langle M_t, e_i \rangle_H^2 < \infty$, we have

$$M_t = \sum_{i=1}^{\infty} \langle M_t, e_i \rangle_H e_i.$$

Therefore,

$$M_t = E \sum_{i=1}^{\infty} \langle M_0, e_i \rangle_H e_i + \sum_{i=1}^{\infty} \sum_{j=1}^{\infty} \int_0^t \lambda_j^{1/2} \phi_j^i(s, \omega) e_i \, dw_j(s). \qquad (2.37)$$

Under the assumptions on M_t, we obtain that $E\|M_0\|_H < \infty$, so that the first term is equal to EM_0. Using the assumptions on M_t and the representations (2.36) and (2.37) above, we obtain

$$\sum_{i=1}^{\infty} E \left(\sum_{j=1}^{\infty} \int_0^t \lambda_j^{1/2} \phi_j^i(s, \omega) \, dw_j(s) \right)^2 = \sum_{j=1}^{\infty} \sum_{i=1}^{\infty} \lambda_j \int_0^t E \left(\phi_j^i(s, \omega) \right)^2 ds.$$

This justifies interchanging the summations in (2.37) to write M_t as

$$M_t = EM_0 + \sum_{j=1}^{\infty} \lambda_j^{1/2} \sum_{i=1}^{\infty} \int_0^t \phi_j^i(s, \omega) e_i \, dw_j(s). \qquad (2.38)$$

Define for $k \in K_Q, h \in H$,

$$\langle \Phi(s, \omega)k, h \rangle_H = \sum_{j=1}^{\infty} \sum_{i=1}^{\infty} \lambda_j \langle h, e_i \rangle_H \langle k, f_j \rangle_{K_Q} \phi_j^i(s, \omega).$$

Then $\Phi(s, \omega) \in \Lambda_2(K_Q, H)$, and, by the definition of the stochastic integral, the second term in (2.38) is equal to

$$\int_0^t \Phi(s, \omega) \, dW(s).$$

We have the following theorem.

Theorem 2.5 (Martingale Representation Theorem I) *Let H and K be separable Hilbert spaces, W_t be a K-valued Q-Wiener process, and M_t an H-valued continuous $\mathscr{F}_t^W$-martingale such that $E\|M_t\|_H^2 < \infty$ for all $t \geq 0$. Then there exists a unique process $\Phi(t) \in \Lambda_2(K_Q, H)$ such that*

$$M_t = E M_0 + \int_0^t \Phi(s)\, dW_s.$$

Remark 2.5 If $E M_0 = 0$, then, by Theorem 2.3, the quadratic variation process corresponding to M_t and the increasing process related to $\|M_t\|_H^2$ are given by (see [57])

$$
\begin{aligned}
\langle\!\langle M \rangle\!\rangle_t &= \int_0^t \left(\Phi(s)Q^{1/2}\right)\left(\Phi(s)Q^{1/2}\right)^* ds, \\
\langle M \rangle_t &= \int_0^t \|\Phi(s)\|_{\mathscr{L}_2(K_Q,H)}^2\, ds = \int_0^t \operatorname{tr}\!\left(\left(\Phi(s)Q^{1/2}\right)\left(\Phi(s)Q^{1/2}\right)^*\right) ds.
\end{aligned}
\tag{2.39}
$$

We shall now prove the converse. We need the following two results.

Theorem 2.6 (Lévy) *Let M_t, $0 \leq t \leq T$, be a K-valued continuous square-integrable martingale with respect to a filtration $\{\mathscr{F}_t\}_{t \leq T}$. Assume that its quadratic variation process is of the form $\langle\!\langle M \rangle\!\rangle_t = tQ$, $t \in [0, T]$. Then M_t is a Q-Wiener process with respect to the filtration $\{\mathscr{F}_t\}_{t \leq T}$.*

Proof Consider $M_t^n = (\frac{1}{\lambda_1^{1/2}}\langle M_t, f_1 \rangle_K, \dots, \frac{1}{\lambda_n^{1/2}}\langle M_t, f_n \rangle_K)$, where $\{f_j\}_{j=1}^\infty$ is an ONB in K such that $Q f_j = \lambda_j f_j$. Then, by the classical Lévy theorem, M_t^n is an n-dimensional Brownian motion with respect to the filtration $\{\mathscr{F}_t\}_{t \leq T}$. This implies that $M_t = \sum_{j=1}^\infty \lambda_j^{1/2} w_j(t) f_j$, where $w_j(t) = \frac{1}{\lambda_j^{1/2}}\langle M_t, f_j \rangle$, $j = 1, 2, \dots$, are independent Brownian motions with respect to $\{\mathscr{F}_t\}_{t \leq T}$. $\qquad\square$

Using a theorem on measurable selectors of Kuratowski and Ryll–Nardzewski [43], we obtain the following lemma in [11]. Below, for a separable Hilbert space H, $\mathscr{L}_1(H)$ denotes the (separable) space of trace-class operators on H.

Lemma 2.9 *Let H be a separable Hilbert space, and Φ be a measurable function from $(\Omega, \mathscr{F})$ to $(\mathscr{L}_1(H), \mathscr{B}(\mathscr{L}_1(H)))$ such that $\Phi(\omega)$ is a nonnegative definite operator for every ω. Then there exists a nonincreasing sequence of real-valued nonnegative measurable functions $\lambda_n(\omega)$ and H-valued measurable functions $g_n(\omega)$ such that for all $h \in H$,*

$$\Phi(\omega)h = \sum_{n=1}^\infty \lambda_n(\omega)\langle g_n(\omega), h \rangle_H g_n(\omega).$$

Moreover, the sequences λ_n and g_n can be chosen to satisfy

$$\|g_n(\omega)\|_H = \begin{cases} 1 & \text{if } \lambda_n(\omega) > 0, \\ 0 & \text{if } \lambda_n(\omega) = 0, \end{cases}$$

and $\langle g_n(\omega), g_m(\omega)\rangle_H = \delta_{m,n}, \forall \omega \in \Omega$.

We know that $M_t = \int_0^t \Phi(s)dW_s$ describes all continuous $\mathscr{F}_t^W$-martingales such that $E\|M_t\|_H^2$ is finite, $\langle\langle M \rangle\rangle_t = \int_0^t (\Phi(s)Q^{1/2})(\Phi(s)Q^{1/2})^* ds$, and $EM_0 = 0$. The related increasing process is given by $\langle M \rangle_t = \int_0^t \operatorname{tr}((\Phi(s)Q^{1/2})(\Phi(s)Q^{1/2})^*) ds$. Now let $(\Omega, \mathscr{F}, \{\mathscr{F}_t\}_{t \leq T}, P)$ be a filtered probability space, and M_t be an H-valued square-integrable martingale relative to $\mathscr{F}_t$. Assume that

$$\langle\langle M \rangle\rangle_t = \int_0^t Q_M(s)\,ds,$$

where the process $Q_M(s, \omega)$ is adapted to $\mathscr{F}_t$ with values in nonnegative definite symmetric trace-class operators on H.

Then we can define a stochastic integral

$$N_t = \int_0^t \Psi(s)\,dM_s \tag{2.40}$$

with respect to M_t exactly as we did for the case of a Wiener process. The integrands are $\mathscr{F}_t$-adapted processes $\Psi(t, \omega)$ with values in linear, but not necessarily bounded, operators from H to a separable Hilbert space G satisfying the condition

$$E \int_0^T \operatorname{tr}\big((\Psi(s)Q_M^{1/2}(s))(\Psi(s)Q_M^{1/2}(s)^*)\big)\,ds < \infty. \tag{2.41}$$

The stochastic integral process $N_t \in \mathscr{M}_T^2(G)$, and its quadratic variation is given by

$$\langle\langle N \rangle\rangle_t = \int_0^t (\Psi(s)Q_M^{1/2}(s))(\Psi(s)Q_M^{1/2}(s))^* ds. \tag{2.42}$$

In particular, we may have

$$Q_M(s, \omega) = (\Phi(s, \omega)Q^{1/2})(\Phi(s, \omega)Q^{1/2})^*$$

with $\Phi(s, \omega) \in \Lambda_2(K_Q, H)$ and M_t adapted to $\mathscr{F}_t^W$ for a Q-Wiener process W_t, and $EM_0 = 0$. In this case

$$M_t = \int_0^t \Phi(s)\,dW_s$$

and

$$N_t = \int_0^t \Psi(s)\,dM_s = \int_0^t \Psi(s)\Phi(s)\,dW_s. \tag{2.43}$$

By (2.39), the quadratic variation process of N_t has the form

$$\langle\!\langle N\rangle\!\rangle_t = \int_0^t \big(\Psi(s)\Phi(s)Q^{1/2}\big)\big(\Psi(s)\Phi(s)Q^{1/2}\big)^* ds. \tag{2.44}$$

Exercise 2.17 Reconcile formulas (2.42) and (2.44). *Hint: use Lemma 2.10 to show that if $L \in \mathcal{L}(H)$, then $(LL^*)^{1/2} = LJ$, where J is a partial isometry on $(\ker L)^{\perp}$.*

Exercise 2.18 Provide details for construction of the stochastic integral (2.40) with respect to square-integrable martingales for the class of stochastic processes satisfying condition (2.41). Prove property (2.42). Show (2.43) for $M_t = \int_0^t \Phi(s)\,dW_s$.

We shall use this integral for representing a square-integrable martingale with its quadratic variation process given by $\langle\!\langle M\rangle\!\rangle_t = \int_0^t (\Phi(s)Q^{1/2})(\Phi(s)Q^{1/2})^*\,ds$, in terms of some Wiener process. This provides the converse of the Martingale Representation Theorem I (Theorem 2.5). The formulation and proof of the theorem we present are taken directly from [11].

Theorem 2.7 (Martingale Representation Theorem II) *Let M_t, $0 \leq t \leq T$, be an H-valued continuous martingale with respect to a filtration $\{\mathcal{F}_t\}_{t=0}^{\infty}$. Assume that its quadratic variation process is given by $\langle\!\langle M\rangle\!\rangle_t = \int_0^t (\Phi(s,\omega)Q^{1/2})(\Phi(s,\omega)Q^{1/2})^*\,ds$, where $\Phi(s,\omega)$ is an adapted $\Lambda_2(K_Q, H)$-valued process. Then there exists a K-valued Q-Wiener process on an extended probability space $(\Omega \times \tilde{\Omega}, \mathcal{F} \times \tilde{\mathcal{F}}, P \times \tilde{P})$ adapted to filtration $\{\mathcal{F}_t \times \tilde{\mathcal{F}}_t\}$ such that*

$$M_t(\omega) = \int_0^t \Phi(s,\omega)\,dW_s(\omega, \tilde{\omega}).$$

In addition, the Wiener process can be constructed so that its increments $W_t - W_s$ are independent of $\mathcal{F}_s$ for $t \geq s$.

Proof To simplify the notation, denote $\Psi(s,\omega) = \Phi(s,\omega)Q^{1/2}$. We shall prove that if M_t is an H-valued continuous $\mathcal{F}_t$-martingale with the quadratic variation process

$$\langle\!\langle M\rangle\!\rangle_t = \int_0^t \Psi(s,\omega)\Psi^*(s,\omega)\,ds$$

such that $E\int_0^T \mathrm{tr}(\Psi(s)\Psi^*(s))\,ds < \infty$, then $M_t = \int_0^t \Phi(s,\omega)\,dW_s$, where W_t is a Q-Wiener process.

Since $\langle\!\langle M\rangle\!\rangle_t = \int_0^t \Psi(s,\omega)\Psi^*(s,\omega)\,ds$, the space $\mathrm{Im}(\Psi(s,\omega)\Psi^*(s,\omega))$ will play a key role in reconstructing M_t.

By Lemma 2.9 we get

$$\big(\Psi(s,\omega)\Psi^*(s,\omega)\big)h = \sum_{n=1}^{\infty} \lambda_n(s,\omega)\langle g_n(s,\omega), h\rangle_H\, g_n(s,\omega),$$

where λ_n and g_n are $\mathscr{F}_t$-adapted. Let $\{f_n\}_{n=1}^\infty$ be an ONB in K and define

$$V(s,\omega)k = \sum_{n=1}^\infty \langle k, f_n \rangle_K g_n(s,\omega).$$

Then

$$V^*(s,\omega)h = \sum_{n=1}^\infty \langle g_n(s,\omega), h \rangle_H f_n,$$

and the $\mathscr{F}_t$-adapted process

$$\Pi(s,\omega) = V(s)V^*(s)$$

is an orthogonal projection on $\mathrm{Im}(\Psi(s,\omega)\Psi^*(s,\omega))$. Thus we can write

$$\begin{aligned} M_t &= \int_0^t \left(\Pi(s) + \Pi^\perp(s) \right) dM_s \\ &= \int_0^t \Pi(s)\, dM_s + \int_0^t \Pi^\perp(s)\, dM_s = M_t' + M_t''. \end{aligned}$$

But $M_0'' = 0$ and $\langle\langle M'' \rangle\rangle_t = \int_0^t \Pi^\perp(s)\Psi(s)\Psi^*(s)\Pi^\perp(s)\, ds = 0$, so that $M_t'' = 0$. In conclusion,

$$M_t = M_t' + M_t'' = \int_0^t V(s)V^*(s)\, dM_s = \int_0^t V(s)\, dN_s$$

with

$$N_t = \int_0^t V^*(s,\omega)\, dM_s,$$

a continuous K-valued square integrable martingale whose quadratic variation is given by

$$\langle\langle N \rangle\rangle_t = \int_0^t V^*(s,\omega)\Psi(s,\omega)\Psi^*(s,\omega)V(s,\omega)\, ds.$$

We define now

$$\Lambda(s,\omega)k = \left(V^*(s,\omega)\Psi(s,\omega)\Psi^*(s,\omega)V(s,\omega) \right)k = \sum_{n=1}^\infty \lambda_n(s,\omega)\langle k, f_n \rangle_K f_n,$$

so that we can represent N_t through its series expansion in K,

$$N_t = \sum_{n=1}^\infty \eta_n(t) f_n,$$

where $\eta_n(t)$ are real-valued martingales with increasing processes $\int_0^t \lambda_n(s,\omega)\, ds$.

We consider now a filtered probability space

$$\left(\Omega \times \Omega', \mathscr{F} \times \mathscr{F}', \{\mathscr{F}_t \times \mathscr{F}'_t\}_{t \leq T}, P \times P'\right),$$

where $(\Omega', \mathscr{F}', \{\mathscr{F}'_t\}_{t \leq T}, P')$ is another filtered probability space. Define

$$\delta_n(s, \omega) = \begin{cases} \lambda_n^{-1/2}(s, \omega) & \text{if } \lambda_n(s, \omega) > 0, \\ 0 & \text{if } \lambda_n(s, \omega) = 0, \end{cases}$$

and

$$\gamma_n(s, \omega) = \begin{cases} 0 & \text{if } \lambda_n(s, \omega) > 0, \\ 1 & \text{if } \lambda_n(s, \omega) = 0. \end{cases}$$

Let $\beta_n(t)$ be independent Brownian motions on $(\Omega', \mathscr{F}', \{\mathscr{F}'_t\}_{t \leq T}, P')$.

We extend processes defined on either Ω or Ω' alone to the product space $\Omega \times \Omega'$, e.g., by $M(t, \omega, \omega') = M(t, \omega)$, and define

$$\hat{w}_n(t) = \int_0^t \delta_n(s) \, d\eta_n(s) + \int_0^t \gamma_n(s) \, d\beta_n(s).$$

Then it is easy to verify that $\langle \hat{w}_n, \hat{w}_m \rangle_t = t \delta_{n,m}$, using the mutual independence of η_n and β_m and the fact that

$$\langle \eta_n, \eta_m \rangle_t = \left(\langle\langle N \rangle\rangle_t f_n, f_m \right)_K = 0.$$

Thus, by Lévy's theorem, $\hat{w}_n(t)$ are independent Brownian motions, and the expression

$$\widehat{W}_t(k) = \sum_{n=1}^{\infty} \left(\hat{w}_n(t) f_n, k \right)_K$$

defines a K-valued cylindrical Wiener process.

Since

$$\int_0^t \lambda_n^{1/2}(s) \, d\hat{w}_n(s) = \int_0^t \lambda_n^{1/2}(s) \delta_n(s) \, d\eta_n(s) = \eta_n(t),$$

we get, using, for example, (2.35), that

$$N_t = \int_0^t \Lambda^{1/2}(s) \, d\widehat{W}_s.$$

Thus we arrive at

$$M_t = \int_0^t V(s) \, dN_s = \int_0^t V(s, \omega) \Lambda^{1/2}(s, \omega) \, d\widehat{W}_s.$$

All that is needed now is a modification of the integrand $\hat{\Psi}(s) = V(s) \Lambda^{1/2}(s)$ to the desired form $\Psi(s)$, and the cylindrical Wiener process $\widehat{W}_t$ needs to be replaced with a Q-Wiener process.

It follows directly from the definitions of $V(s)$ and $\Lambda(s)$ that

$$\Psi(s)\Psi^*(s) = \hat{\Psi}(s)\hat{\Psi}^*(s).$$

Now we need the following general fact from the operator theory (refer to [11], Appendix B).

Lemma 2.10 *Let H be a Hilbert space, $A, B \in \mathcal{L}(H)$, and*

$$AA^* = BB^*.$$

Denote by Π the orthogonal projection onto $(\ker B)^\perp$. Define

$$J = A^{-1}B\Pi : H \to (\ker A)^\perp,$$

where $A^{-1} : \mathcal{R}(A) \to \mathcal{D}(A)$ is the pseudo-inverse operator, i.e., $A^{-1}h$ is defined as the element g of the minimal norm such that $Ag = h$. Then

$$B = AJ,$$

where J is a partial isometry on $(\ker B)^\perp$, and JJ^* is an orthogonal projection onto $(\ker A)^\perp$.

It follows from the formula defining the operator J that there exists an $\mathcal{F}_t$-adapted process $J(t) : (\ker \hat{\Psi}(t))^\perp \to (\ker \Psi(t))^\perp$ such that

$$\hat{\Psi}(t) = \Psi(t)J(t)$$

and such that $J(t)J^*(t)$ is an orthogonal projection onto $(\ker \Psi(t))^\perp$. We need another filtered probability space $(\Omega'', \mathcal{F}'', \{\mathcal{F}_t''\}_{t \leq T}, P'')$ and a cylindrical Wiener process $\widehat{\widehat{W}}_t$ and extend all processes trivially to the product filtered probability space

$$\left(\Omega \times \Omega' \times \Omega'', \mathcal{F} \times \mathcal{F}' \times \mathcal{F}'', \{\mathcal{F}_t \times \mathcal{F}_t' \times \mathcal{F}_t''\}, P \times P' \times P''\right),$$

e.g., by $M_t(\omega, \omega', \omega'') = M_t(\omega, \omega') = M_t(\omega)$. If we define

$$W_t = \int_0^t Q^{1/2}J(s)\,d\widehat{W}_s + \int_0^t Q^{1/2}K(s)\,d\widehat{\widehat{W}}_s$$

with $K(s) = (J(s)J^*(s))^\perp$, the projection onto $\ker \Psi(s)$, then, using Lévy's theorem (Theorem 2.6), W_t is a Q-Wiener process, since by Theorem 2.4

$$\langle\!\langle W \rangle\!\rangle_t = \int_0^t \left(Q^{1/2}J(s)J^*(s)Q^{1/2} + Q^{1/2}K(s)Q^{1/2}\right)ds = tQ.$$

Also

$$\int_0^t \Phi(s)\,dW_s = \int_0^t \Psi(s)J(s)\,d\widehat{W}_s + \int_0^t \Psi(s)K(s)\,d\widehat{\widehat{W}}_s$$

$$= \int_0^t \widehat{\Psi}(s)\,d\widehat{W}_s = M_t. \qquad \square$$

We can now prove the following martingale representation theorem in the cylindrical case.

Corollary 2.2 *Let M_t, $0 \le t \le T$, be an H-valued continuous martingale with respect to a filtration $\{\mathscr{F}_t\}_{t=0}^\infty$. Assume that its quadratic variation process is given by $\langle\!\langle M \rangle\!\rangle_t = \int_0^t \Psi(s, \omega)\Psi^*(s, \omega)\,ds$, where $\Psi(s, \omega)$ is an adapted $\Lambda_2(K, H)$-valued process. Then there exists a cylindrical Wiener process $\tilde{W}_t$ in K on an extended probability space $(\Omega \times \tilde{\Omega}, \mathscr{F} \times \tilde{\mathscr{F}}, P \times \tilde{P})$ adapted to filtration $\{\mathscr{F}_t \times \tilde{\mathscr{F}}_t\}$ such that*

$$M_t(\omega) = \int_0^t \Psi(s, \omega)\,d\tilde{W}_s(\omega, \tilde{\omega}).$$

In addition, the cylindrical Wiener process can be constructed so that for any $k \in K$, the increments $\tilde{W}_t(k) - \tilde{W}_s(k)$ were independent of $\mathscr{F}_s$, if $t \ge s$.

Proof Let Q be a symmetric nonnegative definite trace–class operator on K with strictly positive eigenvalues. Define $\Phi(s) = \Psi(s)Q^{-1/2} \in \Lambda(K_Q, H)$. With the notation from the proof of Theorem 2.7, we have

$$M_t = \int_0^t \Phi(s)\,dW_s = \int_0^t \Psi(s)Q^{-1/2}\,dW_s = \int_0^t \Psi(s)\,d\tilde{W}_s,$$

where the relation $\tilde{W}_t(k) = \langle W_s, Q^{-1/2}k \rangle_K$, $k \in K$, defines the desired cylindrical Wiener process. $\qquad \square$

2.2.6 Stochastic Fubini Theorem

The stochastic version of the Fubini theorem helps calculate deterministic integrals of an integrand that is a stochastic integral process. In literature, this theorem is presented for predictable processes, but there is no need for this restriction if the stochastic integral is relative to a Wiener process.

Theorem 2.8 (Stochastic Fubini Theorem) *Let $(G, \mathscr{G}, \mu)$ be a finite measurable space, and $\Phi : ([0, T] \times \Omega \times G, \mathscr{B}([0, T]) \otimes \mathscr{F}_T \otimes \mathscr{G}) \to (H, \mathscr{B}(H))$ be a measurable map such that for every $x \in G$, the process $\Phi(\cdot, \cdot, x)$ is $\{\mathscr{F}_t\}_{t \le T}$-adapted.*

*Let W_t be a Q-Wiener process on a filtered probability space $(\Omega, \mathcal{F}, \{\mathcal{F}_t\}_{t \leq T}, P)$.
If*

$$\|\|\Phi\|\| = \int_G \|\Phi(\cdot, \cdot, x)\|_{\Lambda_2(K_Q, H)} \mu(dx) < \infty, \tag{2.45}$$

then

(1) $\int_0^T \Phi(t, \cdot, x) \, dW_t$ *has a measurable version as a map from $(\Omega \times G, \mathcal{F}_T \otimes \mathcal{G})$ to $(H, \mathcal{B}(H))$;*
(2) $\int_G \Phi(\cdot, \cdot, x) \mu(dx)$ *is $\{\mathcal{F}_t\}_{t \leq T}$-adapted;*
(3) *The following equality holds P-a.s.:*

$$\int_G \left(\int_0^T \Phi(t, \cdot, x) \, dW_t \right) \mu(dx) = \int_0^T \left(\int_G \Phi(t, \cdot, x) \mu(dx) \right) dW_t. \tag{2.46}$$

Proof Note that condition (2.45) implies that

$$\int_G E \int_0^T \|\Phi(t, \omega, x)\|_{\mathscr{L}_2(K_Q, H)} \, dt \, \mu(dx) < \infty,$$

so that $\Phi \in L^1([0, T] \times \Omega \times G)$. Also, $\Phi(\cdot, \cdot, x) \in \Lambda_2(K_Q, H)$ for almost all $x \in G$.
   We will carry the proof out in two steps.
   (A) We can assume without loss of generality that for all $x \in G$,

$$\|\Phi(\cdot, \cdot, x)\|_{\mathscr{L}_2(K_Q, H)} < M.$$

In order to prove (1)–(3) for an unbounded $\Phi \in L^1([0, T] \times \Omega \times G)$ with $\Phi(\cdot, \cdot, x) \in \Lambda_2(K_Q, H)$ μ-a.e., we only need to know that (1)–(3) hold for a $\|\cdot\|_{\mathscr{L}_2(K_Q, H)}$-norm bounded sequence Φ_n such that $\|\|\Phi_n - \Phi\|\| \to 0$. We define an appropriate sequence by

$$\Phi_n(t, \omega, x) = \begin{cases} n \dfrac{\Phi(t, \omega, x)}{\|\Phi(t, \omega, x)\|_{\mathscr{L}_2(K_Q, H)}} & \text{if } \|\Phi(t, \omega, x)\|_{\mathscr{L}_2(K_Q, H)} > n, \\ \Phi(t, \omega, x) & \text{otherwise.} \end{cases}$$

By the Lebesgue DCT relative to the $\|\cdot\|_{\mathscr{L}_2(K_Q, H)}$-norm, we have that μ-a.e.

$$\|\Phi_n(\cdot, \cdot, x) - \Phi(\cdot, \cdot, x)\|_{\Lambda_2(K_Q, H)} \to 0.$$

By the isometric property (2.19) of the stochastic integral,

$$\lim_{n \to \infty} \int_0^T \Phi_n(t, \cdot, x) \, dW_t = \int_0^T \Phi(t, \cdot, x) \, dW_t \tag{2.47}$$

in $L^2(\Omega)$, and, by selecting a subsequence if necessary, we can assume that the convergence in (2.47) is P-a.s.

We prove (1) by defining an $\mathscr{F}_T \otimes \mathscr{G}$-measurable version of the stochastic integral

$$\int_0^T \Phi(t, \cdot, x)\, dW_t = \begin{cases} \lim_{n \to \infty} \int_0^T \Phi_n(t, \cdot, x)\, dW_t & \text{if the limit exists,} \\ 0 & \text{otherwise.} \end{cases}$$

Again, by the Lebesgue DCT, we have that $\Phi_n \to \Phi$ in $L^1([0, T] \times \Omega \times G)$, so that we can assume, by selecting a subsequence if necessary, that $\Phi_n \to \Phi$ for almost all (t, ω, x), and hence (2) follows for Φ.

To prove (3), we consider

$$E\left\| \int_G \left(\int_0^T \Phi_n(t, \cdot, x)\, dW_t \right) \mu(dx) - \int_G \left(\int_0^T \Phi(t, \cdot, x)\, dW_t \right) \mu(dx) \right\|_H$$

$$\leq \|\|\Phi_n - \Phi\|\| \to 0 \tag{2.48}$$

and

$$E\left\| \int_0^T \left(\int_G \Phi_n(t, \cdot, x)\, \mu(dx) \right) dW_t - \int_0^T \left(\int_G \Phi(t, \cdot, x)\, \mu(dx) \right) dW_t \right\|_H$$

$$\leq \left(E\left\| \int_0^T \left(\int_G (\Phi_n(t, \cdot, x) - \Phi(t, \cdot, x))\, \mu(dx) \right) dW_t \right\|_H^2 \right)^{1/2}$$

$$\leq \left(E \int_0^T \left\| \int_G (\Phi_n(t, \cdot, x) - \Phi(t, \cdot, x))\, \mu(dx) \right\|_{\mathscr{L}_2(K_Q, H)}^2 \right)^{1/2}$$

$$= \left\| \int_G (\Phi_n(\cdot, \cdot, x) - \Phi(\cdot, \cdot, x))\, \mu(dx) \right\|_{\Lambda_2(K_Q, H)}$$

$$\leq \int_G \|\Phi_n(\cdot, \cdot, x) - \Phi(\cdot, \cdot, x)\|_{\Lambda_2(K_Q, H)}\, \mu(dx)$$

$$= \|\|\Phi_n - \Phi\|\| \to 0. \tag{2.49}$$

Now, (3) follows for Φ from (2.48) and (2.49), since it is valid for Φ_n.

(B) If Φ is bounded in the $\| \cdot \|_{\mathscr{L}_2(K_Q, H)}$-norm, then it can be approximated in $\|\| \cdot \|\|$ by bounded elementary processes

$$\Phi(t, \omega, x) = \Phi(0, \omega, x) 1_{\{0\}}(t) + \sum_{j=1}^{n-1} \Phi_j(t_j, \omega, x) 1_{(t_j, t_{j+1}]}(t), \tag{2.50}$$

where $0 \leq t_1 \leq \cdots \leq t_n = T$, and the size of partition of $[0, T]$ converges to 0.

This can be seen by replacing P with $P \otimes \mu$ in Proposition 2.2. Since μ is finite and Φ is bounded, we have

$$\left(\int_G \left(E \int_0^T \| \Phi_n(t, \omega, x) - \Phi(t, \omega, x) \|_{\mathscr{L}_2(K_Q, H)}^2 \, dt \right)^{1/2} \mu(dx) \right)^2$$

$$\leq \mu(G) \int_G \left(E \int_0^T \| \Phi_n(t, \omega, x) - \Phi(t, \omega, x) \|_{\mathscr{L}_2(K_Q, H)}^2 \, dt \right) \mu(dx)$$

$$\leq \int_{G \times \Omega} \int_0^T \| \Phi_n(t, \omega, x) - \Phi(t, \omega, x) \|_{\mathscr{L}_2(K_Q, H)}^2 \, dt \left(P(d\omega) \otimes \mu(dx) \right)$$

with $\Phi(t, \omega, x)$, square integrable with respect to $dt \otimes dP \otimes d\mu$, so that Proposition 2.2 gives the desired approximation.

Clearly, $\Phi_n(\cdot, \cdot, x)$ is $\{\mathscr{F}_t\}_{t \leq T}$-adapted for any $x \in G$, and the stochastic integral $\int_0^T \Phi_n(t, \cdot, x) \, dW_t$ is $\mathscr{F}_T \otimes \mathscr{G} / \mathscr{B}(H)$-measurable.

Since for every $t \in T$ and $A \in \mathscr{L}_2(K_Q, H)$,

$$\langle \Phi_n(t, \cdot, \cdot), A \rangle_{\mathscr{L}_2(K_Q, H)}$$

is $\mathscr{F}_T \otimes \mathscr{G} / \mathscr{B}(\mathbb{R})$-measurable and $P \otimes \mu$-integrable, then by the classical Fubini theorem, the function

$$\int_G \langle \Phi_n(t, \cdot, x), A \rangle_{\mathscr{L}_2(K_Q, H)} \mu(dx)$$

is $\mathscr{F}_t$-measurable, and so is the function

$$\int_G \Phi_n(t, \cdot, x) \mu(dx)$$

by the separability of $\mathscr{L}_2(K_Q, H)$.

Obviously, (3) holds for Φ_n.

Now (B) follows by repeating the arguments in (A), since Φ_n satisfies conditions (1)–(3) and $\|\| \Phi_n - \Phi \|\| \to 0$, so that (2.48) and (2.49) are also valid here. $\square$

Let us now discuss the cylindrical case. In the statement of Theorem 2.8 we can consider $\tilde{W}_t$, a cylindrical Wiener process, and the stochastic integral $\int_0^t \Phi(s) \, d\tilde{W}_s$. Definitions 2.10 and 2.14 differ only by the choice of Q being either a trace-class operator or $Q = I_K$, but in both cases the integrands are in $\Lambda_2(K_Q, H)$. Both stochastic integrals are isometries by either (2.19) or (2.32). We have therefore the following conclusion.

Corollary 2.3 *Under the assumptions of Theorem 2.8, with condition (2.45) replaced with*

$$\|\|\Phi\|\|_1 = \int_G \|\Phi(\cdot, \cdot, x)\|_{\Lambda_2(K,H)} \mu(dx) < \infty, \tag{2.51}$$

conclusions (1)–(3) *of the stochastic Fubini theorem hold for the stochastic integral* $\int_0^T \Phi(t, \cdot, \cdot) d\tilde{W}_t$ *with respect to a standard cylindrical Wiener process* $\{\tilde{W}_t\}_{t\geq 0}$.

2.3 The Itô Formula

We will present a theorem that gives conditions under which a stochastic process $F(t, X(t))$ has a stochastic differential, provided that $X(t)$ has a stochastic differential. First we explain some generally used notation.

2.3.1 The case of a Q-Wiener process

If $\phi \in \mathscr{L}_2(K_Q, H)$ and $\psi \in H$, then $\phi^*\psi \in \mathscr{L}_2(K_Q, \mathbb{R})$, since

$$\|\phi^*\psi\|^2_{\mathscr{L}_2(K_Q,\mathbb{R})} = \sum_{j=1}^\infty \left((\phi^*\psi)(\lambda^{1/2}f_j)\right)^2 = \sum_{j=1}^\infty \langle \psi, \phi(\lambda^{1/2}f_j)\rangle^2_H$$

$$\leq \|\psi\|^2_H \|\phi\|^2_{\mathscr{L}_2(K_Q,H)}.$$

Hence, if $\Phi(s) \in \mathscr{P}(K_Q, H)$ and $\Psi(s) \in H$ are $\mathscr{F}_t$-adapted processes, then the process $\Phi^*(s)\Psi(s)$ defined by

$$\left(\Phi^*(s)\Psi(s)\right)(k) = \langle \Psi(s), \Phi(s)(k)\rangle_H$$

has values in $\mathscr{L}_2(K_Q, \mathbb{R})$. If, in addition, P-a.s., $\Psi(s)$ is bounded as a function of s, then

$$P\left(\int_0^T \|\Phi^*(s)\Psi(s)\|^2_{\mathscr{L}_2(K_Q,\mathbb{R})} ds < \infty\right) = 1,$$

so that $\Phi^*(s)\Psi(s) \in \mathscr{P}(K_Q, \mathbb{R})$, and we can define

$$\int_0^T \langle \Psi(s), \Phi(s) dW_s\rangle_H = \int_0^T \Phi^*(s)\Psi(s) dW_s.$$

Theorem 2.9 (Itô Formula) *Let Q be a symmetric nonnegative trace-class operator on a separable Hilbert space K, and let $\{W_t\}_{0\leq t\leq T}$ be a Q-Wiener process on a filtered probability space $(\Omega, \mathscr{F}, \{\mathscr{F}_t\}_{0\leq t\leq T}, P)$. Assume that a stochastic process $X(t), 0 \leq t \leq T$, is given by*

$$X(t) = X(0) + \int_0^t \Psi(s) ds + \int_0^t \Phi(s) dW_s, \tag{2.52}$$

where $X(0)$ is an $\mathscr{F}_0$-measurable H-valued random variable, $\Psi(s)$ is an H-valued $\mathscr{F}_s$-measurable P-a.s. Bochner-integrable process on $[0, T]$,

$$\int_0^T \|\Psi(s)\|_H \, ds < \infty \quad P\text{-a.s.},$$

and $\Phi \in \mathscr{P}(K_Q, H)$.

Assume that a function $F : [0, T] \times H \to \mathbb{R}$ is such that F is continuous and its Fréchet partial derivatives F_t, F_x, F_{xx} are continuous and bounded on bounded subsets of $[0, T] \times H$. Then the following Itô's formula holds:

$$\begin{aligned}
F(t, X(t)) = F(0, X(0)) &+ \int_0^t \langle F_x(s, X(s)), \Phi(s)dW_s \rangle_H \\
&+ \int_0^t \{ F_t(s, X(s)) + \langle F_x(s, X(s)), \Psi(s) \rangle_H \\
&+ \frac{1}{2} \mathrm{tr}[F_{xx}(s, X(s))(\Phi(s)Q^{1/2})(\Phi(s)Q^{1/2})^*]\} \, ds \quad (2.53)
\end{aligned}$$

P-a.s. for all $t \in [0, T]$.

Proof We will first show that the general statement can be reduced to the case of constant processes $\Psi(s) = \Psi$ and $\Phi(s) = \Phi$, $s \in [0, T]$. For a constant $C > 0$, define the stopping time

$$\begin{aligned}
\tau_C = \inf\Big\{ t \in [0, T] : \max\Big(\|X(t)\|_H, \int_0^t \|\Psi(s)\|_H \, ds, \\
\int_0^t \|\Phi(s)\|_{\mathscr{L}_2(K_Q, H)}^2 \, ds \Big) \geq C \Big\}
\end{aligned}$$

with the convention that the infimum of an empty set equals T.

Then, with the notation $X_C(t) = X(t \wedge \tau_C)$, $\Psi_C(t) = \Psi(t)1_{[0, \tau_C]}(t)$, and $\Phi_C(t) = \Phi(t)1_{[0, \tau_C]}(t)$, we have

$$X_C(t) = X_C(0) + \int_0^t \Psi_C(s) \, ds + \int_0^t \Phi_C(s) \, dW_s, t \in [0, T].$$

By (2.30), it is sufficient to prove Itô's formula for the processes stopped at τ_C. Since

$$P\left(\int_0^T \|\Psi_C(s)\|_H \, ds < \infty \right) = 1$$

and

$$E \int_0^T \|\Phi_C(s)\|_{\mathscr{L}_2(K_Q, H)}^2 \, ds < \infty,$$

by Lemma 2.4 and Corollary 2.1 it follows that Ψ_C and Φ_C can be approximated respectively by sequences of bounded elementary processes $\Psi_{C,n}$ and $\Phi_{C,n}$ for which P-a.s. uniformly in $t \leq T$

$$\int_0^t \|\Psi_{C,n}(s) - \Psi_C(s)\|_H \, ds \to 0$$

and

$$\left\| \int_0^t \Phi_{C,n}(s) \, dW_s - \int_0^t \Phi_C(s) \, dW_s \right\|_H \to 0.$$

Define

$$X_{C,n}(t) = X(0) + \int_0^t \Psi_{C,n}(s) \, ds + \int_0^t \Phi_{C,n}(s) \, dW_s.$$

Then

$$\sup_{t \leq T} \|X_{C,n}(t) - X_C(t)\|_H \to 0$$

with probability one. Assume that we have shown Itô's formula for the process $X_{C,n}(t)$, that is,

$$F\bigl(t, X_{C,n}(t)\bigr) = F\bigl(0, X(0)\bigr) + \int_0^t \bigl\langle F_x\bigl(s, X_{C,n}(s)\bigr), \Phi_{C,n}(s)dW_s \bigr\rangle_H$$

$$+ \int_0^t \Bigl\{ F_t\bigl(s, X_{C,n}(s)\bigr) + \bigl\langle F_x\bigl(s, X_{C,n}(s)\bigr), \Psi_{C,n}(s) \bigr\rangle_H$$

$$+ \frac{1}{2} \operatorname{tr}\bigl[F_{xx}\bigl(s, X_{C,n}(s)\bigr)\bigl(\Phi_{C,n}(s)Q^{1/2}\bigr)\bigl(\Phi_{C,n}(s)Q^{1/2}\bigr)^* \bigr] \Bigr\} \, ds \tag{2.54}$$

P-a.s. for all $t \in [0, T]$. Using the continuity of F and the continuity and local boundedness of its partial derivatives, we will now conclude that

$$F\bigl(t, X_C(t)\bigr) = F\bigl(0, X(0)\bigr) + \int_0^t \bigl\langle F_x\bigl(s, X_C(s)\bigr), \Phi_C(s)dW_s \bigr\rangle_H$$

$$+ \int_0^t \Bigl\{ F_t\bigl(s, X_C(s)\bigr) + \bigl\langle F_x\bigl(s, X_C(s)\bigr), \Psi_C(s) \bigr\rangle_H$$

$$+ \frac{1}{2} \operatorname{tr}\bigl[F_{xx}\bigl(s, X_C(s)\bigr)\bigl(\Phi_C(s)Q^{1/2}\bigr)\bigl(\Phi_C(s)Q^{1/2}\bigr)^* \bigr] \Bigr\} \, ds. \tag{2.55}$$

Clearly, the LHS of (2.54) converges to the LHS of (2.55) a.s.

For the stochastic integrals in (2.54) and (2.55), we have

$$E\left|\int_0^t \langle F_x(s, X_{C,n}(s)), \Phi_{C,n}(s)\, dW_s\rangle_H - \int_0^t \langle F_x(s, X_C(s)), \Phi_C(s)\, dW_s\rangle_H\right|^2$$

$$\leq 2\int_0^t E\left\|\Phi_C^*(s)\big(F_x(s, X_{C,n}(s)) - F_x(s, X_C(s))\big)\right\|_{\mathscr{L}_2(K_Q,\mathbb{R})}^2 ds$$

$$+ 2\int_0^t E\left\|(\Phi_C^*(s) - \Phi_{C,n}^*)F_x(s, X_{C,n}(s))\right\|_{\mathscr{L}_2(K_Q,\mathbb{R})}^2 ds$$

$$\leq 2\int_0^t E\big(\|\Phi_C(s)\|_{\mathscr{L}_2(K_Q,H)}^2 \|F_x(s, X_{C,n}(s)) - F_x(s, X_C(s))\|_H^2\big)\, ds$$

$$+ 2\int_0^t E\big(\|\Phi_C^*(s) - \Phi_{C,n}^*\|_{\mathscr{L}_2(K_Q,H)}^2 \|F_x(s, X_{C,n}(s))\|_H^2\big)\, ds.$$

The first integral converges to zero, since the first factor is an integrable process, and the second factor converges to zero almost surely, so that the Lebesgue DCT applies. The second integral is bounded by $M\|\Phi_C^*(s) - \Phi_{C,n}^*\|_{\Lambda_2(K_Q,H)}^2$ for some constant M, so that it converges to zero, since $\Phi_{C,n}(s) \to \Phi_C$ in the space $\Lambda_2(K_Q, H)$.

In conclusion, the stochastic integrals in (2.54) converge to the stochastic integral in (2.55) in mean square, so that they converge in probability.

We now turn to the nonstochastic integrals.

The first component, involving F_t, of the nonstochastic integral in (2.54) converges P-a.s. to the corresponding component in (2.55) by the continuity and local boundedness of F_t, so that the Lebesgue DCT can be applied.

Note that, P-a.s., $\Psi_{C,n_k} \to \Psi_C$ in $L^1([0, t], H)$, and F_x is locally bounded, so that the functions $s \to F_x(s, X_{C,n_k}(s))$ and $s \to F_x(s, X_C(s))$ are in $L^\infty([0, t], H)$. The convergence with probability one of the second component follows from the duality argument.

To discuss the last nonstochastic integral, we use the fact that

$$\|\Phi_{C,n}(s) - \Phi_C(s)\|_{\Lambda_2(K_Q,H)} \to 0$$

and select a subsequence n_k for which

$$\|\Phi_{C,n_k}(s) - \Phi_C(s)\|_{\mathscr{L}_2(K_Q,H)} \to 0,$$

and therefore, for the eigenvectors f_j of Q,

$$\|\Phi_{C,n_k}(s)f_j - \Phi_C(s)f_j\|_H \to 0 \tag{2.56}$$

a.e. on $[0, T] \times \Omega$. By Exercise 2.19,

$$\mathrm{tr}\big(F_{xx}(s, X_{C,n_k}(s))\Phi_{C,n_k}(s)Q\Phi_{C,n_k}^*(s)\big)$$

$$= \mathrm{tr}\big(\Phi_{C,n_k}^*(s)F_{xx}(s, X_{C,n_k}(s))\Phi_{C,n_k}(s)Q\big)$$

$$= \sum_{j=1}^{\infty} \lambda_j \big\langle F_{xx}\big(s, X_{C,n_k}(s)\big) \Phi_{C,n_k}(s) f_j, \, \Phi_{C,n_k}(s) f_j \big\rangle_H.$$

Since $X_{C,n_k}(s)$ is bounded, the continuity of F_{xx} and (2.56) imply that

$$\big\langle F_{xx}\big(s, X_{C,n_k}(s)\big) \Phi_{C,n_k}(s) f_j, \, \Phi_{C,n_k}(s) f_j \big\rangle_H$$
$$\to \big\langle F_{xx}\big(s, X_C(s)\big) \Phi_C(s) f_j, \, \Phi_C(s) f_j \big\rangle_H.$$

By the Lebesgue DCT (with respect to the counting measure), we get that, a.e. on $[0, T] \times \Omega$,

$$\mathrm{tr}\big(F_{xx}\big(s, X_{C,n_k}(s)\big) \Phi_{C,n_k}(s) Q \Phi^*_{C,n_k}(s)\big) \to \mathrm{tr}\big(F_{xx}\big(s, X_C(s)\big) \Phi_C(s) Q \Phi^*_C(s)\big)$$

and the LHS is bounded by the functions

$$\eta_n(s) = \big\| F_{xx}\big(s, X_{C,n_k}(s)\big) \big\|_{\mathscr{L}(H)} \| \Phi_{C,n_k} \|^2_{\Lambda_2(K_Q, H)}$$

that converge P-a.s. to

$$\eta(s) = \big\| F_{xx}\big(s, X_C(s)\big) \big\|_{\mathscr{L}(H)} \| \Phi_C \|^2_{\Lambda_2(K_Q, H)}.$$

However, by the boundedness of the second derivative of F, $\int_0^t \eta_n(s)\,ds \to \int_0^t \eta(s)\,ds$, so that we can apply the general Lebesgue DCT, Theorem 3.4, to conclude the convergence with probability one of the last nonstochastic integral.[1]

In conclusion, possibly for a subsequence, left- and right-hand sides of (2.54) converge in probability to the left- and right-hand sides of (2.55), so that (2.55) holds P-a.s.

By the additivity of the integrals we can further reduce the proof to the case where

$$X(t) = X(0) + \Psi t + \Phi W_t,$$

where Ψ and Φ are $\mathscr{F}_0$-measurable random variables independent of t.

Now, define

$$u(t, W_t) = F\big(t, X(0) + \Psi t + \Phi W_t\big);$$

then the function u is of the same smoothness order as F. We will now prove Itô's formula for the function u.

---

[1] This elementary proof can be replaced by the following argument. The space of trace-class operators $\mathscr{L}_1(H)$ can be identified with the dual space to the space of compact linear operators on H, the duality between the two spaces is the trace operator ([68], Chap. IV, Sect. 1, Theorem 1). Hence, as a dual separable space, it has the Radon–Nikodym property ([14], Chap. III, Sect. 3, Theorem 1). Thus, $L^1([0, T], \mathscr{L}_1(H))^* = L^\infty([0, T], \mathscr{L}_1(H)^*)$ ([14], Chap. IV, Sect. 1, Theorem 1). But $L^\infty([0, T], \mathscr{L}_1(H)^*) = L^\infty([0, T], \mathscr{L}(H))$ ([68], Chap. IV, Sect. 1, Theorem 2). Thus the convergence of the last nonstochastic integral follows from the duality argument.

Let $0 = t_1 < t_2 < \cdots < t_n = t \le T$ be a partition of an interval $[0, t]$ and denote $\Delta t_j = t_{j+1} - t_j$ and $\Delta W_j = W_{t_{j+1}} - W_{t_j}$. Using Taylor's formula, there exist random variables $\bar{t}_j \in [t_j, t_{j+1}]$ and $\theta_j \in [0, 1]$ such that

$$u(t, W_t) - u(0, 0)$$

$$= \sum_{j=1}^{n-1} \left[u(t_{j+1}, W_{t_{j+1}}) - u(t_j, W_{t_{j+1}}) \right] + \sum_{j=1}^{n-1} \left[u(t_j, W_{t_{j+1}}) - u(t_j, W_{t_j}) \right]$$

$$= \sum_{j=1}^{n-1} u_t(\bar{t}_j, W_{t_{j+1}}) \Delta t_j$$

$$+ \sum_{j=1}^{n-1} \left[\langle u_x(t_j, W_{t_j}), \Delta W_j \rangle_K + \frac{1}{2} \langle u_{xx}(t_j, \bar{W}_j)(\Delta W_j), \Delta W_j \rangle_K \right]$$

$$= \sum_{j=1}^{n-1} u_t(t_j, W_{t_{j+1}}) \Delta t_j + \sum_{j=1}^{n-1} \langle u_x(t_j, W_{t_j}), \Delta W_j \rangle_K$$

$$+ \frac{1}{2} \sum_{j=1}^{n-1} \langle u_{xx}(t_j, W_{t_j})(\Delta W_j), \Delta W_j \rangle_K$$

$$+ \sum_{j=1}^{n-1} \left[u_t(\bar{t}_j, W_{t_{j+1}}) - u_t(t_j, W_{t_{j+1}}) \right] \Delta t_j$$

$$+ \frac{1}{2} \sum_{j=1}^{n-1} \langle \left[u_{xx}(t_j, \bar{W}_j)(\Delta W_j) - u_{xx}(t_j, W_{t_j})(\Delta W_j) \right], \Delta W_j \rangle_K, \qquad (2.57)$$

where $\bar{W}_j = W_{t_j} + \theta_j(W_{t_{j+1}} - W_{t_j})$.

Assuming that $\max\{t_{j+1} - t_j, 1 \le j \le n - 1\} \to 0$ and using the uniform continuity of the mapping $[0, T] \times [0, T] \ni (s, r) \to u_t(s, W_r) \in \mathbb{R}$ (Exercise 2.20) and the continuity of the map $[0, T] \ni t \to u_x(t, W_t) \in K^*$, we get

$$\sum_{j=1}^{n-1} u_t(t_j, W_{t_{j+1}}) \Delta t_j + \sum_{j=1}^{n-1} \langle u_x(t_j, W_{t_j}), \Delta W_j \rangle_K$$

$$\to \int_0^t u_t(s, W_s) \, ds + \int_0^t \langle u_x(s, W_s), dW_s \rangle_K \qquad P\text{-a.s.}$$

by Lemma 2.6. Clearly,

$$\left| \sum_{j=1}^{n-1} \left[u_t(\bar{t}_j, W_{t_{j+1}}) - u_t(t_j, W_{t_{j+1}}) \right] \Delta t_j \right|$$

$$\le T \sup_{j \le n} \left| u_t(\bar{t}_j, W_{t_{j+1}}) - u_t(t_j, W_{t_{j+1}}) \right| \to 0.$$

In view of the continuity of the mapping $[0, T] \times K \ni (t, x) \to u_{xx}(t, x) \in \mathscr{L}(K, K)$, utilizing the result in Exercise 2.5, we have

$$\left| \sum_{j=1}^{n-1} \langle [u_{xx}(t_j, \bar{W}_j) - u_{xx}(t_j, W_{t_j})](\Delta W_j), \Delta W_j \rangle_K \right|$$

$$\leq \sup_{j \leq n-1} \left\| u_{xx}(t_j, \bar{W}_j)(\Delta W_j) - u_{xx}(t_j, W_{t_j})(\Delta W_j) \right\|_{\mathscr{L}(K,K)} \sum_{j=1}^{n-1} \|\Delta W_j\|_K^2 \to 0$$

with probability one as $n \to \infty$ (using arguments as in Exercise 2.20).

It remains to show that

$$\sum_{j=1}^{n-1} \langle u_{xx}(t_j, W_{t_j})(\Delta W_j), \Delta W_j \rangle_K \to \int_0^t \mathrm{tr}[u_{xx}(s, W_s)Q] \, ds \qquad (2.58)$$

in probability P.

Let $1_j^N = 1_{\{\max\{\|W_{t_i}\|_K \leq N, i \leq j\}\}}$. Then 1_j^N is $\mathscr{F}_{t_j}$-measurable, and, using the representation (2.3), we get

$$E\left(\langle 1_j^N u_{xx}(t_j, W_{t_j})(\Delta W_j), \Delta W_j \rangle_K \big| \mathscr{F}_{t_j}\right)$$

$$= E\left(\left\langle 1_j^N u_{xx}(t_j, W_{t_j}) \sum_{k=1}^{\infty} \lambda_k^{1/2}(w_k(t_{j+1}) - w_k(t_j)) f_k, \right.\right.$$

$$\left.\left. \sum_{l=1}^{\infty} \lambda_l^{1/2}(w_l(t_{j+1}) - w_l(t_j)) f_l \right\rangle_K \bigg| \mathscr{F}_{t_j}\right)$$

$$= \sum_{k=1}^{\infty} E\left(\lambda_k \langle 1_j^N u_{xx}(t_j, W_{t_j}) f_k, f_k \rangle_K (w_k(t_{j+1}) - w_k(t_j))^2 \big| \mathscr{F}_{t_j}\right)$$

$$= \mathrm{tr}\left(1_j^N u_{xx}(t_j, W_{t_j})Q\right) \Delta t_j.$$

In view of the above and the fact that u_{xx} is bounded on bounded subsets of $[0, T] \times H$, we obtain

$$E\left(\sum_{j=1}^{n-1} (\langle 1_j^N u_{xx}(t_j, W_{t_j})(\Delta W_j), \Delta W_j \rangle_K - \mathrm{tr}(1_j^N u_{xx}(t_j, W_{t_j})Q)\Delta t_j)\right)^2$$

$$= \sum_{j=1}^{n-1} (E\langle 1_j^N u_{xx}(t_j, W_{t_j})(\Delta W_j), \Delta W_j \rangle_K^2$$

$$- E\left(\mathrm{tr}(1_j^N u_{xx}(t_j, W_{t_j})Q)\right)^2 (\Delta t_j)^2)$$

$$\leq \sup_{s \leq t, \|h\|_H \leq N} |u_{xx}(s,h)|^2_{\mathscr{L}(H)} \sum_{j=1}^{n-1} \left(E\|\Delta W_j\|^4_K - (\operatorname{tr} Q)^2 (\Delta t_j)^2 \right)$$

$$= 2 \sup_{s \leq t, \|h\|_H \leq N} |u_{xx}(s,h)|^2_{\mathscr{L}(H)} \|Q\|^2_{\mathscr{L}_2(K)} \sum_{j=1}^{n-1} (\Delta t_j)^2 \to 0.$$

Also, as $N \to \infty$,

$$P\Bigg(\sum_{j=1}^{n-1} (1 - 1_j^N) \big((u_{xx}(t_j, W_{t_j})(\Delta W_j), \Delta W_j)_K$$

$$- \operatorname{tr}(1_j^N u_{xx}(t_j, W_{t_j}) Q \Delta t_j) \big) \neq 0 \Bigg)$$

$$\leq P\Big(\sup_{s \leq t} \{ \|W_s\| > N \} \Big) \to 0.$$

This proves (2.58). Taking the limit in (2.57), we obtain Itô's formula for the function $u(t, W_t)$,

$$u(t, W_t) = u(0,0) + \int_0^t \left(u_t(s, W_s) + \frac{1}{2} \operatorname{tr}(u_{xx}(s, W_s)Q) \right) ds$$

$$+ \int_0^t (u_x(s, W_s), dW_s)_K. \qquad (2.59)$$

To obtain Itô's formula for $F(t, X(t))$, we calculate the derivatives

$$u_t(t, k) = F_t(t, X(0) + \Psi t + \Phi k) + \langle F_x(t, X(0) + \Psi t + \Phi k), \Psi \rangle_K,$$

$$u_x(t, k) = \Phi^* F_x(t, X(0) + \Psi t + \Phi k),$$

$$u_{xx}(t, k) = \Phi^* F_{xx}(t, X(0) + \Psi t + \Phi k) \Phi.$$

Noting that (see Exercise 2.19)

$$\operatorname{tr}\big[F_{xx}(s, X(s))(\Phi Q^{1/2})(\Phi Q^{1/2})^* \big] = \operatorname{tr}\big[\Phi^* F_{xx}(s, X(s)) \Phi Q \big],$$

we arrive at the desired result (2.53). $\qquad \square$

Exercise 2.19 Show that, for a symmetric operator $T \in \mathscr{L}(H)$ and $\Phi \in \mathscr{L}(K, H)$,

$$\operatorname{tr}(T \Phi Q \Phi^*) = \operatorname{tr}(\Phi^* T \Phi Q).$$

Exercise 2.20 Let $f : ([0, T] \times K) \to \mathbb{R}$ and $g : [0, T] \to K$ be continuous. Show that the mapping $[0, T] \times [0, T] \ni (s, r) \to f(s, g(r)) \in \mathbb{R}$ is uniformly continuous.

2.3.2 *The Case of a Cylindrical Wiener Process*

As in the case of a Q-Wiener process, for $\Phi(s) \in \mathcal{P}(K, H)$ and a P-a.s. bounded H-valued $\mathcal{F}_t$-adapted process $\Psi(s)$, $\Phi^*(s)\Psi(s) \in \mathcal{P}(K, \mathbb{R})$. In addition, since

$$\sum_{j=1}^{\infty} \left(\left(\Phi^*(s)\Psi(s) \right)(f_j) \right)^2 = \sum_{j=1}^{\infty} \langle \Psi(s), \Phi(s)(f_j) \rangle_H^2 \leq \|\Psi(s)\|_H^2 \|\Phi(s)\|_{\mathcal{L}_2(K_Q, H)}^2,$$

the process $\Phi^*(s)\Psi(s)$ can be considered as being K- or K^*-valued, and we can define

$$\int_0^T \langle \Psi(s), \Phi(s)\, d\tilde{W}_s \rangle_H = \int_0^T \langle \Phi^*(s)\Psi(s), d\tilde{W}_s \rangle_K = \int_0^T \Phi^*(s)\Psi(s)\, d\tilde{W}_s.$$

Theorem 2.10 (Itô Formula) *Let H and K be real separable Hilbert spaces, and $\{\tilde{W}_t\}_{0 \leq t \leq T}$ be a K-valued cylindrical Wiener process on a filtered probability space $(\Omega, \mathcal{F}, \{\mathcal{F}_t\}_{0 \leq t \leq T}, P)$. Assume that a stochastic process $X(t), 0 \leq t \leq T$, is given by*

$$X(t) = X(0) + \int_0^t \Psi(s)\, ds + \int_0^t \Phi(s)\, d\tilde{W}_s, \tag{2.60}$$

where $X(0)$ is an $\mathcal{F}_0$-measurable H-valued random variable, $\Psi(s)$ is an H-valued $\mathcal{F}_s$-measurable P-a.s. Bochner-integrable process on $[0, T]$,

$$\int_0^T \|\Psi(s)\|_H\, ds < \infty, \quad P\text{-a.s.},$$

and $\Phi \in \mathcal{P}(K, H)$.

Assume that a function $F : [0, T] \times H \to \mathbb{R}$ is such that F is continuous and its Fréchet partial derivatives F_t, F_x, F_{xx} are continuous and bounded on bounded subsets of $[0, T] \times H$. Then the following Itô's formula holds:

$$\begin{aligned}
F(t, X(t)) = F(0, X(0)) &+ \int_0^t \langle F_x(s, X(s)), \Phi(s)d\tilde{W}_s \rangle_H \\
&+ \int_0^t \Big\{ F_t(s, X(s)) + \langle F_x(s, X(s)), \Psi(s) \rangle_H \\
&+ \frac{1}{2} \text{tr}\big[F_{xx}(s, X(s))\Phi(s)(\Phi(s))^* \big] \Big\}\, ds
\end{aligned} \tag{2.61}$$

P-a.s. for all $t \in [0, T]$.

Proof The proof is nearly identical to the proof of the Itô formula for a Q-Wiener process, and we refer to the notation in the proof of Theorem 2.9. The reduction to the processes $X_C(t) = X(t \wedge \tau_C)$, $\Psi_C(t) = \Psi(t)1_{[0,\tau_C]}(t)$, $\Phi_C(t) = \Phi(t)1_{[0,\tau_C]}(t)$, with

$$X_C(t) = X_C(0) + \int_0^t \Psi_C(s)\,ds + \int_0^t \Phi_C(s)\,d\tilde{W}_s, \quad t \in [0,T],$$

is possible due to (2.33).

A further reduction to bounded elementary processes $\Psi_{C,n}$ and $\Phi_{C,n}$ for which, P-a.s. uniformly in $t \leq T$,

$$\int_0^t \|\Psi_{C,n}(s) - \Psi_C(s)\|_H\,ds \to 0$$

and

$$\left\| \int_0^t \Phi_{C,n}(s)\,d\tilde{W}_s - \int_0^t \Phi_C(s)\,d\tilde{W}_s \right\|_H \to 0$$

is achieved using Lemma 2.4 and Corollary 2.1 with $Q = I_K$, so that we can define

$$X_{C,n}(t) = X(0) + \int_0^t \Psi_{C,n}(s)\,ds + \int_0^t \Phi_{C,n}(s)\,d\tilde{W}_s$$

and claim that

$$\sup_{t \leq T} \|X_{C,n}(t) - X_C(t)\|_H \to 0$$

with probability one. Then, using the isometry property (2.32) and the arguments in the proof of Theorem 2.9 that justify the term-by-term convergence of (2.54) to (2.55), we can reduce the general problem to the case

$$X(t) = X(0) + \Psi t + \Phi\tilde{W}_t, \tag{2.62}$$

where, recalling (2.31) and (2.12),

$$\Phi\tilde{W}_t = \sum_{i=1}^{\infty} \left((\Phi\tilde{W}_t)e_i\right)e_i = \sum_{i=1}^{\infty} \left(\tilde{W}_t(\Phi^*e_i)\right)e_i \in H \tag{2.63}$$

for $\Phi \in \mathscr{L}_2(K,H)$.

From here we proceed as follows. Define

$$u(t,\xi_t) = F\left(t, X(0) + \Psi t + \xi_t\right)$$

with $\xi_t = \Phi\tilde{W}_t \in \mathscr{M}_T^2(H)$. Similarly as in the proof of Theorem 2.9, with $0 = t_1 < t_2 < \cdots < t_n = t \leq T$, $\Delta t_j = t_{j+1} - t_j$, and $\Delta\xi_j = \xi_{t_{j+1}} - \xi_{t_j}$, using Taylor's formula, we obtain

$$u(t, \xi_t) - u(0,0)$$

$$= \sum_{j=1}^{n-1} u_t(t_j, \xi_{t_{j+1}}) \Delta t_j + \sum_{j=1}^{n-1} \langle u_x(t_j, \xi_{t_j}), \Delta \xi_j \rangle_H$$

$$+ \frac{1}{2} \sum_{j=1}^{n-1} \langle u_{xx}(t_j, \xi_{t_j})(\Delta \xi_j), \Delta \xi_j \rangle_H$$

$$+ \sum_{j=1}^{n-1} [u_t(\tilde{t}_j, \xi_{t_{j+1}}) - u_t(t_j, \xi_{t_{j+1}})] \Delta t_j$$

$$+ \frac{1}{2} \sum_{j=1}^{n-1} \langle [u_{xx}(t_j, \tilde{\xi}_j)(\Delta \xi_j) - u_{xx}(t_j, \xi_{t_j})(\Delta \xi_j)], \Delta \xi_j \rangle_H,$$

$$= S_1 + S_2 + S_3 + S_4 + S_5,$$

where $\Delta \xi_j = \Phi(\tilde{W}_{t_{j+1}} - \tilde{W}_{t_j})$, $\tilde{\xi}_j = \Phi \tilde{W}_{t_j} + \theta_j \Phi(\tilde{W}_{t_{j+1}} - \tilde{W}_{t_j})$, and $\tilde{t}_j \in [t_j, t_{j+1}]$, $\theta_j \in [0, 1]$ are random variables.

Using the smoothness of the function u, we conclude that S_4 and S_5 converge to zero with probability one as $n \to \infty$ and that

$$S_1 + S_2 \to \int_0^t u_t(s, \xi_s) \, ds + \int_0^t \langle \Phi^* u_x(s, \xi_s), d\tilde{W}_s \rangle_K$$

$$= \int_0^t u_t(s, \xi_s) \, ds + \int_0^t \langle u_x(s, \xi_s), \Phi \, d\tilde{W}_s \rangle_H.$$

To show that

$$\sum_{j=1}^{n-1} \langle u_{xx}(t_j, \xi_{t_j})(\Delta \xi_j), \Delta \xi_j \rangle_H \to \int_0^t \mathrm{tr}[u_{xx}(s, \tilde{W}_s) \Phi \Phi^*] \, ds$$

in probability P, we let $1_j^N = 1_{\{\max\{\|\xi_{t_i}\|_H \le N, i \le j\}\}}$. Then 1_j^N is $\mathscr{F}_{t_j}$-measurable, and, using the representation (2.63), we get

$$E(\langle 1_j^N u_{xx}(t_j, \xi_{t_j})(\Delta \xi_j), \Delta \xi_j \rangle_H | \mathscr{F}_{t_j})$$

$$= E\left(\left\langle 1_j^N u_{xx}(t_j, \xi_{t_j}) \sum_{i=1}^\infty (\tilde{W}_{t_{j+1}}(\Phi^* e_i) - \tilde{W}_{t_j}(\Phi^* e_i)) e_i, \right.\right.$$

$$\left.\left. \sum_{l=1}^\infty (\tilde{W}_{t_{j+1}}(\Phi^* e_l) - \tilde{W}_{t_j}(\Phi^* e_l)) e_l \right\rangle_H \bigg| \mathscr{F}_{t_j} \right)$$

$$= \sum_{i=1}^{\infty} E\left(\langle 1_j^N u_{xx}(t_j, \xi_{t_j}) e_i, e_i \rangle_H \left(\tilde{W}_{t_{j+1}}(\Phi^* e_i) - \tilde{W}_{t_j}(\Phi^* e_i)\right)^2 \Big| \mathscr{F}_{t_j}\right)$$

$$= \operatorname{tr}\left(1_j^N u_{xx}(t_j, \xi_{t_j}) \Phi \Phi^*\right) \Delta t_j.$$

Now, to complete the proof, we can now follow the arguments in the proof of Theorem 2.9 with $\Phi\Phi^*$ replacing Q. $\qquad\square$

Chapter 3
Stochastic Differential Equations

3.1 Stochastic Differential Equations and Their Solutions

Let K and H be real separable Hilbert spaces, and W_t be a K-valued Q-Wiener process on a complete filtered probability space $(\Omega, \mathscr{F}, \{\mathscr{F}_t\}_{t \leq T}, P)$ with the filtration $\mathscr{F}_t$ satisfying the *usual conditions*. We consider semilinear SDEs (SSDEs for short) on $[0, T]$ in H. The general form of such SSDE is

$$\begin{cases} dX(t) = (AX(t) + F(t, X)) \, dt + B(t, X) \, dW_t, \\ X(0) = \xi_0. \end{cases} \tag{3.1}$$

Here, $A : \mathscr{D}(A) \subset H \to H$ is the generator of a C_0-semigroup of operators $\{S(t), \ t \geq 0\}$ on H. Recall from Chap. 1 that for a C_0-semigroup $S(t)$, we have $\|S(t)\|_{\mathscr{L}(H)} \leq M \exp\{\alpha t\}$ and if $M = 1$, then $S(t)$ is called a pseudo-contraction semigroup.

The coefficients F and B are, in general, nonlinear mappings,

$$F : \Omega \times [0, T] \times C\big([0, T], H\big) \to H,$$

$$B : \Omega \times [0, T] \times C\big([0, T], H\big) \to \mathscr{L}_2(K_Q, H).$$

The initial condition ξ_0 is an $\mathscr{F}_0$-measurable H-valued random variable.

We will study the existence and uniqueness problem under various regularity assumptions on the coefficients of (3.1) that include:

(A1) F and B are jointly measurable, and for every $0 \leq t \leq T$, they are measurable with respect to the product σ-field $\mathscr{F}_t \otimes \mathscr{C}_t$ on $\Omega \times C([0, T], H)$, where $\mathscr{C}_t$ is a σ-field generated by cylinders with bases over $[0, t]$.

(A2) F and B are jointly continuous.

(A3) There exists a constant ℓ such that for all $x \in C([0, T], H)$,

$$\big\|F(\omega, t, x)\big\|_H + \big\|B(\omega, t, x)\big\|_{\mathscr{L}_2(K_Q, H)} \leq \ell\Big(1 + \sup_{0 \leq s \leq T} \|x(s)\|_H\Big)$$

for $\omega \in \Omega$ and $0 \leq t \leq T$.

L. Gawarecki, V. Mandrekar, *Stochastic Differential Equations in Infinite Dimensions*,   73
Probability and Its Applications,
DOI 10.1007/978-3-642-16194-0_3, © Springer-Verlag Berlin Heidelberg 2011

For every $t \in [0, T]$, we define the following operator θ_t on $C([0, T], H)$:

$$\theta_t x(s) = \begin{cases} x(s), & 0 \le s \le t, \\ x(t), & t < s \le T. \end{cases}$$

Assumption (A1) implies that

$$F(\omega, t, x) = F(\omega, t, x_1) \quad \text{and} \quad B(\omega, t, x) = B(\omega, t, x_1)$$

if $x = x_1$ on $[0, t]$. Because $\theta_t x$ is a Borel function of t with values in $C([0, T], H)$, $F(\omega, t, \theta_t x)$ and $B(\omega, t, \theta_t x)$ also are Borel functions in t. With this notation, (3.1) can be rewritten as

$$\begin{cases} dX(t) = (AX(t) + F(t, \theta_t X)) \, dt + B(t, \theta_t X) \, dW_t, \\ X(0) = \xi_0. \end{cases}$$

We will say that F and B satisfy the Lipschitz condition if

(A4) For all $x, y \in C([0, T], H)$, $\omega \in \Omega$, $0 \le t \le T$, there exists $\mathscr{K} > 0$ such that

$$\left\| F(\omega, t, x) - F(\omega, t, y) \right\|_H + \left\| B(\omega, t, x) - B(\omega, t, y) \right\|_{\mathscr{L}_2(K_Q, H)}$$

$$\le \mathscr{K} \sup_{0 \le s \le T} \left\| x(s) - y(s) \right\|_H.$$

To simplify the notation, we will not indicate the dependence of F and B on ω whenever this does not lead to confusion.

There exist different notions of a solution to the semilinear SDE (3.1), and we now define *strong, weak, mild,* and *martingale* solutions.[1]

Definition 3.1 A stochastic process $X(t)$ defined on a filtered probability space $(\Omega, \mathscr{F}, \{\mathscr{F}_t\}_{t \le T}, P)$ and adapted to the filtration $\{\mathscr{F}_t\}_{t \le T}$

(a) is a *strong solution* of (3.1) if
   (1) $X(\cdot) \in C([0, T], H)$;
   (2) $X(t, \omega) \in \mathscr{D}(A)$ $dt \otimes dP$-almost everywhere;
   (3) the following conditions hold:

$$P\left(\int_0^T \| AX(t) \|_H \, dt < \infty \right) = 1,$$

$$P\left(\int_0^T \left(\| F(t, X) \|_H + \| B(t, X) \|_{\mathscr{L}_2(K_Q, H)}^2 \right) dt < \infty \right) = 1;$$

---

[1] A *weak (mild)* solution is called *mild* (respectively *mild integral*) solution in [9], where also a concept of a *weakened solution* is studied.

(4) for every $t \leq T$, P-a.s.,

$$X(t) = \xi_0 + \int_0^t \left(AX(s) + F(s, X)\right) ds + \int_0^t B(s, X) \, dW_s; \qquad (3.2)$$

(b) is a *weak solution* of (3.1) (in the sense of duality) if
  (1) the following conditions hold:

$$P\left(\int_0^T \|X(t)\|_H \, dt < \infty\right) = 1, \qquad (3.3)$$

$$P\left(\int_0^T \left(\|F(t, X)\|_H + \|B(t, X)\|_{\mathcal{L}_2(K_Q, H)}^2\right) dt < \infty\right) = 1; \qquad (3.4)$$

  (2) for every $h \in \mathcal{D}(A^*)$ and $t \leq T$, P-a.s.,

$$\langle X(t), h \rangle_H = \langle \xi_0, h \rangle_H + \int_0^t \left(\langle X(s), A^*h \rangle_H + \langle F(s, X), h \rangle_H\right) ds$$

$$+ \int_0^t \langle h, B(s, X) \, dW_s \rangle_H; \qquad (3.5)$$

(c) is a *mild solution* of (3.1) if
  (1) conditions (3.3) and (3.4) hold;
  (2) for all $t \leq T$, P-a.s.,

$$X(t) = S(t)\xi_0 + \int_0^t S(t - s)F(s, X) \, ds + \int_0^t S(t - s)B(s, X) \, dW_s. \qquad (3.6)$$

We say that a process X is a *martingale* solution of the equation

$$\begin{cases} dX(t) = (AX(t) + F(t, X)) \, dt + B(t, X) \, dW_t, \\ X(0) = x \in H \quad \text{(deterministic)}, \end{cases} \qquad (3.7)$$

if there exists a filtered probability space $(\Omega, \mathcal{F}, \{\mathcal{F}_t\}_{t \in [0, T]}, P)$ and, on this probability space, a Q-Wiener process W_t, relative to the filtration $\{\mathcal{F}_t\}_{t \leq T}$, such that X_t is a mild solution of (3.7).

Unlike the strong solution, where the filtered probability space and the Wiener process are given, a martingale solution is a system $((\Omega, \mathcal{F}, \{\mathcal{F}_t\}_{t \leq T}, P), W, X)$ where the filtered probability space and the Wiener process are part of the solution.

If $A = 0$, $S(t) = I_H$ (identity on H), we obtain the SDE

$$\begin{cases} dX(t) = F(t, X) \, dt + B(t, X) \, dW_t, \\ X(0) = x \in H \quad \text{(deterministic)}, \end{cases} \qquad (3.8)$$

and a martingale solution of (3.8) is called a *weak solution* (in the stochastic sense, see [77]).

Remark 3.1 In the presence or absence of the operator A, there should be no confusion between a weak solution of (3.1) in the sense of duality and a weak solution
of (3.8) in the stochastic context.

Obviously, a strong solution is a weak solution (either meaning) and a mild solution is a martingale solution.

We will first study solutions to an SDE corresponding to the deterministic abstract inhomogeneous Cauchy problem (1.32),

$$X(t) = \int_0^t AX(s)\,ds + \int_0^t \Phi(s)\,dW_s. \tag{3.9}$$

The role of the deterministic convolution $\int_0^t S(t-s)f(s)\,ds$ will now be played by
the stochastic process

$$S \star \Phi(t) = \int_0^t S(t-s)\Phi(s)\,dW_s, \quad \Phi \in \mathscr{P}(K_Q, H), \tag{3.10}$$

which will be called *stochastic convolution*. Let $\|\cdot\|_{\mathscr{D}(A)}$ be the graph norm on
$\mathscr{D}(A)$,

$$\|h\|_{\mathscr{D}(A)} = \left(\|h\|_H^2 + \|Ah\|_H^2\right)^{1/2}.$$

The space $(\mathscr{D}(A), \|\cdot\|_{\mathscr{D}(A)})$ is a separable Hilbert space (Exercise 1.2). If $f:$
$[0, T] \to \mathscr{D}(A)$ is a measurable function and $\int_0^T \|f(s)\|_{\mathscr{D}(A)} < \infty$, then for any
$t \in [0, T]$,

$$\int_0^t f(s)\,ds \in \mathscr{D}(A) \quad \text{and} \quad \int_0^t Af(s)\,ds = A\int_0^t f(s)\,ds.$$

We have the following stochastic analogue of this fact.

Proposition 3.1 *Assume that A is the infinitesimal generator of a C_0-semigroup
on H and that W_t is a K-valued Q-Wiener process. If $\Phi(t) \in \mathscr{D}(A)$ P-a.s. for all
$t \in [0, T]$ and*

$$P\left(\int_0^T \|\Phi(t)\|_{\mathscr{L}_2(K_Q, H)}^2\,dt < \infty\right) = 1,$$

$$P\left(\int_0^T \|A\Phi(t)\|_{\mathscr{L}_2(K_Q, H)}^2\,dt < \infty\right) = 1,$$

then $P(\int_0^T \Phi(t)\,dW_t \in \mathscr{D}(A)) = 1$ and

$$A\int_0^T \Phi(t)\,dW_t = \int_0^T A\Phi(t)\,dW_t \quad P\text{-a.s.} \tag{3.11}$$

Proof Equality (3.11) is true for bounded elementary processes in $\mathscr{E}(\mathscr{L}(K, \mathscr{D}(A)))$. Let $\Phi_n \in \mathscr{E}(\mathscr{L}(K, \mathscr{D}(A)))$ be bounded elementary processes approximating Φ as in Lemma 2.3,

$$\int_0^T \|\Phi(t, \omega) - \Phi_n(t, \omega)\|_{\mathscr{L}_2(K_Q, \mathscr{D}(A))}^2 \to 0 \quad \text{as } n \to \infty$$

P-a.s. and hence,

$$\int_0^t \Phi_n(s)\, dW_s \to \int_0^t \Phi(s)\, dW_s,$$

$$A \int_0^t \Phi_n(s)\, dW_s = \int_0^t A\Phi_n(s)\, dW_s \to \int_0^t A\Phi(s)\, dW_s$$

in probability as $n \to \infty$. Now (3.11) follows since the infinitesimal generator A is a closed operator. □

Theorem 3.1 *Assume that A is an infinitesimal generator of a C_0-semigroup of operators $S(t)$ on H and that W_t is a K-valued Q-Wiener process.*

(a) *If $\Phi \in \mathscr{P}(K_Q, H)$ and, for $h \in \mathscr{D}(A^*)$ and every $0 \le t \le T$,*

$$\langle X(t), h\rangle_H = \int_0^t \langle X(s), A^* h\rangle_H\, ds + \left\langle \int_0^t \Phi(s)\, dW_s, h\right\rangle_H \quad P\text{-a.s.,} \quad (3.12)$$

then $X(t) = S \star \Phi(t)$.

(b) *If $\Phi \in \Lambda_2(K_Q, H)$, then for every $0 \le t \le T$, $S \star \Phi(t)$ satisfies (3.12).*

(c) *If $\Phi \in \Lambda_2(K_Q, H)$, $\Phi(K_Q) \subset \mathscr{D}(A)$, and $A\Phi \in \Lambda_2(K_Q, H)$, then $S \star \Phi(t)$ is a strong solution of (3.9).*

Proof (a) The proof in [11] relies on the fact which we make a subject of Exercise 3.2. Another method is presented in [9]. We choose to use an Itô formula type of proof which is consistent with the deterministic approach (see [63]).

Assume that (3.12) holds and let

$$u(s, x) = \langle x, S^*(t - s)h\rangle_H,$$

where $h \in \mathscr{D}(A^*)$ is arbitrary but fixed, $x \in H$, and $0 \le s \le t \le T$. The problem is to determine the differential of $u(s, X(s))$.

Since the adjoint semigroup $S^*(t)$ is a C_0-semigroup whose infinitesimal generator is A^* (see Theorem 1.2), we have

$$u_s(s, x) = \langle x, -A^* S^*(t - s)h\rangle_H,$$

$$u_x(s, x) = \langle \cdot, S^*(t - s)h\rangle_H,$$

$$u_{xx}(s, x) = 0.$$

Let $0 = s_1 \leq s_2 \leq \cdots \leq s_n = s$ be a partition of an interval $[0, s]$ and denote $\Delta s_j = s_{j+1} - s_j$ and $\Delta X_j = X(s_{j+1}) - X(s_j)$. According to (2.57), there exist random variables $\tilde{s}_j \in [s_j, s_{j+1}]$ such that

$$u\big(s, X(s)\big) - u\big(0, X(0)\big)$$

$$= \sum_{j=1}^{n-1} u_s\big(s_j, X(s_{j+1})\big)\Delta s_j + \sum_{j=1}^{n-1}\big\langle u_x\big(s_j, X(s_j)\big), \Delta X_j\big\rangle_H$$

$$+ \sum_{j=1}^{n-1}\big[u_s\big(\tilde{s}_j, X(s_{j+1})\big) - u_s\big(s_j, W_{s_{j+1}}\big)\big]\Delta s_j. \tag{3.13}$$

Due to the continuity of $u_s(s, X(s))$,

$$\sum_{j=1}^{n-1} u_s\big(s_j, X(s_{j+1})\big)\Delta s_j \to \int_0^s u_s\big(r, X(r)\big)\, dr = \int_0^s \big\langle X(r), -A^*S^*(t-r)h\big\rangle_H\, dr.$$

We consider the second sum,

$$\sum_{j=1}^{n-1}\big\langle u_x\big(s_j, X(s_j)\big), \Delta X_j\big\rangle_H = \sum_{j=1}^{n-1}\big\langle S^*(t-s_j)h, X(s_{j+1}) - X(s_j)\big\rangle_H$$

$$= \sum_{j=1}^{n-1}\left(\left\langle \int_0^{s_{j+1}} X(r)\, dr, A^*S^*(t-s_j)h\right\rangle_H + \left\langle \int_0^{s_{j+1}} \Phi(r)\, dW_r, S^*(t-s_j)h\right\rangle_H\, dr\right.$$

$$\left. - \left\langle \int_0^{s_j} X(r)\, dr, S^*(t-s_j)h\right\rangle_H - \left\langle \int_0^{s_j} \Phi(r)\, dW_r, S^*(t-s_j)h\right\rangle_H\right)$$

$$= \sum_{j=1}^{n-1}\left(\left\langle \int_{s_j}^{s_{j+1}} X(r)\, dr, A^*S^*(t-s_j)h\right\rangle_H + \left\langle \int_{s_j}^{s_{j+1}} \Phi(r)\, dW_r, S^*(t-s_j)h\right\rangle_H\right).$$

Due to the continuity of $A^*S^*(t-s)h = S^*(t-s)A^*h$, the fist sum converges to

$$\int_0^s \big\langle X(r), A^*S^*(t-r)h\big\rangle_H\, dr.$$

Denote $M_s = \int_0^s \Phi(r)\, dW_r \in \mathscr{M}_T^2(H)$. Then, $\int_{s_j}^{s_{j+1}} \Phi(r)\, dW_r = M_{s_{j+1}} - M_{s_j}$. By (2.43), the second sum converges in $L^2(\Omega, \mathbb{R})$ to

$$\left\langle \int_0^s S(t-r)\, dM_r, h\right\rangle_H = \left\langle \int_0^s S(t-r)\Phi(r)\, dW_r, h\right\rangle_H.$$

The last sum in (3.13) converges to zero, since it is bounded by

$$t \sup_{0 \leq j \leq n-1}\big|\big\langle X(s_{j+1}), A^*S^*(t-\tilde{s}_j)h\big\rangle_H - \big\langle X(s_{j+1}), A^*S^*(t-s_j)h\big\rangle_H\big| \to 0$$

due to the uniform continuity of $S(s)h$ on finite intervals and commutativity of A^* and S^* on the domain of A^*. We have proved that

$$u(s, X(s)) - u(0, X(0)) = \langle X(s), S^*(t-s)h \rangle_H$$

$$= \left\langle \int_0^s S(t-r)\Phi(r)\, dW_r, h \right\rangle_H. \qquad (3.14)$$

For $s = t$, we have

$$\langle X(t), h \rangle_H = \left\langle \int_0^t S(t-r)\Phi(r)\, dW_r, h \right\rangle_H.$$

Since $\mathscr{D}(A^*)$ is dense in H, (a) follows.

(b) For $h \in \mathscr{D}(A^*)$ and $k \in K$, consider the process defined by

$$\Psi(s, \omega, t)(k) = \left(1_{\{(0,t]\}}(s)\big(S(t-s)\Phi(s)\big)^* A^* h\right)(k)$$

$$= \langle 1_{\{(0,t]\}}(s) S(t-s)\Phi(s)(k), A^* h \rangle_H.$$

Then $\Psi : [0, T] \times \Omega \times [0, T] \to \mathscr{L}_2(K_Q, \mathbb{R})$.

For every $0 \le t \le T$, $\Psi(\cdot, \cdot, t)$ is $\{\mathscr{F}_s\}_{0 \le s \le T}$-adapted, and

$$|||\Psi||| = \int_0^T \|\Psi(\cdot, \cdot, t)\|_{\Lambda_2(K_Q, \mathbb{R})}\, dt$$

$$= \int_0^T \left(E \int_0^T \|1_{\{(0,t]\}}(s)\big(S(t-s)\Phi(s)\big)^* A^* h\|_{\Lambda_2(K_Q, \mathbb{R})}^2\, ds \right)^{1/2} dt$$

$$\le T \|A^* h\|_H M e^{\alpha t} \left(E \int_0^T \|\Phi(s)\|_{\Lambda_2(K_Q, H)}^2\, ds \right)^{1/2}$$

$$= C \|\Phi\|_{\Lambda_2(K_Q, H)} < \infty,$$

so that the assumptions of the stochastic Fubini theorem, Theorem 2.8, are satisfied. We obtain

$$\left\langle \int_0^t S \star \Phi(s)\, ds, A^* h \right\rangle_H = \int_0^t \int_0^s \langle S(s-u)\Phi(u)\, dW_u, A^* h \rangle_H\, ds$$

$$= \int_0^t \left(\int_0^t \Psi(u, \omega, s)\, dW_u \right) ds$$

$$= \int_0^t \left(\int_0^t \Psi(u, \omega, s)\, ds \right) dW_u$$

$$= \int_0^t \left(\int_0^t 1_{\{(0,s]\}}(u) \langle S(s-u)\Phi(u)(\cdot), A^* h \rangle_H\, ds \right) dW_u$$

$$= \int_0^t \left\langle \left(A \int_u^t S(s-u)\Phi(u)(\cdot)\, ds \right), h \right\rangle_H\, dW_u$$

$$= \int_0^t \langle (S(t-u)\varPhi(u) - \varPhi(u))(\cdot), h\rangle_H \, dW_u$$

$$= \left\langle \int_0^t (S(t-u)\varPhi(u) - \varPhi(u)) \, dW_u, h\right\rangle_H,$$

where we have used the fact that for $x \in H$, the integral $\int_0^t S(r)x \, dr \in \mathscr{D}(A)$, and

$$A\left(\int_0^t S(r)x \, dr \right) = S(t)x - x$$

(Theorem 1.2). Thus we conclude that

$$\left\langle \int_0^t S \star \varPhi(s) \, ds, A^* h \right\rangle_H = \langle S \star \varPhi(t), h\rangle_H - \left\langle \int_0^t \varPhi(s) \, dW_s, h\right\rangle_H,$$

proving (b).

(c) Recall from (1.22), Chap. 1, the Yosida approximation $A_n = AR_n$ of A, and let $S_n(s) = e^{sA_n}$ be the corresponding semigroups. Then part (b) implies that

$$S_n \star \varPhi(t) = \int_0^t A_n S_n \star \varPhi(s) \, ds + \int_0^t \varPhi(s) \, dW_s. \qquad (3.15)$$

Part (b) of Lemma 2.2, Chap. 2, implies that

$$\sup_{0 \le t \le T} E \|S_n \star \varPhi(t) - S \star \varPhi(t)\|_H^2 \to 0. \qquad (3.16)$$

Recall the commutativity property (1.16) from Chap. 1 that for $x \in \mathscr{D}(A)$, $AR_n x = R_n Ax$. In addition, $AS_n(t)x = S_n(t)Ax$ for $x \in \mathscr{D}(A)$, see Exercise 3.1. Using Proposition 3.1, we obtain

$$A_n S_n \star \varPhi(t) = AR_n \int_0^t S_n(t-s)\varPhi(s) \, dW_s$$

$$= R_n \int_0^t S_n(t-s)A\varPhi(s) \, dW_s$$

$$= R_n S_n \star A\varPhi(t).$$

Hence,

$$\sup_{0 \le t \le T} E \left\| \int_0^t (A_n S_n \star \varPhi(s) - AS \star \varPhi(s)) \, ds \right\|_H^2$$

$$\le T^2 \sup_{0 \le t \le T} E \int_0^t \|A_n S_n \star \varPhi(s) - AS \star \varPhi(s)\|_H^2 \, ds$$

$$\le T^2 E \int_0^T \|R_n S_n \star A\varPhi(s) - S \star A\varPhi(s)\|_H^2 \, ds$$

$$\leq T^2 E \int_0^T \left\| R_n\big(S_n \star A\Phi(s) - S \star A\Phi(s)\big) \right\|_H^2 ds$$

$$+ T^2 E \int_0^T \left\| (R_n - I) S \star A\Phi(s) \right\|_H^2 ds$$

$$\leq C \bigg(E \int_0^T \left\| S_n \star A\Phi(s) - S \star A\Phi(s) \right\|_H^2 ds$$

$$+ E \int_0^T \left\| (R_n - I) S \star A\Phi(s) \right\|_H^2 ds \bigg).$$

The first summand converges to zero by (c) of Lemma 2.2, Chap. 2.
Since $R_n x \to x$ for $x \in H$, we have

$$\left\| (R_n - I) S \star A\Phi(s) \right\|_H \to 0$$

and

$$\left\| (R_n - I) S \star A\Phi(s) \right\|_H^2 \leq C_1 \left\| S \star A\Phi(s) \right\|_H^2$$

with

$$E \int_0^T \left\| S \star A\Phi(s) \right\|_H^2 ds = \int_0^T E \int_0^s \left\| S(s-u) A\Phi(u) \right\|_{\mathscr{L}_2(K_Q, H)}^2 du\, ds$$

$$\leq C_2 \| A\Phi \|_{\Lambda_2(K_Q, H)}^2 < \infty,$$

and the second summand converges to zero by the Lebesgue DCT.
Summarizing,

$$\sup_{0 \leq t \leq T} E \left\| \int_0^t \big(A_n S_n \star \Phi(s) - A S \star \Phi(s)\big) ds \right\|_H^2 \to 0. \tag{3.17}$$

Combining (3.16) and (3.17), we obtain that both terms in (3.15) converge uniformly in mean square to the desired limits, so that (3.9) is satisfied by $S \star \Phi(t)$. This concludes the proof. $\qquad \square$

Exercise 3.1 Show that $A S_n(t)x = S_n(t) Ax$, for $x \in \mathscr{D}(A)$.

After the preliminary discussion concerning stochastic convolution, we turn to a general problem of the relationship among different types of solutions to the semilinear SDE (3.1).

Theorem 3.2 *A weak solution to (3.1) is a mild solution. Conversely, if X is a mild solution of (3.1) and*

$$E \int_0^T \| B(t, X) \|_{\mathscr{L}_2(K_Q, H)}^2 dt < \infty,$$

then $X(t)$ is a weak solution of (3.1). If, in addition $X(t) \in \mathcal{D}(A)$, $dP \otimes dt$ almost everywhere, then $X(t)$ is a strong solution of (3.1).

Proof The techniques for proving parts (a) and (b) of Theorem 3.1 are applicable to a more general case. Consider the process $X(t)$ satisfying the equation

$$\langle X(t), h \rangle_H = \langle \xi_0, h \rangle_H + \int_0^t \left(\langle X(s), A^*h \rangle_H + \langle f(s), h \rangle_H \right) ds$$

$$+ \int_0^t \langle h, \Phi(s) \, dW_s \rangle_H \tag{3.18}$$

with an adapted process $f(\cdot) \in L^1(\Omega, H)$, $\Phi \in \mathcal{P}(K_Q, H)$, and $h \in \mathcal{D}(A^*)$.

As in (a) of Theorem 3.1, we let

$$u(s, x) = \langle x, S^*(t - s)h \rangle_H,$$

where $h \in \mathcal{D}(A^*)$ is arbitrary but fixed, $x \in H$, and $0 \le s \le t \le T$. Then, formula (3.14) takes the form

$$u(s, X(s)) - u(0, X(0)) = \langle X(s), S^*(t - s)h \rangle_H - \langle X(0), S^*(t)h \rangle_H$$

$$= \left\langle \int_0^s S(t - r) \Phi(r) \, dW_r, h \right\rangle_H + \lim_{n \to \infty} \sum_{j=1}^{n-1} \left\langle S^*(t - s_j)h, \int_{s_j}^{s_{j+1}} f(r) \, dr \right\rangle_H$$

$$= \left\langle \int_0^s S(t - r) \Phi(r) \, dW_r, h \right\rangle_H + \left\langle \int_0^s S(t - r) f(r) \, dr, h \right\rangle_H.$$

For $s = t$, we have

$$\langle X(t), h \rangle_H = \langle S(t)\xi_0, h \rangle_H + \left\langle \int_0^t S(t - r) \Phi(r) \, dW_r, h \right\rangle_H$$

$$+ \left\langle \int_0^s S(t - r) f(r) \, dr, h \right\rangle_H.$$

Now it follows that $X(t)$ is a mild solution if we substitute $f(t) = F(t, X)$ and $\Phi(t) = B(t, X)$ and use the fact that $\mathcal{D}(A^*)$ is dense in H.

To prove the converse statement, consider the process

$$X(t) = S(t)\xi_0 + \int_0^t S(t - s) f(s) \, ds + S \star \Phi(t),$$

where $f(t)$ is as in the first part, and $\Phi \in \Lambda_2(K_Q, H)$. We need to show that

$$\langle X(t), h \rangle_H = \langle \xi_0, h \rangle_H$$

$$+ \int_0^t \left\langle S(s)\xi_0 + \int_0^s S(s - u) f(u) \, du + S \star \Phi(s), A^*h \right\rangle_H ds$$

$$+ \int_0^t \langle f(s), h \rangle_H \, ds + \left\langle \int_0^t \Phi(s) \, dW_s, h \right\rangle_H.$$

Using the result in (b) of Theorem 3.1, we have that

$$\langle S \star \Phi(t), h \rangle_H = \left\langle \int_0^t S \star \Phi(s) \, ds, A^* h \right\rangle_H + \left\langle \int_0^t \Phi(s) \, dW_s, h \right\rangle_H.$$

Since (see Theorem 1.2) for any $\xi \in H$, $\int_0^t S(s)\xi \, ds \in \mathscr{D}(A)$ and

$$A \int_0^t S(s)\xi \, ds = S(t)\xi - \xi,$$

we get

$$\langle S(t)\xi_0, h \rangle_H = \langle \xi_0, h \rangle_H + \int_0^t \langle S(s)\xi_0, A^* h \rangle_H \, ds.$$

Finally, using (deterministic) Fubini's theorem,

$$\left\langle \int_0^t \int_0^s S(s-u) f(u) \, du \, ds, A^* h \right\rangle_H = \left\langle \int_0^t \int_u^t S(s-u) f(u) \, ds \, du, A^* h \right\rangle_H$$

$$= \left\langle \int_0^t A \int_u^t S(s-u) f(u) \, ds \, du, h \right\rangle_H$$

$$= \left\langle \int_0^t A \int_0^{t-u} S(v) f(u) \, dv \, du, h \right\rangle_H$$

$$= \left\langle \int_0^t \left(S(t-u) f(u) - f(u) \right) du, h \right\rangle_H,$$

completing the calculations.

The last statement of the theorem is now obvious. $\qquad\square$

The following existence and uniqueness result for linear SDEs is a direct application of Theorem 3.2.

Corollary 3.1 *Let $\{W_t, 0 \le t \le T\}$ be a Q-Wiener process defined on a filtered probability space $(\Omega, \mathscr{F}, \{\mathscr{F}_t\}_{t \le T}, P)$, and A be the infinitesimal generator of a C_0-semigroup $\{S(t), t \ge 0\}$. Assume that $B \in \mathscr{L}(K, H)$, $f(\cdot) \in L^1(\Omega, H)$ is an $\{\mathscr{F}_t\}_{t \le T}$-adapted process, and ξ_0 is an H-valued $\mathscr{F}_0$-measurable random variable. Then the linear equation*

$$\begin{cases} dX(t) = (AX(t) + f(t)) \, dt + B \, dW_t, \\ X(0) = \xi_0, \end{cases} \tag{3.19}$$

has a unique weak solution given by

$$X(t) = S(t)\xi_0 + \int_0^t S(t-s)f(s)\,ds + \int_0^t S(t-s)B\,dW_s, \quad 0 \le t \le T.$$

Exercise 3.2 Prove that if $X(t)$ is a weak solution of the equation

$$\begin{cases} dX(t) = AX(t)\,dt + B\,dW_t, \\ X(0) = \xi_0, \end{cases}$$

with $B \in \mathscr{L}(K, H)$, then for an arbitrary function $\zeta(\cdot) \in C^1([0, T], (\mathscr{D}(A^*),$ $\|\cdot\|_{\mathscr{D}(A^*)}))$ and $t \in [0, T]$,

$$\langle X(t), \zeta(t) \rangle_H = \int_0^t \langle X(s), \zeta'(s) + A^*\zeta(s) \rangle_H \,ds + \int_0^t \langle \zeta(s), B\,dW_s \rangle_H. \tag{3.20}$$

Hint: Prove the result for a linearly dense subset of $C^1([0, T], (\mathscr{D}(A^*), \|\cdot\|_{\mathscr{D}(A^*)}))$ consisting of functions $\zeta(s) = \zeta_0 \varphi(s)$, where $\varphi(s) \in C^1([0, T], \mathbb{R})$.

Exercise 3.3 Apply (3.20) to a function $\zeta(s) = S^*(t-s)\zeta_0$ with $\zeta_0 \in \mathscr{D}(A^*)$ to show that if $X(t)$ is a weak solution of the linear SDE (3.19) with $\xi_0 = 0$ and $f(t) \equiv 0$, then, P-a.s., $X(t) = S \star B(t)$. Extend this result to a general case of ξ_0 and $f(t)$.

3.2 Solutions Under Lipschitz Conditions

We first prove the uniqueness and existence of a mild solution to (3.1) in the case of Lipschitz-type coefficients. This result is known (see Ichikawa [32]) if the coefficients $F(t, \cdot)$ and $B(t, \cdot)$ depend on $x \in C([0, T], H)$ through $x(t)$ only. We follow a technique extracted from the work of Gikhman and Skorokhod, [25] and extend it from R^n to H-valued processes.

Note that conditions (A3) and (A4) imply, respectively, that

$$\left\| \int_a^b F(t, x)\,dt \right\|_H \le \ell \int_a^b \left(1 + \sup_{s \le T} \|(\theta_t x)(s)\|_H \right) dt$$

and

$$\left\| \int_a^b \big(F(t, x) - F(t, y)\big)\,dt \right\|_H \le \mathscr{K} \int_a^b \sup_{s \le T} \|(\theta_t(x-y))(s)\|\,dt.$$

We will now state inequalities useful for proving the existence, uniqueness, and properties of solutions to the SDE (3.1). We begin with well-known inequalities (refer to (7.8), (7.9) in [11] and (24) in [34]).

Lemma 3.1 *Let $\Phi \in \Lambda_2(K_Q, H)$ and $p \geq 1$. Then*

$$E\left(\sup_{0 \leq t \leq T}\left\|\int_0^t \Phi(s)\,dW_s\right\|_H^{2p}\right) \leq c_{1,p} E\left(\left\|\int_0^T \Phi(s)\,dW_s\right\|_H^{2p}\right)$$

$$\leq c_{2,p} E\left(\int_0^T \|\Phi(s)\|_{\mathscr{L}_2(K_Q,H)}^2\,ds\right)^p$$

$$\leq c_{2,p} T^{p-1} E\left(\int_0^T \|\Phi(s)\|_{\mathscr{L}_2(K_Q,H)}^{2p}\,ds\right) \quad (3.21)$$

with the constants

$$c_{1,p} = \left(\frac{2p}{2p-1}\right)^{2p},$$

$$c_{2,p} = \left(p(2p-1)\right)^p (c_{1,p})^{2p^2}.$$

Proof The first inequality follows from the fact that the stochastic integral is an $L^p(\Omega)$-martingale and from Doob's maximal inequality, Theorem 2.2. The third is just Hölder's inequality. We now prove the second. For $p = 1$, it is the isometry property of the stochastic integral.

Assume now that $p > 1$. Let $F(\cdot) = \|\cdot\|_H^{2p} : H \to \mathbb{R}$. Then F is continuous, and its partial derivatives

$$\left(F_x(x)\right)(h) = 2p\|x\|_H^{2(p-1)}\langle x, h\rangle_H, \quad h \in H,$$

$$\left(F_{xx}(x)\right)(h, g) = 4p(p-1)\|x\|_H^{2(p-2)}\langle x, h\rangle_H \langle x, g\rangle_H$$

$$+ 2p\|x\|_H^{2(p-1)}\langle h, g\rangle_H, \quad h, g \in H,$$

are continuous and bounded on bounded subsets of H, with

$$\left\|F_{xx}(x)\right\|_{\mathscr{L}(H \times H, \mathbb{R})} \leq 2p(2p-1)\|x\|_H^{2(p-1)}. \quad (3.22)$$

Let $M(t) = \int_0^t \Phi(s)\,dW_s$. Applying Itô's formula (2.53) to $F(M(t))$ and taking expectations, we obtain, using (3.22), Hölder's inequality, and Doob's maximal inequality,

$$E\|M(s)\|^{2p} = E\int_0^s \left|\frac{1}{2}\,\mathrm{tr}\left[F_{xx}(M(u))\left(\Phi(u)Q\Phi^*(u)\right)\right]\right|\,du$$

$$\leq p(2p-1)E\left(\int_0^s \|M(u)\|_H^{2(p-1)}\|\Phi(u)\|_{\mathscr{L}_2(K_Q,H)}^2\,du\right)$$

$$\leq p(2p-1)E\left(\sup_{0 \leq u \leq s}\|M(u)\|_H^{2(p-1)}\int_0^s \|\Phi(u)\|_{\mathscr{L}_2(K_Q,H)}^2\,du\right)$$

$$\leq p(2p-1)\left[E\left(\sup_{0\leq u\leq s}\|M(u)\|_H^{2p}\right)\right]^{\frac{p-1}{p}}\left[E\left(\int_0^s\|\Phi(u)\|_{\mathscr{L}_2(K_Q,H)}^2\,du\right)^p\right]^{\frac{1}{p}}$$

$$\leq p(2p-1)\left[\left(\frac{2p}{2p-1}\right)^{2p}E\|M(s)\|_H^{2p}\right]^{\frac{p-1}{p}}\left[E\left(\int_0^s\|\Phi(u)\|_{\mathscr{L}_2(K_Q,H)}^2\,du\right)^p\right]^{\frac{1}{p}}.$$

Dividing both sides by $(E\|M(s)\|_H^{2p})^{\frac{p-1}{p}}$, we obtain

$$E\|M(s)\|_H^{2p}\leq\frac{c_{2,p}}{c_{1,p}}E\left(\int_0^s\|\Phi(u)\|_{\mathscr{L}_2(K_Q,H)}^2\,du\right)^p,$$

and (3.21) follows.   □

Corollary 3.2 *Let $\{S(t),\,0\leq t\leq T\}$ be a C_0-semigroup and $p\geq 1$. For $\Phi\in\Lambda_2(K_Q,H)$ and $t\in[0,T]$,*

$$E\left\|\int_0^t S(t-s)\Phi(s)\,dW_s\right\|_H^{2p}\leq C_{p,\alpha,M,T}^1 E\left(\int_0^t\|\Phi(s)\|_{\mathscr{L}_2(K_Q,H)}^2\,ds\right)^p$$

$$\leq C_{p,\alpha,M,T}^2 E\int_0^t\|\Phi(s)\|_{\mathscr{L}_2(K_Q,H)}^{2p}\,ds. \qquad (3.23)$$

The constants $C_{p,\alpha,M,T}^1$ and $C_{p,\alpha,M,T}^2$ depend only on the indicated parameters.

Proof We define $G(s)=S(t-s)\Phi(s)$. Then, for $u\in[0,t]$, we have by Lemma 3.1

$$E\left\|\int_0^u S(t-s)\Phi(s)\,dW_s\right\|_H^{2p}=E\left\|\int_0^u G(s)\,dW_s\right\|_H^{2p}$$

$$\leq\frac{c_{2,p}}{c_{1,p}}E\left(\int_0^u\|G(s)\|_{\mathscr{L}_2(K_Q,H)}^2\,ds\right)^p$$

$$=\frac{c_{2,p}}{c_{1,p}}E\left(\int_0^u\|S(t-s)\Phi(s)\|_{\mathscr{L}_2(K_Q,H)}^2\,ds\right)^p$$

$$\leq\frac{c_{2,p}}{c_{1,p}}M^{2p}e^{2p\alpha T}E\left(\int_0^u\|\Phi(s)\|_{\mathscr{L}_2(K_Q,H)}^2\,ds\right)^p.$$

In particular, for $u=t$, we get the first inequality in (3.23), the second is the Hölder inequality.   □

We will need inequalities of Burkholder type for the process of stochastic convolution. We begin with a supporting lemma [10].

Lemma 3.2 *Let $0<\alpha\leq 1$ and $p>1$ be numbers such that $\alpha>1/p$. Then, for an arbitrary $f\in L^p([0,T],H)$, the function*

$$G_\alpha f(t)=\int_0^t(t-s)^{\alpha-1}S(t-s)f(s)\,ds,\quad 0\leq t\leq T, \qquad (3.24)$$

is continuous, and there exists a constant $C > 0$ such that

$$\sup_{0 \leq t \leq T} \|G_\alpha f(t)\|_H \leq C \|f\|_{L^p([0,T],H)}.$$

Exercise 3.4 Prove Lemma 3.2.

Hint: use the Hölder inequality to show the bound for $G_\alpha f(t)$. To show the continuity, start with a smooth function $f \in C^\infty([0, T], H)$ vanishing near $t = 0$ and show that

$$\frac{d}{dt} G_\alpha f(t) = \int_0^t s^{\alpha-1} S(s) \frac{d}{dt} f(t-s) \, ds$$

is bounded on $[0, T]$. For general $f(t)$, use the fact that $C^\infty([0, T]) \hookrightarrow L^p([0, T], H)$ densely.

The following Burkholder-type inequalities concern two cases. The first allows a general C_0-semigroup but is restricted only to the powers strictly greater than two. Its proof relies on a *factorization technique* developed in [10], and it is a consequence of (3.21) and (3.23). The second inequality allows the power of two but is restricted to pseudo-contraction semigroups only. Curiously, the general case of power two is still an open problem.

Lemma 3.3 *Let W_t be a K-valued Wiener process with covariance Q, and $\Phi \in \Lambda_2(K_Q, H)$.*

(a) *Let $S(t)$ be a general C_0-semigroup and $p > 1$. If*

$$E \left(\int_0^T \|\Phi(t)\|_{\mathcal{L}_2(K_Q,H)}^{2p} \, dt \right) < \infty,$$

then there exists a continuous modification of the stochastic convolution $S \star \Phi(t) = \int_0^t S(t-s)\Phi(s) \, dW_s$.

For this continuous version, there exists a constant $C_{p,\alpha,M,T}$, depending only on the indicated parameters, such that for any stopping time τ,

$$E \sup_{0 \leq t \leq T \wedge \tau} \|S \star \Phi(t)\|_H^{2p} \leq C_{p,\alpha,M,T} E \int_0^{T \wedge \tau} \|\Phi(t)\|_{\mathcal{L}_2(K_Q,H)}^{2p} \, dt. \tag{3.25}$$

Let $A_n = A R_n$ be the Yosida approximations, and $S_n(t) = e^{A_n t}$. Then a continuous version of $S \star \Phi(t)$ can be approximated by the (continuous) processes $S_n \star \Phi(t)$ in the following sense:

$$\lim_{n \to \infty} E \sup_{0 \leq t \leq T} \|S \star \Phi(t) - S_n \star \Phi(t)\|_H^{2p} = 0. \tag{3.26}$$

(b) *Let $S(t)$ be a C_0-pseudo-contraction semigroup and $p \geq 1$. If*

$$E \left(\int_0^T \|\Phi(t)\|_{\mathcal{L}_2(K_Q,H)}^2 \, dt \right)^p < \infty,$$

then there exists a continuous modification of the stochastic convolution $S \star \Phi(t)$. For this continuous version, there exists a constant $C_{p,\alpha,T}$, depending only on the indicated parameters, such that for any stopping time τ,

$$E \sup_{0 \le t \le T \wedge \tau} \|S \star \Phi(t)\|_H^{2p} \le C_{p,\alpha,T} E \left(\int_0^{T \wedge \tau} \|\Phi(t)\|_{\mathscr{L}_2(K_Q,H)}^2 \, dt \right)^p. \tag{3.27}$$

Let $\Phi_n(t) = R_n \Phi(t)$. Then a continuous version of $S \star \Phi(t)$ can be approximated by the (continuous) processes $S \star \Phi_n(t)$ in the following sense:

$$\lim_{n \to \infty} E \sup_{0 \le t \le T} \|S \star \Phi(t) - S \star \Phi_n(t)\|_H^{2p} = 0. \tag{3.28}$$

Proof (a) We follow the proof of Proposition 7.3 in [11], which uses the factorization method introduced in [10]. Let us begin with the following identity (see Exercise 3.6):

$$\int_\sigma^t (t-s)^{\alpha-1}(s-\sigma)^{-\alpha} \, ds = \frac{\pi}{\sin \pi \alpha}, \quad 0 < \alpha < 1, \ \sigma < t. \tag{3.29}$$

Using this identity and the stochastic Fubini theorem 2.8, we obtain

$$\int_0^t S(t-s)\Phi(s) \, dW_s$$

$$= \frac{\sin \pi \alpha}{\pi} \int_0^t \left(\int_\sigma^t (t-s)^{\alpha-1}(s-\sigma)^{-\alpha} \, ds \right) S(t-\sigma)\Phi(\sigma) \, dW_\sigma$$

$$= \frac{\sin \pi \alpha}{\pi} \int_0^t (t-s)^{\alpha-1} S(t-s) \left(\int_0^s (s-\sigma)^{-\alpha} S(s-\sigma)\Phi(\sigma) \, dW_\sigma \right) ds$$

$$= \frac{\sin \pi \alpha}{\pi} \int_0^t (t-s)^{\alpha-1} S(t-s) Y(s) \, ds \quad \text{P-a.s.}$$

with

$$Y(s) = \int_0^s (s-\sigma)^{-\alpha} S(s-\sigma)\Phi(\sigma) \, dW_\sigma, \quad 0 \le s \le T.$$

Hence, we have the modification

$$S \star \Phi(t) = \frac{\sin \pi \alpha}{\pi} \int_0^t (t-s)^{\alpha-1} S(t-s) Y(s) \, ds, \tag{3.30}$$

for which we need to prove the assertions in (a).

Let $\frac{1}{2p} < \alpha < \frac{1}{2}$. Applying Hölder's inequality to the integral in (3.30), we obtain, for some constant $C^1_{p,\alpha,M,T}$,

$$\sup_{0 \le t \le T} \left\| \int_0^t S(t-s)\Phi(s) \, dW_s \right\|_H^{2p}$$

$$\leq \left(\frac{M \sin \pi \alpha}{\pi}\right)^{2p} e^{2p\alpha T} \left(\int_0^T (t-s)^{(\alpha-1)(\frac{2p}{2p-1})} ds\right)^{2p-1} \int_0^T \|Y(s)\|_H^{2p} ds$$

$$\leq C_{p,\alpha,M,T}^1 \int_0^T \|Y(s)\|_H^{2p} ds,$$

since $\alpha > 1/(2p)$. By Corollary 3.2, there exists a constant $C_{p,\alpha,M,T}^2 > 0$ such that

$$\int_0^T E\|Y(s)\|_H^{2p} ds$$

$$\leq C_{p,\alpha,M,T}^2 E \int_0^T \left(\int_0^s (s-\sigma)^{-2\alpha} \|\Phi(\sigma)\|_{\mathcal{L}_2(K_Q,H)}^2 d\sigma\right)^p ds$$

$$\leq C_{p,\alpha,M,T}^2 \left(\int_0^T \sigma^{-2\alpha} d\sigma\right)^p E\left(\int_0^T \|\Phi(\sigma)\|_{\mathcal{L}_2(K_Q,H)}^{2p} d\sigma\right)$$

$$\leq C_{p,\alpha,M,T}^3 E\left(\int_0^T \|\Phi(\sigma)\|_{\mathcal{L}_2(K_Q,H)}^{2p} d\sigma\right) \qquad (3.31)$$

with some constant $C_{p,\alpha,M,T}^3 > 0$, by the theorem about convolution in $L^p(\mathbb{R}^d)$, see Exercise 3.7. Now (3.25), in case $\tau = T$, follows with $C_{p,\alpha,M,T} = C_{p,\alpha,M,T}^1 C_{p,\alpha,M,T}^3$.

We will consider (3.25) with a stopping time τ. Let $\tau_n \uparrow \tau$ P-a.s. be an increasing sequence of stopping times approximating τ, each τ_n taking k_n values $0 \leq t_1 \leq \cdots \leq t_{k_n} \leq T$. Then

$$E \sup_{0 \leq s \leq \tau_n \wedge t} \left\|\int_0^s S(s-r)\Phi(r) dW_r\right\|_H^{2p}$$

$$= \sum_{i=1}^{k_n} E\left(1_{\{\tau_n=t_i\}} \sup_{0 \leq s \leq t_i \wedge t} \left\|\int_0^s S(s-r)\Phi(r) dW_r\right\|_H^{2p}\right)$$

$$= \sum_{i=1}^{k_n} E \sup_{0 \leq s \leq t_i \wedge t} \left\|\int_0^s 1_{\{\tau_n=t_i\}} S(s-r)\Phi(r) dW_r\right\|_H^{2p}$$

$$\leq \sum_{i=1}^{k_n} C_{p,\alpha,M,T} E \int_0^{t_i \wedge t} \left\|1_{\{\tau_n=t_i\}}\Phi(r)\right\|_{\mathcal{L}_2(K_Q,H)}^{2p} dr$$

$$= C_{p,\alpha,M,T} E \int_0^{\tau_n \wedge t} \|\Phi(r)\|_{\mathcal{L}_2(K_Q,H)}^{2p} dr,$$

and (3.25) is obtained by the monotone convergence theorem.

Note that by (3.31), the process $Y(t)$ has almost surely $2p$-integrable paths, so that by Lemma 3.2,

$$S \star \Phi(t) = \frac{\sin \pi \alpha}{\pi} G_\alpha Y(t) \qquad (3.32)$$

has a continuous version. Equation (3.32) is referred to as the *factorization formula*.

Finally, we justify (3.26). As in (3.30),

$$S_n \star \Phi(t) = \frac{\pi}{\sin \pi \alpha} \int_0^t (t-s)^{\alpha-1} S_n(t-s) Y_n(s) \, ds$$

with

$$Y_n(s) = \int_0^s (s-\sigma)^{-\alpha} S_n(s-\sigma) \Phi(\sigma) \, dW_\sigma, \quad 0 \le s \le T.$$

Hence,

$$S \star \Phi(t) - S_n \star \Phi(t) = \frac{\pi}{\sin \pi \alpha} \int_0^t (t-s)^{\alpha-1} \big(S(t-s) - S_n(t-s)\big) Y(s) \, ds$$

$$+ \frac{\pi}{\sin \pi \alpha} \int_0^t (t-s)^{\alpha-1} S_n(t-s) \big(Y(s) - Y_n(s)\big) \, ds$$

$$= I_n(t) + J_n(t).$$

Let us analyze the terms $I_n(t)$ and $J_n(t)$ separately. By the Hölder inequality,

$$\sup_{0 \le t \le T} \|I_n(t)\|_H^{2p} \le C \int_0^T \big\| \big(S(t-s) - S_n(t-s)\big) Y(s) \big\|_H^{2p} \, ds,$$

with the expression on the right-hand side converging to zero and being bounded by a P-integrable function, so that

$$\lim_{n \to \infty} E \sup_{0 \le t \le T} \|I_n(t)\|_H^{2p} = 0$$

by the Lebesgue DCT.

The expression for J_n is an integral of the type in (3.30), so that, by applying the Hölder inequality, we obtain

$$\sup_{0 \le t \le T} \|J_n(t)\|_H^{2p}$$

$$= \sup_{0 \le t \le T} \left\| \frac{\sin \pi \alpha}{\pi} \int_0^t (t-s)^{\alpha-1} S_n(t-s) \big(Y(s) - Y_n(s)\big) \, ds \right\|_H^{2p}$$

$$\le C_{p,\alpha,M,T}^3 \int_0^T \|Y(t) - Y_n(t)\|_H^{2p} \, dt.$$

Similarly to (3.31), using the convolution inequality in Exercise 3.7 (with $r = 1$ and $s = p$), we have

$$\int_0^T E \|Y(t) - Y_n(t)\|_H^{2p} \, dt$$

$$= E \int_0^T \left\| \int_0^t (t-s)^{-\alpha} \big(S(t-s) - S_n(t-s)\big) \Phi(s) \, dW_s \right\|_H^{2p} \, dt$$

$$\leq C_p^4 E \int_0^T \left(\int_0^t (t-s)^{-2\alpha} \left\| \big(S(t-s) - S_n(t-s)\big)\Phi(s) \right\|_{\mathscr{L}_2(K_Q,H)}^2 ds \right)^p dt$$

$$\leq C_p^4 E \int_0^T \left(\int_0^t (t-s)^{-2\alpha} \sup_{0 \leq u \leq T} \left\| \big(S(u) - S_n(u)\big)\Phi(s) \right\|_{\mathscr{L}_2(K_Q,H)}^2 ds \right)^p dt$$

$$\leq C_p^4 E \left\{ \left(\int_0^T t^{-2\alpha} dt \right)^p \left(\int_0^T \sup_{0 \leq u \leq T} \left\| \big(S(u) - S_n(u)\big)\Phi(t) \right\|_{\mathscr{L}_2(K_Q,H)}^{2p} dt \right) \right\}$$

$$\leq C_{p,\alpha}^5 E \int_0^T \sup_{0 \leq u \leq T} \left\| \big(S(u) - S_n(u)\big)\Phi(t) \right\|_{\mathscr{L}_2(K_Q,H)}^{2p} dt \to 0$$

by Lemma 2.2, part (c).

(b) We follow the idea in [72] for pseudo-contraction semigroups. Let $A_n = A R_n$ be the Yosida approximations of A; then $\Phi_n = R_n \Phi \in \Lambda_2(K_Q, H)$, $\Phi_n(K_Q) \subset \mathscr{D}(A)$ (see Chap. 1, (1.15)), and $A\Phi_n = A_n \Phi \in \Lambda_2(K_Q, H)$. By Theorem 3.1(c),

$$X_n = S \star \Phi_n(s)$$

is a strong solution of the equation

$$X(t) = \int_0^t AX(s)\,ds + \int_0^t \Phi_n(s)\,dW_s. \tag{3.33}$$

Applying Itô's formula to $F(x) = \|x\|_H^{2p}$, we get, similarly as in Lemma 3.1,

$$\|X_n(s)\|_H^{2p} \leq \int_0^s \big\langle 2p\|X_n(u)\|_H^{2(p-1)} X_n(u),\, \Phi_n(u)\,dW_u \big\rangle_H$$

$$+ \int_0^s \big\langle 2p\|X_n(u)\|_H^{2(p-1)} X(u),\, AX(u) \big\rangle_H du$$

$$+ \frac{1}{2} \int_0^s 2p(2p-1)\|X_n(u)\|_H^{2(p-1)} \|\Phi_n(u)\|_{\mathscr{L}_2(K_Q,H)}^2 du.$$

Since $S(t)$ is a pseudo-contraction semigroup,

$$\langle Ax, x \rangle_H \leq \alpha \|x\|_H^2, \quad x \in \mathscr{D}(A)$$

(see Exercise 3.5). Thus,

$$\sup_{0 \leq s \leq t} \|X_n(s)\|_H^{2p} \leq 2p \sup_{0 \leq s \leq t} \left| \int_0^s \big\langle \|X_n(u)\|_H^{2(p-1)} X_n(u),\, \Phi_n(u)\,dW_u \big\rangle_H \right|$$

$$+ 2p\alpha \int_0^t \sup_{0 \leq u \leq s} \|X_n(u)\|_H^{2p} du$$

$$+ p(2p-1) \sup_{0 \leq s \leq t} \|X_n(s)\|_H^{2(p-1)} \int_0^t \|\Phi_n(s)\|_{\mathscr{L}_2(K_Q,H)}^2 ds.$$

To simplify the notation, denote

$$X_n^*(t) = \sup_{0 \le s \le t} \|X_n(s)\|_H, \quad \text{and} \quad \phi_n(t) = \int_0^t \|\Phi_n(s)\|_{\mathscr{L}_2(K_Q, H)}^2 \, ds.$$

Let $\tau_k = \inf_{t \le T}\{X_n^*(t) > k\}$, $k = 1, 2, \ldots$, with the *infimum* over an empty set being equal T. By the Burkholder inequality for real-valued martingales, we have

$$E\left|\int_0^{s \wedge \tau_k} \langle \|X_n(u)\|_H^{2(p-1)} X_n(u), \Phi_n(u) \, dW_u\rangle_H\right|$$

$$\le E\left(\int_0^{s \wedge \tau_k} \|X_n(u)\|_H^{2(2p-1)} \|\Phi_n(u)\|_{\mathscr{L}_2(K_Q, H)}^2 \, du\right)^{1/2}$$

$$\le E\left((X_n^*(s \wedge \tau_k))^{2p-1}(\phi_n(s \wedge \tau_k))^{1/2}\right).$$

Now, by Hölder's inequality,

$$E\left((X_n^*(s \wedge \tau_k))^{2p-1}(\phi_n(s \wedge \tau_k))^{1/2}\right)$$
$$\le \left(E(X_n^*(s \wedge \tau_k))^{2p}\right)^{1-1/2p}\left(E(\phi_n(s \wedge \tau_k))^p\right)^{1/2p}$$

and

$$E\left((X_n^*(s \wedge \tau_k))^{2(p-1)}\phi_n(s \wedge \tau_k)\right)$$
$$\le \left(E(X_n^*(s \wedge \tau_k))^{2p}\right)^{1-1/p}\left(E(\phi_n(s \wedge \tau_k))^p\right)^{1/p}.$$

We arrive at the following estimate

$$E(X_n^*(t \wedge \tau_k))^{2p} \le 2p\left(E(X_n^*(t \wedge \tau_k))^{2p}\right)^{1-1/2p}\left(E(\phi_n(t \wedge \tau_k))^p\right)^{1/2p}$$
$$+ p(2p-1)\left(E(X_n^*(t \wedge \tau_k))^{2p}\right)^{1-1/p}\left(E(\phi_n(t \wedge \tau_k))^p\right)^{1/p}$$
$$+ 2p\alpha E\int_0^t (X_n^*(s \wedge \tau_k))^{2p} \, ds,$$

since

$$\int_0^{t \wedge \tau_k} (X_n^*(s))^{2p} \, ds \le \int_0^t (X_n^*(s \wedge \tau_k))^{2p} \, ds.$$

This is an expression of the form

$$g(t) \le u(t) + p\alpha \int_0^t g(s) \, ds$$

with constants $p, \alpha > 0$ and an integrable function $u : [0, T] \to \mathbb{R}$. Now we use Gronwall's lemma and the fact that $\sup_{0 \le s \le t} u(s) = u(t)$:

$$g(t) \le u(t) + p\alpha \int_0^t u(s) e^{p\alpha(t-s)} \, ds$$

$$\le u(t) + \sup_{0 \le s \le t} u(s) \, p\alpha \int_0^t e^{p\alpha(t-s)} \, ds = u(t) e^{p\alpha t}.$$

Multiplying the obtained inequality by $(E(X_n^*(t \wedge \tau_k))^{2p})^{1/(2p)-1}$ we can see that

$$\left(E\left(X_n^*(t \wedge \tau_k)^{2p}\right)\right)^{1/2p}$$
$$\le e^{2p\alpha t} \left(p(2p-1)\left(E\left(X_n^*(t \wedge \tau_k)\right)^{2p}\right)^{-1/2p}\left(E\left(\phi_n(t \wedge \tau_k)^p\right)\right)^{1/p}\right.$$
$$+ 2p\left(E\left(\phi_n(t \wedge \tau_k)^p\right)\right)^{1/2p}\bigg).$$

Let $z = (E(X_n^*(t \wedge \tau_k))^{2p})^{1/2p}$, we have

$$z^2 \le e^{2p\alpha t} p(2p-1)(E\left(\phi_n(t \wedge \tau_k)^p\right))^{1/p} + e^{2p\alpha t} 2p(E\left(\phi_n(t \wedge \tau_k)^p\right))^{1/2p} z,$$

giving

$$z \le \frac{1}{2}\left(2pe^{2p\alpha t} + \left(4p^2 e^{4p\alpha t} + 4p(2p-1)e^{2p\alpha t}\right)^{1/2}\right)\left(E\phi_n(t \wedge \tau_k)^p\right)^{1/2p}$$

$$\le C_{p,T} e^{2p\alpha t} \left(E\phi_n(t \wedge \tau_k)^p\right)^{1/2p}.$$

Thus, we proved that

$$E \sup_{0 \le s \le t} \left\| X_n(s \wedge \tau_k) \right\|_H^{2p} \le C_{p,T} e^{2p\alpha t} E\left(\int_0^t \left\| \Phi_n(s) \right\|_{\mathcal{L}_2(K_Q,H)}^2\right)^p$$

(dropping the stopping time in the RHS does not decrease its value).

Since $\sup_{0 \le s \le t \wedge \tau_k} \|X_n(s)\|_H \uparrow \sup_{0 \le s \le t} \|X_n(s)\|_H$, P-a.s., as $k \to \infty$, by the continuity of $X_n(t)$ as a solution of (3.33), we get by the monotone convergence that

$$E \sup_{0 \le s \le t} \left\| X_n(s) \right\|_H^{2p} \le C_{p,T} e^{2p\alpha t} E\left(\int_0^t \left\| \Phi_n(s) \right\|_{\mathcal{L}_2}^2\right)^p.$$

In conclusion, note that $\Phi_n \to \Phi$ in $\Lambda_2(K_Q, H)$ by the Lebesgue DCT, so that $X_n(t) = S \star \Phi_n(t) \to S \star \Phi(t) = X(t)$, in $L^2(\Omega, H)$, for any $0 \le t \le T$. In addition, note that

$$E\left(\int_0^t \left\| \Phi_n(s) - \Phi(s) \right\|_{\mathcal{L}_2(K_Q,H)}^2\right)^p \to 0$$

by the Lebesgue DCT. By applying inequality (3.27) to $X_n(t) - X_m(t)$, we deduce that X_n is a Cauchy sequence in the norm

$$\| \cdot \|_{L^{2p}(\Omega, C([0,T],H))} = \left(E \sup_{0 \leq t \leq T} \| \cdot \|^{2p} \right)^{1/2p},$$

so that

$$E \sup_{0 \leq t \leq T} \| X_n(t) - \tilde{X}(t) \|^{2p} \to 0, \qquad (3.34)$$

where $\tilde{X}(t)$ is a continuous modification of $X(t)$, proving simultaneously (3.27) for $S \star \Phi(s)$ and (3.28).

The argument that (3.27) holds with a stopping time τ is the same as that for (3.25).   □

Remark 3.2 Under the assumptions of Lemma 3.3, part (a), the continuous modification of the stochastic convolution $S \star \Phi(s)$ defined by (3.30) can be approximated, as in (3.28), by the processes $X_n(t) = S \star \Phi_n(t)$ defined in the proof of part (b). This is because for $\tilde{X}$ defined in the proof of part (b),

$$P\big(\tilde{X}(t) = S \star \Phi(t), \ 0 \leq t \leq T\big) = 1$$

for a continuous version of $S \star \Phi(t)$.

Thus, for a general C_0-semigroup $S(t)$, $p > 1$, and $\Phi \in \Lambda_2(K_Q, H)$, we have two approximations

$$S_n \star \Phi(t) \to S \star \Phi(t),$$
$$S \star \Phi_n(t) \to S \star \Phi(t),$$

both converging in $L^{2p}(\Omega, C([0, T], H))$.

Exercise 3.5 (Lumer–Phillips) Prove that if A generates a pseudo-contraction semigroup $S(t)$ on H, then

$$\langle Ax, x \rangle_H \leq \alpha \|x\|_H^2, \quad x \in \mathscr{D}(A).$$

Hint: this is Theorem 4.3 in [63] if $\alpha = 0$. See also [78].

Exercise 3.6 Prove (3.29).

Exercise 3.7 Prove that if $f, g \in L^1(\mathbb{R}^d)$, then $f * g$ exists almost everywhere in $\mathbb{R}^d$, $f * g \in L^1(\mathbb{R}^d)$, and

$$\|f * g\|_{L^1(\mathbb{R}^d)} \leq \|f\|_{L^1(\mathbb{R}^d)} \|g\|_{L^1(\mathbb{R}^d)}.$$

More general, prove that if $f \in L^r(\mathbb{R}^d)$ and $g \in L^s(\mathbb{R}^d)$ with $r, s \geq 1$, $\frac{1}{r} + \frac{1}{s} \geq 1$, $\frac{1}{p} = \frac{1}{r} + \frac{1}{s} - 1$, then $f * g(s)$ exists almost everywhere in $\mathbb{R}^d$, and

$$\|f * g\|_{L^p(\mathbb{R}^d)} \leq \|f\|_{L^r(\mathbb{R}^d)} \|g\|_{L^s(\mathbb{R}^d)}.$$

For an adapted process $\xi(\cdot) \in C([0, T], H)$, denote

$$I(t, \xi) = \int_0^t S(t - s) F(s, \xi) \, ds + \int_0^t S(t - s) B(s, \xi) \, dW_s. \tag{3.35}$$

Lemma 3.4 *If $F(t, x)$ and $B(t, x)$ satisfy conditions (A1) and (A3), $S(t)$ is either a pseudo-contraction semigroup and $p \geq 1$ or a general C_0-semigroup and $p > 1$, then, for a stopping time τ,*

$$E \sup_{0 \leq s \leq t \wedge \tau} \|I(s, \xi)\|_H^{2p} \leq C \left(t + \int_0^t E \sup_{0 \leq u \leq s \wedge \tau} \|\xi(u)\|_H^{2p} \, ds \right) \tag{3.36}$$

with the constant C depending only on p, M, α, T and the constant ℓ.

Proof We note that

$$\sup_{0 \leq s \leq t \wedge \tau} \|I(s, \xi)\|_H^{2p}$$

$$\leq 2^{2p-1} \sup_{0 \leq s \leq t \wedge \tau} \left(\left\| \int_0^s S(s - u) F(u, \xi) \, du \right\|_H^{2p} + \left\| \int_0^s S(s - u) B(u, \xi) \, dW_u \right\|_H^{2p} \right).$$

We can find a bound for the expectation of the first term,

$$E \sup_{0 \leq s \leq t \wedge \tau} \left\| \int_0^s S(s - u) F(u, \xi) \, du \right\|_H^{2p}$$

$$\leq E \sup_{0 \leq s \leq t \wedge \tau} \left(\ell C_{M,\alpha,t} \int_0^s \left(1 + \sup_{0 \leq r \leq u} \|\xi(r)\|_H \right) du \right)^{2p}$$

$$\leq 2^{2p-1} (\ell C_{M,\alpha,t})^{2p} \left(t^{2p} + t E \sup_{0 \leq s \leq t} \int_0^{s \wedge \tau} \sup_{0 \leq r \leq u} \|\xi(r)\|_H^{2p} \, du \right)$$

$$\leq C_{p,M,\alpha,T,\ell} \left(t + \int_0^t E \sup_{0 \leq u \leq s \wedge \tau} \|\xi_u\|_H^{2p} \, ds \right),$$

and, using (3.25) or (3.27), a bound for the expectation of the second term,

$$E \sup_{0 \leq s \leq t \wedge \tau} \left\| \int_0^s S(s - u) B(u, \xi) \, dW_u \right\|_H^{2p} \leq C_{p,M,\alpha,t} E \int_0^{t \wedge \tau} \|B(s, \xi)\|_{\mathscr{L}_2(K_Q, H)}^{2p} \, ds.$$

The latter is dominated by

$$C_{p,M,\alpha,t}\ell^{2p} E \int_0^{t\wedge\tau} \left(1 + \sup_{0\le u\le s} \|\xi(u)\|_H\right)^{2p} ds$$

$$\le 2^{2p-1} C_{p,M,\alpha,T}\ell^{2p} \left(t + E \int_0^{t\wedge\tau} \sup_{0\le u\le s} \|\xi(u)\|_H^{2p} ds\right)$$

$$= C'_{p,M,\alpha,T,\ell} \left(t + \int_0^t E \sup_{0\le u\le s\wedge\tau} \|\xi(u)\|_H^{2p} ds\right).$$

We complete the proof by combining the inequalities for both terms.   □

Lemma 3.5 *Let conditions* (A1) *and* (A4) *be satisfied, and* $S(t)$ *be either a pseudo-contraction semigroup and* $p \ge 1$ *or a general* C_0-*semigroup and* $p > 1$. *Then,*

$$E \sup_{0\le s\le t} \|I(s,\xi_1) - I(s,\xi_2)\|_H^{2p} \le C_{p,M,\alpha,T,\mathscr{K}} \int_0^t E \sup_{0\le u\le s} \|\xi_1(u) - \xi_2(u)\|_H^{2p} ds$$

with the constant $C_{p,M,\alpha,T,\mathscr{K}}$ *depending only on the indicated parameters.*

Proof We begin with the following estimate:

$$E \sup_{0\le s\le t} \|I(s,\xi_1) - I(s,\xi_2)\|_H^{2p}$$

$$\le 2^{2p-1} E \sup_{0\le s\le t} \left(\left\|\int_0^s S(s-u)\big(F(u,\xi_1) - F(u,\xi_2)\big) du\right\|_H^{2p}\right.$$

$$\left. + \left\|\int_0^s S(s-u)\big(B(u,\xi_1) - B(u,\xi_2)\big) dW_u\right\|_H^{2p}\right).$$

Considering the two terms separately, we obtain

$$E \sup_{0\le s\le t} \left\|\int_0^s S(s-u)\big(F(u,\xi_1) - F(u,\xi_2)\big) du\right\|_H^{2p}$$

$$\le M^{2p} e^{2p\alpha t} \mathscr{K}^{2p} t^{2p-1} E \sup_{0\le s\le t} \int_0^s \sup_{0\le r\le u} \|\xi_1(r) - \xi_2(r)\|_H^{2p} du$$

$$= C_{p,M,\alpha,T,\mathscr{K}} \int_0^t E \sup_{0\le u\le s} \|\xi_1(u) - \xi_2(u)\|_H^{2p} ds,$$

by Hölder's inequality, and similarly, using (3.25) or (3.27),

$$E \sup_{0\le s\le t} \left\|\int_0^s S(s-u)\big(B(u,\xi_1) - B(u,\xi_2)\big) dW_u\right\|_H^{2p}$$

$$\leq C_{p,M,\alpha,T} E \int_0^t \left\| B(s,\xi_1) - B(s,\xi_2) \right\|_{\mathscr{L}_2(K_Q,H)}^{2p} ds$$

$$\leq C_{p,M,\alpha,T,\mathscr{K}} \int_0^t E \sup_{0 \leq u \leq s} \left\| \xi_1(u) - \xi_2(u) \right\|_H^{2p} ds.$$

This completes the proof. □

Let $\mathscr{H}_{2p}$ denote the space of $C([0,T],H)$-valued random variables ξ such that the process $\xi(t)$ is jointly measurable, adapted to the filtration $\{\mathscr{F}_t\}_{t \in [0,T]}$, with $E \sup_{0 \leq s \leq T} \|\xi(s)\|_H^{2p} < \infty$. Then $\mathscr{H}_{2p}$ is a Banach space with the norm

$$\|\xi\|_{\mathscr{H}_{2p}} = \left(E \sup_{0 \leq s \leq T} \|\xi(s)\|_H^{2p} \right)^{\frac{1}{2p}}.$$

Theorem 3.3 *Let the coefficients F and B satisfy conditions* (A1), (A3), *and* (A4). *Assume that $S(t)$ is either a pseudo-contraction semigroup and $p \geq 1$ or a general C_0-semigroup and $p > 1$. Then the semilinear equation* (3.1) *has a unique continuous mild solution. If, in addition, $E\|\xi_0\|_H^{2p} < \infty$, then the solution is in $\mathscr{H}_{2p}$.*

If $A = 0$, then (3.8) *has unique strong solution. If, in addition, $E\|\xi_0\|_H^{2p} < \infty$, then the solution is in $\mathscr{H}_{2p}$, $p \geq 1$.*

Proof We first assume that $E\|\xi_0\|_H^{2p} < \infty$. Let $I(t,X)$ be defined as in (3.35), and consider $I(X)(t) = I(t,X)$. Then, by Lemma 3.4, $I : \mathscr{H}_{2p} \to \mathscr{H}_{2p}$. The solution can be approximated by the following sequence:

$$X_0(t) = S(t)\xi_0,$$
$$X_{n+1}(t) = S(t)\xi_0 + I(t,X_n), \quad n = 0,1,\ldots. \tag{3.37}$$

Indeed, let $v_n(t) = E \sup_{0 \leq s \leq t} \|X_{n+1}(s) - X_n(s)\|_H^{2p}$. Then $v_0(t) = E \sup_{0 \leq s \leq t} \|X_1(s) - X_0(s)\|_H^{2p} \leq v_0(T) \equiv V_0$, and, using Lemma 3.5, we obtain

$$v_1(t) = E \sup_{0 \leq s \leq t} \|X_2(s) - X_1(s)\|_H^{2p} = E \sup_{0 \leq s \leq t} \|I(s,X_1) - I(s,X_0)\|_H^{2p}$$

$$\leq C \int_0^t E \sup_{0 \leq u \leq s} \|X_1(u) - X_0(u)\|_H^{2p} ds \leq C V_0 t$$

and, in general,

$$v_n(t) \leq C \int_0^t v_{n-1}(s) ds \leq \frac{V_0(Ct)^n}{n!}.$$

Next, similarly to the proof of Gikhman and Skorokhod in [25], we show that

$$\sup_{0 \leq t \leq T} \|X_n(t) - X(t)\|_H \to 0 \quad \text{a.s.}$$

for some $X \in \mathcal{H}_{2p}$. If we let $\varepsilon_n = (V_0(CT)^n/n!)^{1/(1+2p)}$, then, using Chebychev's inequality, we arrive at

$$P\left(\sup_{0 \le t \le T} \|X_{n+1}(t) - X_n(t)\|_H > \varepsilon_n\right) = P\left(\sup_{0 \le t \le T} \|X_{n+1}(t) - X_n(t)\|_H^{2p} > \varepsilon_n^{2p}\right)$$

$$\le \left(\frac{V_0(CT)^n}{n!}\right) \Big/ \left(\frac{[V_0(CT)^n]^{2p/(1+2p)}}{n!}\right) = \varepsilon_n.$$

Because $\sum_{n=1}^{\infty} \varepsilon_n < \infty$, by the Borel–Cantelli lemma, $\sup_{0 \le t \le T} \|X_{n+1}(t) - X_n(t)\|_H < \varepsilon_n$ P-a.s. Thus, the series

$$\sum_{n=1}^{\infty} \sup_{0 \le t \le T} \|X_{n+1}(t) - X_n(t)\|_H$$

converges P-a.s., showing that X_n converges to some X a.s. in $C([0, T], H)$.
  Moreover,

$$E \sup_{0 \le t \le T} \|X(t) - X_n(t)\|_H^{2p} = E \lim_{m \to \infty} \sup_{0 \le t \le T} \|X_{n+m}(t) - X_n(t)\|_H^{2p}$$

$$= E \lim_{m \to \infty} \sup_{0 \le t \le T} \left\| \sum_{k=n}^{n+m-1} (X_{k+1}(t) - X_k(t)) \right\|_H^{2p}$$

$$\le E \lim_{m \to \infty} \left(\sum_{k=n}^{n+m-1} \sup_{0 \le t \le T} \|X_{k+1}(t) - X_k(t)\|_H \right)^{2p}$$

$$= \lim_{m \to \infty} E \left(\sum_{k=n}^{n+m-1} \sup_{0 \le t \le T} \|X_{k+1}(t) - X_k(t)\|_H k \frac{1}{k} \right)^{2p}$$

$$\le \sum_{k=n}^{\infty} E \sup_{0 \le t \le T} \|X_{k+1}(t) - X_k(t)\|_H^{2p} k^{2p} \left(\sum_{k=n}^{\infty} k^{-2q} \right)^{p/q}$$

with $1/2p + 1/2q = 1$. Note that $q > 1/2$; hence, the second series converges. The first series is bounded by: $\sum_{k=n}^{\infty} v_k(T)k^{2p} \le \sum_{k=n}^{\infty} V_0(CT)^k k^{2p}/k! \to 0$ as $n \to \infty$.
  To justify that $X(t)$ is a mild solution to (3.1), we note that, a.s., $F(s, X_n) \to F(s, X)$ uniformly in s. Therefore,

$$\int_0^t S(t - s)F(s, X_n)\,ds \to \int_0^t S(t - s)F(s, X)\,ds \quad \text{a.s.}$$

Using the fact, proved above, that $E \sup_t \|X(t) - X_n(t)\|_H^{2p} \to 0$, we obtain

$$E \left\| \int_0^t S(t - s)\big(B(s, X) - B(s, X_n)\big)\,dW_s \right\|_H^{2p}$$

$$\leq C_{p,M,\alpha,T} E \int_0^t \left\| (B(s, X) - B(s, X_n)) \right\|_{\mathscr{L}(K_Q,H)}^{2p} ds$$

$$\leq C_{p,M,\alpha,T,\mathscr{K}} E \sup_{0 \leq t \leq T} \left\| X(t) - X_n(t) \right\|_H^{2p} \to 0.$$

Now, if $E \|\xi_0\|_H^{2p} \leq \infty$, take the $\mathscr{F}_0$-measurable random variable $\chi_k = 1_{\{\xi_0 < k\}}$ and let

$$\xi_k = \xi_0 \chi_k.$$

Let $X^k(t)$ be a mild solution of (3.1) with the initial condition ξ_k. We will first show that

$$X^k \chi_k = X^{k+1} \chi_k.$$

Let X_n^k and X_n^{k+1} be the approximations of mild solutions X^k and X^{k+1} defined by (3.37). Since

$$X_0^k(t) = S(t)\xi_0 \chi_k = S(t)\xi_0 \chi_{k+1} \chi_k = X_0^{k+1}(t) \chi_k,$$

we deduce that

$$X_0^k \chi_k = X_0^{k+1} \chi_k,$$
$$F\left(t, X_0^k\right) \chi_k = F\left(t, X_0^{k+1}\right) \chi_k,$$
$$B\left(t, X_0^k\right) \chi_k = B\left(t, X_0^{k+1}\right) \chi_k,$$

so that

$$X_1^k(t)\chi_k = S(t)\xi_0 \chi_k + \chi_k \int_0^t S(t-s)F\left(s, X_0^k\right) ds + \chi_k \int_0^t S(t-s)B\left(s, X_0^k\right) dW_s$$

$$= S(t)\xi_0 \chi_k + \chi_k \int_0^t S(t-s)F\left(s, X_0^{k+1}\right) ds$$

$$+ \chi_k \int_0^t S(t-s)B\left(s, X_0^{k+1}\right) dW_s$$

$$= \left(S(t)\xi_0 \chi_{k+1}\right) \chi_k + I\left(t, X_0^{k+1}\right) \chi_k$$

$$= X_1^{k+1}(t)\chi_k.$$

This, by induction, leads to $X_n^k \chi_k = X_n^{k+1} \chi_k$. Since

$$X_n^k \to X^k \quad \text{and} \quad X_n^{k+1} \to X^{k+1}$$

in $\mathscr{H}_{2p}$, we also have by the generalized Lebesgue DCT, Theorem 3.4, that

$$X_n^k \chi_k \to X^k \chi_k \quad \text{and} \quad X_n^{k+1} \chi_k \to X^{k+1} \chi_k$$

in $\mathcal{H}_{2p}$, so that P-a.s., for all $t \in [0, T]$, $X^k(t)\chi_k = X^{k+1}(t)\chi_k$. The limit

$$\lim_{k \to \infty} X^k(t) = X(t)$$

exists P-a.s. and is as an element of $C([0, T], H)$. $X(t)$ satisfies (3.1), since

$$X(t)\chi_k = X^k(t) \quad P\text{-a.s.} \tag{3.38}$$

and $X^k(t)$ satisfies (3.1), so that

$$X(t)\chi_k = S(t)\xi_0\chi_k + \chi_k \int_0^t S(t-s)F(s, X)\,ds + \chi_k \int_0^t S(t-s)B(s, X)\,dW_s$$

and $P(\bigcup_k \{\chi_k = 1\}) = 1$.

The obtained solution is unique. If $X(t), Y(t)$ are two solutions to (3.1), then consider the processes $X^k(t) = X(t)\chi_k$ and $Y^k(t) = Y(t)\chi_k$, $k \geq 1$. We define $V(t) = E \sup_{s \leq t} \|X^k(s) - Y^k(s)\|_H^{2p}$. By Lemma 3.5,

$$V(t) \leq C \int_0^t V(s)\,ds \leq \cdots \leq E \sup_{0 \leq s \leq T} \|X^k(s) - Y^k(s)\|_H^{2p} \frac{(Ct)^n}{n!} \to 0$$

as $n \to \infty$, giving $V(t) = 0$. Consequently,

$$X(t)\chi_k = X^k(t) = Y^k(t) = Y(t)\chi_k \quad P\text{-a.s.} \qquad \square$$

We have used the following facts in the proof of Theorem 3.3.

Exercise 3.8 Let $A \in \mathcal{F}_0$, $\Phi \in \mathcal{P}(K_Q, H)$. Prove that

$$\int_0^t 1_A \Phi(s)\,dW_s = 1_A \int_0^t \Phi(s)\,dW_s. \tag{3.39}$$

Exercise 3.9 Prove the following theorem.

Theorem 3.4 (Generalized Lebesgue DCT) *Let (E, μ) be a measurable space, and g_n be a sequence of nonnegative real-valued integrable functions such that $g_n(x) \to g(x)$ for μ-a.e. $x \in E$ and*

$$\int_E g_n(x)\mu(dx) \to \int_E g(x)\mu(dx).$$

Let f_n be another sequence of functions such that $|f_n| \leq g_n$ and $f_n(x) \to f(x)$ for μ-a.e. $x \in E$, then f_n and f are integrable functions, and

$$\int_E f_n(x)\mu(dx) \to \int_E f(x)\mu(dx).$$

The Itô formula (2.53) is applicable to calculate differentials of functions of processes of the form (2.52) (in particular, to strong solutions), but not of the form in which a mild solution is provided. Hence, the following proposition will be essential, since the Itô formula can be applied to strong solutions (if they exist) of the following approximating problems, where the unbounded operator A is replaced by its Yosida approximations $A_n = A R_n$ with $R_n = n R(n, A)$ and $R(n, A)$ being the resolvent operator of A:

$$\begin{cases} dX(t) = (A_n X(t) + F(t, X)) \, dt + B(t, X) \, dW_t, \\ X(0) = \xi_0. \end{cases} \qquad (3.40)$$

Proposition 3.2 *Let the coefficients F and B satisfy conditions* (A1), (A3), *and* (A4) *and* $E \|\xi_0\|_H^{2p} < \infty$. *Let $X(t)$ be a mild solution to the semilinear equation* (3.1), *and $X_n(t)$ be strong solutions of the approximating problems* (3.40). *If $S(t)$ is a pseudo-contraction semigroup and $p \geq 1$ or a general C_0-semigroup and $p > 1$, then the mild solution $X(t)$ of* (3.1) *is approximated in $\mathscr{H}_{2p}$ by the sequence of strong solutions $X_n(t)$ to* (3.40), *that is,*

$$\lim_{n \to \infty} E \sup_{0 \leq t \leq T} \|X_n(t) - X(t)\|_H^{2p} = 0$$

Proof First note that under the assumption on the coefficients of (3.40), by Theorem 3.3, strong solutions X_n exist, and they are unique and coincide with mild solutions. Moreover,

$$E \sup_{0 \leq s \leq t} \|X_n(s) - X(s)\|_H^{2p} \leq C \Bigg(E \sup_{0 \leq s \leq t} \|(S_n(s) - S(s))\xi_0\|_H^{2p}$$

$$+ E \int_0^t \|(S_n(t - s) - S(t - s))F(s, X_n)\|_H^{2p} \, ds$$

$$+ E \int_0^t \|(S_n(t - s) - S(t - s))B(s, X_n)\|_L^{2p} \, ds$$

$$+ E \sup_{0 \leq s \leq t} \|I(s, X_n) - I(s, X)\|_H^{2p} \Bigg).$$

Since for $x \in H$ and $n \to \infty$, $(S_n(t) - S(t))(x) \to 0$ uniformly in $t \in [0, T]$, the first three summands converge to zero by the Lebesgue DCT. Thus first three terms are bounded by $\varepsilon(n) \to 0$ as $n \to \infty$.

Lemma 3.5 implies that the last summand is bounded by $C \int_0^t E \sup_{s \leq t} \|X_n(s) - X(s)\|_H^{2p} \, ds$. By the Gronwall lemma, we deduce that

$$E \sup_{0 \leq t \leq T} \|X_n(s) - X(s)\|_H^{2p} \, ds \leq \varepsilon(n) e^{Ct} \to 0$$

as $n \to \infty$. $\square$

The following estimate on moments for mild solutions of SSDEs is available.

Lemma 3.6 *Let $X(t) = \xi_0 + \int_0^t S(t-s)F(s, X)\,ds + \int_0^t S(t-s)B(s, X)\,dW_s$ with the coefficients F and B satisfying conditions (A1) and (A3) and an $\mathscr{F}_0$-measurable H-valued random variable ξ_0. Let $S(t)$ be either a pseudo-contraction semigroup and $p \geq 1$ or a general C_0-semigroup and $p > 1$. Then*

$$E \sup_{0 \leq s \leq t} \|X(s)\|_H^{2p} < C_{p,M,\alpha,\ell,T}\left(1 + E\|\xi_0\|_H^{2p}\right)e^{C_{p,M,\alpha,\ell,T}t}$$

with $C_{p,M,\alpha,\ell,T}$ depending only on the indicated constants.

Proof Let $\tau_n = \inf\{t : \|X(t)\|_H > n\}$. Lemma 3.4 implies that

$$E \sup_{0 \leq s \leq t} \|X(s \wedge \tau_n)\|_H^{2p} = E \sup_{0 \leq s \leq t} \|S(t \wedge \tau_n)\xi_0 + I(s \wedge \tau_n, X)\|_H^{2p}$$

$$\leq 2^{2p-1}\left(E \sup_{0 \leq s \leq t \wedge \tau_n} \|S(s)\xi_0\|_H^{2p} + E \sup_{0 \leq s \leq t \wedge \tau_n} \|I(s, X)\|_H^{2p}\right)$$

$$\leq 2^{2p-1}\left(E\|\xi_0\|_H^{2p}Me^{\alpha t} + E \sup_{0 \leq s \leq t \wedge \tau_n} \|I(s, X)\|_H^{2p}\,ds\right)$$

$$\leq C\left(E\|\xi_0\|_H^{2p} + t + \int_0^t E \sup_{0 \leq u \leq s} \|X(u \wedge \tau_n)\|_H^{2p}\,ds\right).$$

By the Gronwall lemma, we conclude that

$$E \sup_{0 \leq s \leq t} \|X(s \wedge \tau_n)\|_H^{2p} \leq C\left(E\|\xi_0\|_H^{2p} + t\right) + C\int_0^t \left(E\|\xi_0\|_H^{2p} + t\right)e^{C(t-s)}\,ds$$

$$\leq C_{p,M,\alpha,\ell,T}\left(1 + E\|\xi_0\|_H^{2p}\right)e^{C_{p,M,\alpha,\ell,T}t}.$$

Also, because $\tau_n \to \infty$ a.s., we have $\sup_{s \leq t} \|X(s \wedge \tau_n)\|_H \to \sup_{s \leq t} \|X(s)\|_H$ a.s. Therefore,

$$E \sup_{s \leq t} \|X(s)\|_H^{2p} = E \lim_{n \to \infty} \sup_{s \leq t} \|X(s \wedge \tau_n)\|_H^{2p}$$

$$\leq \liminf_{n \to \infty} E \sup_{s \leq t} \|X(s \wedge \tau_n)\|_H^{2p}$$

$$\leq C_{p,M,\alpha,\ell,T}\left(1 + E\|\xi_0\|_H^{2p}\right)e^{C_{p,M,\alpha,\ell,T}t}. \qquad \square$$

Example 3.1 Consider a K-valued Q-Wiener process W_t with sample paths in $C([0, t], K)$. For $B \in \mathscr{L}(K, H)$, the process $W_t^B = BW_t$ is an H-valued Wiener process with covariance BQB^*, and it can be realized in $C([0, T], H)$. Consider now the equation

$$\begin{cases} dX(t) = (AX(t) + F(t, X))\,dt + dW_t^B, \\ X(0) = x \in H \quad \text{(deterministic)}, \end{cases} \qquad (3.41)$$

with F satisfying conditions (A1), (A2), and (A4) and A generating a C_0-semigroup of operators on H. Then, Theorem 3.3 guarantees the existence of a unique mild solution to (3.41) in $C([0, T], H)$, which is given by

$$X(t) = S(t)x + \int_0^t S(t-s)F(s, X)\, ds + \int_0^t S(t-s)B\, dW_t.$$

In case $F \equiv 0$, this process is called an H-valued Ornstein–Uhlenbeck process.

In Sect. 3.3 we will consider a special case where the coefficients F and B of an SSDE (3.1) depend on $X(t)$ rather than on the entire past of the solution. It is known that even in the deterministic case where $A \equiv 0$, $B \equiv 0$, and $F(t, X) = f(t, X(t))$ with a continuous function $f : \mathbb{R} \times H \to H$, a solution to a Peano differential equation

$$\begin{cases} X'(t) = f(t, X(t)), \\ X(0) = x \in H, \end{cases} \tag{3.42}$$

may not exist (see [26] for a counterexample), unless H is finite-dimensional. Thus either one needs an additional assumption on A, or one has to seek a solution in a larger space. These topics will be discussed in Sects. 3.8 and 3.9.

3.3 A Special Case

We proved the existence and uniqueness theorem for mild solutions to (3.1), Theorem 3.3, for general C_0-semigroups if $p > 1$, and for pseudo-contraction semigroups if $p = 1$. This defect is due to the fact that it is still an open problem if the maximum inequality (3.25) is valid for $p = 1$ and a general strongly continuous semigroup $S(t)$. In this section we include the case $p = 1$ and a general C_0-semigroup. Assume that the coefficients of (3.1) depend on the value of the solution at time t alone rather than on the entire past, so that $F(\omega, t, x) = F(\omega, t, x(t))$ and $B(\omega, t, x) = B(\omega, t, x(t))$ for $x \in C([0, T], H)$. We note that if $\tilde{F}(\omega, t, h) = F(\omega, t, x(t))$ and $\tilde{B}(\omega, t, h) = B(\omega, t, x(t))$ for $x(t) \equiv h$, a constant function, then conditions (A1), (A3), and (A4) on F and B in Sect. 3.1 imply the following conditions on $\tilde{F}$ and $\tilde{B}$:

(A1') $\tilde{F}$ and $\tilde{B}$ are jointly measurable on $\Omega \times [0, T] \times H$, and for every $0 \le t \le T$, and $x \in H$, they are measurable with respect to the σ-field $\mathscr{F}_t$ on Ω.

(A3') There exists a constant ℓ such that for all $x \in H$,

$$\left\| \tilde{F}(\omega, t, x) \right\|_H + \left\| \tilde{B}(\omega, t, x) \right\|_{\mathscr{L}_2(K_Q, H)} \le \ell \big(1 + \|x\|_H \big).$$

(A4') For $x_1, x_2 \in H$,

$$\left\| \tilde{F}(\omega, t, x_1) - \tilde{F}(\omega, t, x_2) \right\|_H + \left\| \tilde{B}(\omega, t, x_1) - \tilde{B}(\omega, t, x_2) \right\|_{\mathscr{L}_2(K_Q, H)}$$
$$\le \mathscr{K} \|x_1 - x_2\|_H.$$

On the other hand, if we define $F(\omega, t, x) = \tilde{F}(\omega, t, x(t))$ and $B(\omega, t, x) = \tilde{B}(\omega, t, x(t))$, then conditions (A1'), (A3'), and (A4') imply conditions (A1), (A3), and (A4).

We will remain consistent with our general approach and use the notation and conditions developed for $F(t, x) = F(t, x(t))$ and $B(t, x) = B(t, x(t))$.

Let $\tilde{\mathscr{H}}_{2p}$ denote the class of H-valued stochastic processes X that are measurable as mappings from $([0, T] \times \Omega, \mathscr{B}([0, T]) \otimes \mathscr{F})$ to $(H, \mathscr{B}(H))$, adapted to the filtration $\{\mathscr{F}_t\}_{t \leq T}$, and satisfying $\sup_{0 \leq s \leq T} E \|\xi(s)\|_H^{2p} < \infty$. Then $\tilde{\mathscr{H}}_{2p}$ is a Banach space with the norm

$$\|X\|_{\tilde{\mathscr{H}}_{2p}} = \left(\sup_{0 \leq t \leq T} E \|X(t)\|_H^{2p} \right)^{\frac{1}{2p}}.$$

Let $I(t, \xi)$ be as in (3.35). By repeating the proofs of Lemmas 3.4 and 3.5, with all suprema dropped and with (3.23) replacing (3.25) and (3.27), we obtain the following inequalities for $\xi_1, \xi_2 \in \tilde{\mathscr{H}}_{2p}$, $p \geq 1$, and a general C_0-semigroup:

$$E \|I(t, \xi)\|_H^{2p} \leq C_{p,M,\alpha,T,\ell} \left(t + \int_0^t E \|\xi(s)\|_H^{2p} \, ds \right), \quad p \geq 1, \quad (3.43)$$

$$E \|I(t, \xi_1) - I(t, \xi_2)\|_H^{2p}$$

$$\leq C_{p,M,\alpha,T,\mathscr{K}} \int_0^t E \|\xi_1(s) - \xi_2(s)\|_H^{2p} \, ds, \quad p \geq 1. \quad (3.44)$$

Inequality (3.44) implies

$$E \|I(t, \xi_1) - I(t, \xi_2)\|_H^{2p} \leq C_{p,M,\alpha,T,\mathscr{K}} t \sup_{0 \leq s \leq t} E \|\xi_1(s) - \xi_2(s)\|_H^{2p}$$

$$\leq C_{p,M,\alpha,T,\mathscr{K}} t \|\xi_1 - \xi_2\|_{\tilde{\mathscr{H}}_{2p}}^{2p}, \quad p \geq 1. \quad (3.45)$$

Hence, we have the following corollary to the two lemmas.

Corollary 3.3 *Let $\xi, \xi_1, \xi_2 \in \tilde{\mathscr{H}}_{2p}$, $p \geq 1$. If $F(t, x) = F(t, x(t))$ and $B(t, x) = B(t, x(t))$ satisfy conditions (A1) and (A3), then there exists a constant $C_{p,M,\alpha,T,\ell}$, such that*

$$\|I(\cdot, \xi)\|_{\tilde{\mathscr{H}}_{2p}}^{2p} \leq C_{p,M,\alpha,T,\ell} \left(1 + \|\xi\|_{\tilde{\mathscr{H}}_{2p}}^{2p} \right). \quad (3.46)$$

If $F(t, x) = F(t, x(t))$ and $B(t, x) = B(t, x(t))$ satisfy conditions (A1) and (A4), then there exists a constant $C_{p,M,\alpha,T,\mathscr{K}}$ such that

$$\|I(\cdot, \xi_1) - I(\cdot, \xi_2)\|_{\tilde{\mathscr{H}}_{2p}}^{2p} \leq C_{p,M,\alpha,T,\mathscr{K}} \|\xi_1 - \xi_2\|_{\tilde{\mathscr{H}}_{2p}}^{2p}. \quad (3.47)$$

The constants depend only on the indicated parameters.

Now we can prove the existence and uniqueness result.

Theorem 3.5 *Let the coefficients $F(t, x) = F(t, x(t))$ and $B(t, x) = B(t, x(t))$ satisfy conditions (A1), (A3), and (A4). Assume that $S(t)$ is a general C_0-semigroup. Then the semilinear equation (3.1) has a unique continuous mild solution. If, in addition, $E\|\xi_0\|_H^{2p} < \infty$, $p \geq 1$, then the solution is in $\mathscr{H}_{2p}$.*

In this case, either for $p > 1$ and a general C_0-semigroup or for $p = 1$ and a pseudo-contraction semigroup, the solution is in $\mathscr{H}_{2p}$.

Proof We follow the proof of the existence and uniqueness for deterministic Volterra equations, which uses the Banach contraction principle. The idea is to change a given norm on a Banach space to an equivalent norm, so that the integral transformation related to the Volterra equation becomes contractive.

We first assume that

$$E\|\xi_0\|_H^{2p} < \infty.$$

Let $\mathfrak{B}$ be the Banach space of processes $X \in \mathscr{H}_{2p}$, equipped with the norm

$$\|X\|_{\mathfrak{B}} = \left(\sup_{0 \leq t \leq T} e^{-Lt} E \|X(t)\|_H^{2p} \right)^{\frac{1}{2p}},$$

where $L = C_{p,M,\alpha,T,\mathscr{K}}$, the constant in Corollary 3.3. The norms $\| \cdot \|_{\tilde{\mathscr{H}}_{2p}}$ and $\| \cdot \|_{\mathfrak{B}}$ are equivalent since

$$e^{-LT/2p}\| \cdot \|_{\tilde{\mathscr{H}}_{2p}} \leq \| \cdot \|_{\mathfrak{B}} \leq \| \cdot \|_{\tilde{\mathscr{H}}_{2p}}.$$

With $I(t, \xi)$ as in (3.35), define

$$\tilde{I}(X)(t) = S(t)\xi_0 + I(t, X). \tag{3.48}$$

Note that $\tilde{I} : \mathfrak{B} \to \mathfrak{B}$ by (3.43). We will find a fixed point of the transformation $\tilde{I}$. Let $X, Y \in \mathfrak{B}$. We use (3.44) and calculate

$$\left\| \tilde{I}(X) - \tilde{I}(Y) \right\|_{\mathfrak{B}}^{2p} = \sup_{0 \leq t \leq T} e^{-Lt} E \left\| I(t, X) - I(t, Y) \right\|_H^{2p}$$

$$\leq \sup_{0 \leq t \leq T} e^{-Lt} L \int_0^t E \|X(s) - Y(s)\|_H^{2p} \, ds$$

$$= \sup_{0 \leq t \leq T} e^{-Lt} L \int_0^t e^{Ls} e^{-Ls} E \|X(s) - Y(s)\|_H^{2p} \, ds$$

$$\leq L \|X - Y\|_{\mathfrak{B}}^{2p} \sup_{0 \leq t \leq T} e^{-Lt} \int_0^t e^{Ls} \, ds$$

$$= L \|X - Y\|_{\mathfrak{B}}^{2p} \sup_{0 \leq t \leq T} e^{-Lt} \frac{e^{Lt} - 1}{L}$$

$$\leq \left(1 - e^{-LT} \right) \|X - Y\|_{\mathfrak{B}}^{2p},$$

proving that $\tilde{I} : \mathfrak{B} \to \mathfrak{B}$ is a contraction, and we can use Banach's contraction principle to find its unique fixed point.

We need to show that the fixed point is a process having a continuous modification. The only problem is when $p = 1$, since for $p > 1$, the continuity is ensured by Theorem 3.3 (note that we allow $S(t)$ to be a general C_0-semigroup). Similarly as in the proof of Theorem 3.3, let

$$\chi_k = 1_{\{\xi_0 < k\}} \quad \text{and} \quad \xi_k = \xi_0 \chi_k,$$

so that $E\|\xi_k\|_H^{2p} < \infty$ for any $p > 1$. Let $X^k(\cdot) \in \mathscr{H}_{2p}$ be mild solutions of (3.1) with the initial condition ξ_k. The processes $X^k(t)$ and the limit

$$\lim_{k \to \infty} X^k(t) = \tilde{X}(t)$$

are continuous (see (3.38)), and $\tilde{X}(t)$ is a mild solution of (3.1) with the initial condition ξ_0. By the uniqueness, $\tilde{X}(t)$ is the sought continuous modification of $X(t)$. The proof is complete for ξ_0 satisfying $E\|\xi_0\|_H^{2p} < \infty$ and $p \geq 1$.

If $E\|\xi\|_H^{2p} \leq \infty$, then we apply the corresponding part of the proof of Theorem 3.3.

The uniqueness is justified as in the proof of Theorem 3.3.

The final assertion of the theorem is a direct consequence of Theorem 3.3.　□

Denote $\mathscr{F}_t^X = \sigma\{X(s), s \leq t\}$ and

$$\mathscr{F}_s^{W,\xi_0} = \sigma\left(\mathscr{F}_s^W \cup \sigma(\xi_0)\right).$$

Remark 3.3 In case $E\|\xi_0\|_H^{2p} < \infty$, the unique mild solution X constructed in Theorem 3.5 can be approximated in $\mathscr{H}_{2p}$ by the sequence

$$X_0(t) = S(t)\xi_0,$$

$$X_{n+1}(t) = \tilde{I}(X_n)(t)$$

$$= S(t)\xi_0 + \int_0^t S(t-s)F\left(s, X_n(s)\right) ds + \int_0^t S(t-s)B\left(s, X_n(s)\right) dW_s.$$

Then $X_n(t)$ and its limit $X(t)$ are measurable with respect to $\mathscr{F}_t^{W,\xi_0}$.

If $E\|\xi_0\|_H^{2p} \leq \infty$, the mild solution X is obtained as a P-a.e. limit of mild solutions adapted to $\mathscr{F}_t^{W,\xi_0}$, so that it is also adapted to that filtration.

We conclude with the following corollary to Proposition 3.2. It follows by dropping suprema and using (3.44) in the proof of Proposition 3.2.

Corollary 3.4 *Let the coefficients* $F(t, x) = F(t, x(t))$ *and* $B(t, x) = B(t, x(t))$ *satisfy conditions* (A1), (A3), *and* (A4) *and* $E\|\xi_0\|_H^{2p} < \infty$. *Let* $X(t)$ *be a mild* *solution to the semilinear equation* (3.1), *and* $X_n(t)$ *be strong solutions of the approximating problems* (3.40). *If* $S(t)$ *is a* C_0-*semigroup and* $p \geq 1$, *then the mild solution* $X(t)$ *of* (3.1) *is approximated in* $\mathcal{H}_{2p}$ *by the sequence of strong solutions* $X_n(t)$ *to* (3.40),

$$\lim_{n \to \infty} \sup_{0 \leq t \leq T} E \| X_n(t) - X(t) \|_H^{2p} = 0.$$

3.4 Markov Property and Uniqueness

We examine the Markov property of mild solutions obtained under the Lipschitz condition on the coefficients of the SSDE (3.1) in the setting of Sect. 3.3, that is, we assume that the coefficients of (3.1) depend on t and the value of the solution at time t alone, so that $F(t, x) = F(t, x(t))$ and $B(t, x) = B(t, x(t))$ for $x \in C([0, T], H)$. In addition, let F and B not depend on ω, so that we can think of F and B as $F : [0, T] \times H \to H$ and $B : [0, T] \times H \to \mathcal{L}_2(K_Q, H)$.

Definition 3.2 An H-valued stochastic process $\{X(t), t \geq 0\}$ defined on a probability space $(\Omega, \mathcal{F}, P)$ is called a *Markov process* if it satisfies the following *Markov property*:

$$E\big(\varphi\big(X(t + h)\big)\big|\mathcal{F}_t^X\big) = E\big(\varphi\big(X(t + h)\big)\big|X(t)\big) \tag{3.49}$$

for all t, $h \geq 0$ and any real-valued bounded measurable function φ.

Exercise 3.10 Prove that if (3.49) holds true for any real-valued bounded measurable function φ, then it is also valid for any φ, such that $\varphi(X(t + h)) \in L^1(\Omega, \mathbb{R})$.

We now want to consider mild solutions to (3.1) on the interval $[s, T]$. To that end, let $\{W_t\}_{t \geq 0}$, be a Q-Wiener process with respect to the filtration $\{\mathcal{F}_t\}_{t \leq T}$ and consider $\bar{W}_t = W_{t+s} - W_s$, the increments of W_t. The process $\bar{W}_t$ is a Q-Wiener process with respect to $\bar{\mathcal{F}}_t = \mathcal{F}_{t+s}$, $t \geq 0$. Its increments on $[0, T - s]$ are identical to the increments of W_t on $[s, T]$.

Consider (3.1) with $\bar{W}_t$ replacing W_t and $\bar{\mathcal{F}}_0 = \mathcal{F}_s$ replacing $\mathcal{F}_0$. Under the assumptions of Theorem 3.5, there exists a mild solution $X(t)$ of (3.1), and it is unique, so that for any $0 \leq s \leq T$ and an $\mathcal{F}_s$-measurable random variable ξ, there exists a unique process $X(\cdot, s; \xi)$ such that

$$X(t, s; \xi) = S(t - s)\xi + \int_s^t S(t - r)F\big(r, X(r, s; \xi)\big)\, dr$$

$$+ \int_s^t S(t - r)B\big(r, X(r, s; \xi)\big)\, dW_r. \tag{3.50}$$

Let φ be a real bounded measurable function on H. For $x \in H$, define

$$(P_{s,t}\varphi)(x) = E\big(\varphi\big(X(t, s; x)\big)\big). \tag{3.51}$$

This definition can be extended to functions φ such that $\varphi(X(t, s; x)) \in L^1(\Omega, \mathbb{R})$ for arbitrary $s \le t$. Note that for any random variable η,

$$(P_{s,t}\varphi)(\eta) = E\left(\varphi\left(X(t, s; x)\right)\right)\big|_{x=\eta}.$$

Due to the uniqueness of the solution, we have the following theorem.

Theorem 3.6 *Let the coefficients F and B satisfy conditions (A1), (A3), and (A4). Assume that $S(t)$ is a general C_0-semigroup. Then, for $u \le s \le t \le T$, the solutions $X(t, u; \xi)$ of (3.50) are Markov processes, i.e., they satisfy the following Markov property*:

$$E\left(\varphi\left(X(t, u; \xi)\right)\big|\mathscr{F}_s^{W,\xi}\right) = (P_{s,t}\varphi)\left(X(s, u; \xi)\right) \tag{3.52}$$

for any real-valued function φ, such that $\varphi(X(t, s; \xi)) \in L^1(\Omega, \mathbb{R})$ for arbitrary $s \le t$.

Proof Using the uniqueness, we have, for $0 \le u \le s \le t \le T$,

$$X(t, u; \xi) = X\left(t, s; X(s, u; \xi)\right)   \text{P-a.e.,}$$

so that we need to prove

$$E\left(\varphi\left(X\left(t, s; X(s, u; \xi)\right)\right)\big|\mathscr{F}_s^{W,\xi}\right) = (P_{s,t}\varphi)\left(X(s, u; \xi)\right).$$

We will prove that

$$E\left(\varphi\left(X(t, s; \eta)\right)\big|\mathscr{F}_s^{W,\xi}\right) = P_{s,t}(\varphi)(\eta) \tag{3.53}$$

for all $\sigma(X(s, u; \xi))$-measurable random variables η. By the monotone class theorem (functional form) it suffices to prove (3.53) for φ bounded continuous on H.

Note that if $\eta = x \in H$, then clearly the solution $X(t, s; x)$ obtained in Theorem 3.3 is measurable with respect to $\sigma\{W_t - W_s, t \ge s\}$ and hence independent of $\mathscr{F}_s^{W,\xi}$, by the fact that the increments $W_t - W_s$, $t \ge s$, are independent of $\mathscr{F}_s^{W,\xi}$. This implies

$$E\left(\varphi(X(t, s; x))\big|\mathscr{F}_s^{W,\xi}\right) = (P_{s,t}\varphi)(x). \tag{3.54}$$

Consider a simple function

$$\eta = \sum_{j=1}^{n} x_j 1_{A_j}\left(X(s, u; \xi)\right),$$

where $\{A_j, j = 1, 2, \ldots, n\}$ is a measurable partition of H, and $x_1, x_2, \ldots, x_n \in H$. Then, P-a.e.,

$$X(t, s; \eta) = \sum_{j=1}^{n} X(t, s; x_j) 1_{A_j}\left(X(s, u; \xi)\right)$$

and

$$E\big(\varphi\big(X(t,s;\eta)\big)\big|\mathscr{F}_s^{W,\xi}\big) = \sum_{j=1}^n E\big(\varphi\big(X(t,s;x_j)\big)1_{A_j}\big(X(s,u;\xi)\big)\big|\mathscr{F}_s^{W,\xi}\big).$$

Now $\varphi(X(t,s;x_j))$, $j = 1,2,\ldots,n$ are independent of $\mathscr{F}_s^{W,\xi}$, and $1_{A_j}(X(s,u;\xi))$ is $\mathscr{F}_s^{W,\xi}$-measurable, giving that, P-a.e.,

$$E\big(\varphi\big(X(t,s;\eta)\big)\big|\mathscr{F}_s^{W,\xi}\big) = \sum_{j=1}^n (P_{s,t}\varphi)(x_j)1_{A_j}\big(X(s,u;\xi)\big)$$

$$= (P_{s,t}\varphi)(\eta). \tag{3.55}$$

If $E\|\eta\|_H^2 < \infty$, then there exists a sequence of simple functions η_n of the above form such that $E\|\eta_n\|^2 < \infty$ and $E\|\eta_n - \eta\|^2 \to 0$. Lemma 3.7, in Sect. 3.5, yields

$$E\big\|X(t,s;\eta_n) - E\big(X(t,s;\eta)\big)\big\|_H^2 \to 0.$$

By selecting a subsequence if necessary, we can assume that $X(t,s;\eta_n) \to X(t,s;\eta)$ P-a.e. Since φ is continuous and bounded, (3.55) implies that

$$E\big(\varphi\big(X(t,s;\eta)\big)\big|\mathscr{F}_s^{W,\xi}\big) = \lim_n E\big(\varphi(X(t,s;\eta_n)|\mathscr{F}_s^{W,\xi}\big)$$

$$= \lim_n (P_{s,t}\varphi)(\eta_n) = (P_{s,t}\varphi)(\eta). \tag{3.56}$$

For a general η, we consider the solutions $X(t,s,\eta_n)$ with initial conditions $\eta_n = \eta\chi_n$, where $\chi_n = 1_{\{\eta<n\}}$ as we did in the final step of the proof of Theorem 3.3. Then, $X(t,s,\eta_n) \to X(t,s,\eta)$ P-a.e., and we can repeat the argument in (3.56) to conclude (3.53). $\qquad\square$

Corollary 3.5 *Under the assumptions of Theorem 3.5, for $u \le s \le t \le T$, the solutions $X(t,u;\xi)$ of (3.50) satisfy the following Markov property:*

$$E\big(\varphi\big(X(t,u;\xi)\big)\big|\mathscr{F}_s^X\big) = (P_{s,t}\varphi)\big(X(s,u;\xi)\big) \tag{3.57}$$

with $\mathscr{F}_s^X = \sigma\{X(r,u;\xi), u \le r \le s\}$ for every real-valued function φ such that $\varphi(X(t,s;\xi)) \in L^1(\Omega,\mathbb{R})$ for arbitrary $s \le t$.

Proof Since the RHS of (3.52) is $\mathscr{F}_s^X$-measurable and $\mathscr{F}_s^X \subset \mathscr{F}_s^{W,\xi}$, which is a consequence of Remark 3.3, it is enough to take conditional expectation with respect to $\mathscr{F}_s^X$ in (3.52). $\qquad\square$

Exercise 3.11 Show that if $X(t) = X(t,0;\xi_0)$ is a mild solution to (3.1) as in Theorem 3.5, then the Markov property (3.52) implies (3.49).

We now consider the case where F and B are independent of t, and assume that $x \in H$. Then we get

$$X(t+s,t;x) = S(s)x + \int_t^{t+s} S(t+s-u)F\big(X(u,t;x)\big)\,du$$

$$+ \int_t^{t+s} S(t+s-u)B\big(X(u,t,x)\big)\,dW_u$$

$$= S(s)x + \int_0^s S(s-u)F\big(X(t+u,t;x)\big)\,du$$

$$+ \int_0^s S(s-u)B\big(X(t+u,t;x)\big)\,d\bar{W}_u,$$

where $\bar{W}_u = W_{t+u} - W_t$.

Since $\bar{W}_u$ is a Wiener process with the same distribution as W_u, the processes $X(t+s,t;x)$ and $X(s,0;x)$ are mild solutions to the same SDE (3.1) but with different Q-Wiener processes. By the uniqueness of the solution we have that

$$\big\{X(t+s,t;x),\ s \ge 0\big\} \stackrel{d}{=} \big\{X(s,0;x),\ s \ge 0\big\},$$

i.e., the solution is a homogeneous Markov process. In particular, we get

$$P_{s,t}(\varphi) = P_{0,t-s}(\varphi), \quad 0 \le s \le t,$$

for all bounded measurable functions φ on H. Let us denote by

$$P_t = P_{0,t}.$$

Note that for $\varphi \in C_b(H)$, the space of bounded continuous functions on H, $P_t(\varphi) \in C_b(H)$, due to the continuity of the solution with respect to the initial condition, which we will prove in Lemma 3.7. This property is referred to as the *Feller property* of P_t. In addition, by the Markov property,

$$(P_t \circ P_s)(\varphi) = P_{t+s}(\varphi), \tag{3.58}$$

so that P_t is a *Feller semigroup*. A Markov process X_t whose corresponding semigroup has Feller property is called a *Feller process*.

Exercise 3.12 Show (3.58).

Denote $X^x(t) = X(t,0;x)$. In the case of time-independent coefficients F and B and with $x \in H$, the Markov property (3.57) takes the form

$$E\big(\varphi\big(X^x(t+s)\big)|\mathscr{F}_t^{X^x}\big) = (P_s\varphi)\big(X^x(t)\big) \tag{3.59}$$

with $\mathscr{F}_t^{X^x} = \sigma(X^x(u),\ u \le t)$.

In Sect. 3.5 we examine the dependence of the solution on the initial condition in detail.

3.5 Dependence of the Solution on the Initial Value

As in Sect. 3.4, we consider the semilinear SDE (3.1) with the coefficients $F(t, x) = F(t, x(t))$ and $B(t, x) = B(t, x(t))$ for $x \in C([0, T], H)$ such that F and B do not depend on ω.

Before we study the dependence on the initial condition, we need the following lemma.

Lemma 3.7 *Let $\{S(t), t \geq 0\}$ be a C_0-semigroup and for $\xi \in L^2(\Omega, H)$ and $X \in C([0, T], H)$, extend the operator $\tilde{I}$ defined in (3.48) to*

$$\tilde{I}(\xi, X)(t) = S(t)\xi + \int_0^t S(t-s)F(s, X(s)) \, ds + \int_0^t S(t-s)B(s, X(s)) \, dW_s.$$

$$(3.60)$$

Let F and B satisfy conditions (A1), (A3), and (A4). Then, for $0 \leq t \leq T$,

(1) $E\|\tilde{I}(\xi, X)(t) - \tilde{I}(\eta, X)(t)\|_H^2 \leq C_{1,T} E\|\xi - \eta\|_H^2$.
(2) $E\|\tilde{I}(\xi, X)(t) - \tilde{I}(\xi, Y)(t)\|_H^2 \leq C_{2,T} \int_0^T E\|X(t) - Y(t)\|_H^2 \, dt$.

Proof Condition (2) follows from (3.44) in Lemma 3.5. To prove (1), we let $X^\xi(t)$ and $X^\eta(t)$ be mild solutions of (3.1) with initial conditions ξ and η, respectively. Then,

$$E\|X^\xi(t) - X^\eta(t)\|_H^2$$

$$\leq 3C_T \left(E\|\xi - \eta\|_H^2 + E\left\{ \int_0^t \|F(s, X^\xi(s)) - F(s, X^\eta(s))\|_H^2 \, ds \right.\right.$$

$$\left.\left. + \int_0^t \|B(s, X^\xi(s)) - B(s, X^\eta(s))\|_{\mathscr{L}_2(K_Q, H)}^2 \, ds \right\} \right)$$

$$\leq 3C_T \left(E\|\xi - \eta\|_H^2 + \mathscr{K}^2 \int_0^t E\|X^\xi(s) - X^\eta(s)\|_H^2 \, ds \right).$$

We now obtain the result using Gronwall's lemma.   □

We now prove the continuity of the solution with respect to the initial value.

Theorem 3.7 *Let X_n be mild solutions to the sequence of stochastic semilinear equations (3.1) with coefficients F_n, B_n and initial conditions ξ_n, so that the following equations hold:*

$$X_n(t) = S(t)\xi_n + \int_0^t S(t-r)F_n(r, X_n(r)) \, dr + \int_0^t S(t-r)B_n(r, X_n(r)) \, dW_r.$$

Assume that $F_n(t, x)$ and $B_n(t, x)$ satisfy conditions (A1), (A3), and (A4) of Sect. 3.1, and in addition, let the following conditions hold:

(IV1) $\sup_n E\|\xi_n\|^2 < \infty$.

(IV2) *With* $n \to \infty$, $\|F_n(t, x) - F_0(t, x)\|_H^2 + \|B_n(t, x) - B_0(t, x)\|_{\mathscr{L}_2(K_Q, H)} \to 0$, *and* $E\|\xi_n - \xi_0\|_H^2 \to 0$.

Then $X_n(t) \to X_0(t)$ *in* $L^2(\Omega, H)$ *uniformly in* t.

Proof For any $t \le T$,

$$E\|X_n(t) - X_0(t)\|_H^2$$

$$\le 3\left\{ \|S(t)(\xi_n - \xi_0)\|_H^2 \right.$$

$$+ 2E\left(\int_0^t \|S(t - s)\big(F_n(s, X_n(s)) - F_n(s, X_0(s))\big)\|_H\, ds \right)^2$$

$$+ 2E\left\| \int_0^t S(t - s)\big(B_n(s, X_n(s)) - B_n(s, X_0(s))\big)\, dW_s \right\|_H^2$$

$$+ 2E\left\| \int_0^t S(t - s)\big(F_n(s, X_0(s)) - F_0(s, X_0(s))\big)\, ds \right\|_H^2$$

$$+ 2E\left\| \int_0^t S(t - s)\big(B_n(s, X_0(s)) - B_0(s, X_0(s))\big)\, dW_s \right\|_H^2 \right\}$$

$$\le 3e^{2\alpha t}\left\{ \|\xi_n - \xi_0\|_H^2 + 2\mathscr{K}^2 \int_0^t \|X_n(s) - X_0(s)\|_H^2\, ds + 2\alpha_1^{(n)}(t) + 2\alpha_2^{(n)}(t) \right\}$$

by the Lipschitz condition (A4). Now,

$$\alpha_1^{(n)}(t) = E\int_0^t \|S(t - s)\big(F_n(s, X_0(s)) - F_0(s, X_0(s))\big)\|_H^2\, ds$$

$$\le e^{2\alpha T} E\int_0^T \|F_n(s, X_0(s)) - F_0(s, X_0(s))\|_H^2\, ds.$$

As $F_n(s, X_0(s)) \to F_0(s, X_0(s))$, by condition (A3) and Lemma 3.6, we have, for all s,

$$E\|F_n(s, X_0(s))\|_H^2 \le 2\ell^2 E\big(1 + \|X_0(s)\|_H^2\big) \le C\big(1 + E\|\xi_0\|_H^2\big)$$

with the constant C independent of n. Using the uniform integrability, we obtain

$$\alpha_1^{(n)} \to 0 \quad \text{uniformly in } t.$$

Similarly,

$$\alpha_2^{(n)}(t) = \left\| E\int_0^t S(t - s)\big(B_n(s, X_0(s)) - B_0(x, X_0(s))\big)\, dW_s \right\|_H^2$$

$$\leq E \int_0^T \left\| S(t-s)\big(B_n\big(s, X_0(s)\big) - B_0\big(s, X_0(s)\big)\big) \right\|^2_{\mathscr{L}_2(K_Q, H)} ds \to 0$$

uniformly in $t \leq T$.

We obtain the result using Gronwall's lemma. □

We discuss differentiability of the solution with respect to the initial value in the case where the coefficients $F : [0, T] \times H \to H$ and $B : [0, T] \times H \to \mathscr{L}_2(K_Q, H)$ of (3.1) are Fréchet differentiable in the second (Hilbert space) variable.[2]

Theorem 3.8 *Assume that $F : [0, T] \times H \to H$ and $B : [0, T] \times H \to \mathscr{L}_2(K_Q, H)$ satisfy conditions (A1), (A3), and (A4).*

(a) If Fréchet derivatives $DF(t, \cdot)$ and $DB(t, \cdot)$ are continuous in H and bounded,

$$\left\| DF(t, x)y \right\|_H + \left\| DB(t, x)y \right\|_{\mathscr{L}_2(K_Q, H)} \leq M_1 \|y\|_H \tag{3.61}$$

for $x, y \in H$, $0 \leq t \leq T$, with the constant $M_1 \geq 0$, then $\tilde{I} : H \times \tilde{\mathscr{H}}_2 \to \tilde{\mathscr{H}}_2$ is continuously Fréchet differentiable, and its partial derivatives are

$$\left(\frac{\partial \tilde{I}(x, \xi)}{\partial x} y \right)(t) = S(t)y,$$

$$\left(\frac{\partial \tilde{I}(x, \xi)}{\partial \xi} \eta \right)(t) = \int_0^t S(t-s)DF\big(s, \xi(s)\big)\eta(s)\, ds \tag{3.62}$$

$$+ \int_0^t S(t-s)DB\big(s, \xi(s)\big)\eta(s)\, dW_s$$

P-a.s., with $\xi, \eta \in \tilde{\mathscr{H}}_2$, $x, y \in H$, $0 \leq t \leq T$.

(b) If in addition, second-order Fréchet derivatives $D^2 F(t, \cdot)$ and $D^2 B(t, \cdot)$ are continuous in H and bounded,

$$\left\| D^2 F(t, x)(y, z) \right\|_H + \left\| D^2 B(t, x)(y, z) \right\|_{\mathscr{L}_2(K_Q, H)} \leq M_2 \|y\|_H \|z\|_H \tag{3.63}$$

for $x, y, z \in H$, $0 \leq t \leq T$, with the constant $M_2 \geq 0$, then $\tilde{I} : H \times \tilde{\mathscr{H}}_2 \to \tilde{\mathscr{H}}_2$ is twice continuously Fréchet differentiable, and its second partial derivative is

$$\left(\frac{\partial^2 \tilde{I}(x, \xi)}{\partial \xi^2}(x, \xi)(\eta, \zeta) \right)(t)$$

---

[2]A reader interested in the theory of reaction–diffusion systems in a bounded domain $\mathcal{O}$ of $\mathbb{R}^d$, perturbed by a Gaussian random field, and the related stochastic evolution equations in an infinite-dimensional Hilbert space H, is referred to the work of Cerrai [8]. The author considers the nonlinear case with the coefficient F defined on $H = L^2(\mathcal{O}, \mathbb{R}^d)$, which is (necessarily) not Fréchet differentiable but can be assumed Gateaux differentiable.

$$= \int_0^t S(t-s)D^2F\big(s,\xi(s)\big)\big(\eta(s),\zeta(s)\big)\,ds$$

$$+ \int_0^t S(t-s)D^2B\big(s,\xi(s)\big)\big(\eta(s),\zeta(s)\big)\,dW_s \qquad (3.64)$$

P-a.s., with $\xi,\eta,\zeta \in \tilde{\mathscr{H}}_2$, $x \in H$, $0 \le t \le T$.

Proof Consider

$$\frac{\tilde{I}(x+h,\xi)(t) - \tilde{I}(x,\xi)(t) - S(t)h}{\|h\|_H} = \frac{S(t)(x+h) - S(t)x - S(t)h}{\|h\|_H} = 0,$$

proving the first equality in (3.62). To prove the second equality, let

$$r_F(t,x,h) = F(t,x+h) - F(t,x) - DF(t,x)h,$$

$$r_B(t,x,h) = B(t,x+h) - B(t,x) - DB(t,x)h.$$

By Exercise 3.13,

$$\big\|r_F(t,x,h)\big\|_H \le 2M_1\|h\|_H \quad \text{and} \quad \big\|r_B(t,x,h)\big\|_{\mathscr{L}_2(K_Q,H)} \le 2M_1\|h\|_H.$$

Now with $\frac{\partial \tilde{I}(x,\xi)}{\partial \xi}$ as defined in (3.62), we have

$$r_{\tilde{I}}(x,\xi,\eta)(t) = \tilde{I}(x,\xi+\eta)(t) - \tilde{I}(x,\xi)(t) - \left(\frac{\partial \tilde{I}(x,\xi)}{\partial \xi}\eta\right)(t)$$

$$= \int_0^t S(t-s)r_F\big(s,\xi(s),\eta(s)\big)\,ds + \int_0^t S(t-s)r_B\big(s,\xi(s),\eta(s)\big)\,dW_s$$

$$= I_1 + I_2.$$

We need to show that, as $\|\eta\|_{\tilde{\mathscr{H}}_2} \to 0$,

$$\frac{\|r_{\tilde{I}}(x,\xi,\eta)\|_{\tilde{\mathscr{H}}_2}}{\|\eta\|_{\tilde{\mathscr{H}}_2}} \to 0.$$

Consider

$$\frac{\big(\sup_{0 \le t \le T} E\big\|\int_0^t S(t-s)r_F(s,\xi(s),\eta(s))\,ds\big\|_H^2\big)^{1/2}}{\|\eta\|_{\tilde{\mathscr{H}}_2}}$$

$$\le C\left(E\int_0^T \frac{\|r_F(s,\xi(s),\eta(s))\|_H^2}{\|\eta(s)\|_H^2}\frac{\|\eta(s)\|_H^2}{\|\eta\|_{\tilde{\mathscr{H}}_2}^2}1_{\{\|\eta(s)\|_H \ne 0\}}\,ds\right)^{1/2}.$$

Since F is Fréchet differentiable, the factor

$$\frac{\|r_F(s,\xi(s),\eta(s))\|_H^2}{\|\eta(s)\|_H^2} \to 0,$$

and, as noted earlier, it is bounded by $4M_1^2$. In addition, the factor

$$\frac{\|\eta(s)\|_H^2}{\|\eta\|_{\tilde{\mathscr{H}}_2}^2} 1_{\{\|\eta(s)\|_H \neq 0\}} \le 1.$$

Consequently, by the Lebesgue DCT, $\|I_1\|_{\tilde{\mathscr{H}}_2}/\|\eta\|_{\tilde{\mathscr{H}}_2} \to 0$ as $\|\eta\|_{\tilde{\mathscr{H}}_2} \to 0$.
Because

$$\|I_2\|_{\tilde{\mathscr{H}}_2} = \left(\sup_{0 \le t \le T} E \int_0^t \left\| S(t-s) r_B(s,\xi(s),\eta(s)) \right\|_{\mathscr{L}_2(K_Q,H)}^2 ds \right)^{1/2},$$

we obtain that $\|I_2\|_{\tilde{\mathscr{H}}_2}/\|\eta\|_{\tilde{\mathscr{H}}_2} \to 0$ as $\|\eta\|_{\tilde{\mathscr{H}}_2} \to 0$, similarly as for I_1.

This concludes the proof of part (a), and the proof of part (b) can be carried out using similar arguments.   □

Exercise 3.13 Let H_1, H_2 be two Hilbert spaces. For a Fréchet differentiable function $F : H_1 \to H_2$, define $r_F(x,h) = F(x+h) - F(x) - DF(x)h$. Show that

$$\left\| r_F(x,h) \right\|_{H_2} \le 2 \sup_{x \in H_1} \left\| DF(x) \right\|_{\mathscr{L}(H_1,H_2)} \|h\|_{H_1}.$$

We will use the following lemma on contractions depending on a parameter.

Lemma 3.8 *Let X, U be Banach spaces, and $f : X \times U \to U$ be a contraction with respect to the second variable, i.e., for some $0 \le \alpha < 1$,*

$$\left\| f(x,u) - f(x,v) \right\|_U \le \alpha \|u - v\|_U, \quad x \in X, \ u, v \in V \tag{3.65}$$

and let for every $x \in X$, $\varphi(x)$ denote the unique fixed point of the contraction $f(x,\cdot) : U \to U$. Then the unique transformation $\varphi : X \to U$ defined by

$$f(x,\varphi(x)) = \varphi(x) \quad \text{for every } x \in X \tag{3.66}$$

is of class $C^k(X)$ whenever $f \in C^k(X \times U)$, $k = 0, 1, \ldots$. The derivatives of φ can be calculated using the chain rule; in particular,

$$D\varphi(x)y = \left[\frac{\partial f(x, \varphi(x))}{\partial u} - I\right]^{-1} \left(\frac{\partial f(x, \varphi(x))}{\partial x} y\right),$$

$$D^2\varphi(x)(y, z) = \left[\frac{\partial f(x, \varphi(x))}{\partial u} - I\right]^{-1} \left(\frac{\partial^2 f(x, \varphi(x))}{\partial x^2}(y, z)\right.$$

$$+ \frac{\partial^2 f(x, \varphi(x))}{\partial x \partial u}\left(D\varphi(x)y, z\right) + \frac{\partial^2 f(x, \varphi(x))}{\partial u \partial x}\left(y, D\varphi(x)z\right)$$

$$\left. + \frac{\partial^2 f(x, \varphi(x))}{\partial^2 u}\left(D\varphi(x)y, D\varphi(x)z\right)\right).$$

(3.67)

Let $\{f_n\}_{n=1}^{\infty}$ be a sequence of mappings in $C^l(X \times U)$ satisfying condition (3.65), denote by $\varphi_n : X \to U$ the unique transformations satisfying condition (3.66), and assume that for all $x, x_1, \ldots, x_k \in X$, $u, u_1, \ldots, u_j \in U$, $0 \le k + j \le l$,

$$\lim_{n \to \infty} \frac{\partial^{k+j} f_n(x, u)}{\partial x^k \partial u^j}(u_1, \ldots, u_j, x_1, \ldots, x_k) = \frac{\partial^{k+j} f(x, u)}{\partial x^k \partial u^j}(u_1, \ldots, u_j, x_1, \ldots, x_k).$$

(3.68)

Then

$$\lim_{n \to \infty} D^l \varphi_n(x)(x_1, \ldots, x_l) = D^l \varphi(x)(x_1, \ldots, x_l).$$

(3.69)

Proof Let $F(x, u) = u - f(x, u)$. Then $F(x, u) = 0$ generates the implicit function $\varphi(x)$ defined in (3.66). In addition, $F_u(x, u) = I - f_u(x, u)$ is invertible, since $\|f_u(x, u)\|_{\mathscr{L}(U)} \le \alpha < 1$. The differentiability and the form of the derivatives of $\varphi(x)$ follow from the *implicit function theorem* (see VI.2 in [44]). The last statement follows from the convergence in (3.68) and the form of the derivatives of $\varphi(x)$ given in (3.67).   □

We are now ready to prove a result on differentiability of the solution with respect to the initial condition.

Theorem 3.9 *Assume that $F : [0, T] \times H \to H$ and $B : [0, T] \times H \to \mathscr{L}_2(K_Q, H)$ satisfy conditions (A1), (A3), and (A4). Let Fréchet derivatives $DF(t, \cdot)$, $DB(t, \cdot)$, $D^2 F(t, \cdot)$, and $D^2 B(t, \cdot)$ be continuous in H and satisfy conditions (3.61) and (3.63). Then the solution X^x of (3.1) with initial condition $x \in H$, viewed as a mapping $X^\cdot : H \to \tilde{\mathscr{H}}_2$, is twice continuously differentiable in x and for any $y, z \in H$, the first and second derivative processes $DX^x(\cdot)y$ and $D^2 X^x(\cdot)(y, z)$ are mild solutions of the equations*

$$\begin{cases} dZ(t) = (AZ(t) + DF(t, X^x(t))Z(t)) \, dt + DB(t, X^x(t))Z(t) \, dW_t, \\ Z(0) = y, \end{cases}$$

(3.70)

and

$$\begin{cases} dZ(t) = (AZ(t) + DF(t, X^x(t))Z(t) + D^2F(t, X^x(t))(DX^x(t)y, DX^x(t)z)) \, dt \\ \qquad + (DB(X^x(t))Z(t) + D^2B(t, X^x(t))(DX^x(t)y, DX^x(t)z)) \, dW_t, \\ Z(0) = 0. \end{cases}$$

$$(3.71)$$

If X_n is the solution to (3.40) with deterministic initial condition $x \in H$, then for $y, z \in H$, we have the following approximations for the first and second derivative processes:

$$\lim_{n \to \infty} \left\| (DX_n^x(\cdot) - DX^x(\cdot))y \right\|_{\tilde{\mathscr{H}}_2} = 0,$$
$$\lim_{n \to \infty} \left\| (D^2X_n^x(\cdot) - D^2X^x(\cdot))(y, z) \right\|_{\tilde{\mathscr{H}}_2} = 0.$$

$$(3.72)$$

Proof Consider the operator $\tilde{I} : \mathfrak{B} \to \mathfrak{B}$ with $\tilde{I}$ defined in (3.60) and the Banach space $\mathfrak{B}$ defined in the proof of Theorem 3.5 in the case $p = 1$. Since $\mathfrak{B}$ is just $\tilde{\mathscr{H}}_2$ renormed with the norm $\| \cdot \|_{\mathfrak{B}}$ that is equivalent to $\| \cdot \|_{\tilde{\mathscr{H}}_2}$, we can as well prove the theorem in $\mathfrak{B}$, and the result will remain valid in $\tilde{\mathscr{H}}_2$.

Since $\tilde{I}$ is a contraction on $\mathfrak{B}$, as shown in the proof of Theorem 3.5, the solution X^x of (3.1) is the unique fixed point in $\mathfrak{B}$ of the transformation $\tilde{I}$, so that

$$X^x = \tilde{I}(X^x). \tag{3.73}$$

By Theorem 3.8, $\tilde{I} \in C^2(H \times \mathfrak{B})$ satisfies the conditions of Lemma 3.8, so that $X^x \in C^2(H)$ and formulas (3.70) and (3.71) follow from the chain rule and from (3.62) and (3.64). The last part of the assertion follows from the approximation (3.69). $\square$

3.6 Kolmogorov's Backward Equation

We put an additional restriction on the coefficients in this section and assume that F and B depend only on $x \in H$. We will now discuss analytical properties of the transition semigroup P_t. Recall that for a bounded measurable function φ on H,

$$P_t\varphi(x) = P_{0,t}\varphi(x) = E(\varphi(X^x(t))),$$

where $X^x(t) = X(t, 0; x)$ is a solution of (3.50) with deterministic initial condition $\xi_0 = x \in H$ and $s = 0$, or simply, a solution of (3.1) with $\xi_0 = x$. The smooth dependence of the solution with respect to the initial condition results in smooth dependence of the function

$$u(t, x) = P_t\varphi(x)$$

on t and x, and formulas (3.62) and (3.64) allow to establish a specific form of a parabolic-type PDE for $u(t, x)$, which is called *Kolmogorov's backward equation*,

$$
\begin{cases}
\dfrac{\partial u(t,x)}{\partial t} = \left\langle Ax + F(x), \dfrac{\partial u(t,x)}{\partial x} \right\rangle_H \\[2mm]
\qquad + \dfrac{1}{2} \operatorname{tr}\left(\dfrac{\partial^2 u(t,x)}{\partial x^2} \big(B(x)Q^{1/2}\big)\big(B(x)Q^{1/2}\big)^* \right), \\[2mm]
0 < t < T, \ x \in \mathcal{D}(A), \\[2mm]
u(0,x) = \varphi(x).
\end{cases} \tag{3.74}
$$

We follow the presentation in [11] and begin with the case where A is a bounded linear operator on H.

Theorem 3.10 (Kolmogorov's Backward Equation I) *Assume that $A \in \mathscr{L}(H)$ and F, B do not depend on t, $F : H \to H$, and $B : H \to \mathscr{L}_2(K_Q, H)$. Let Fréchet derivatives $DF(x)$, $DB(x)$, $D^2 F(x)$, and $D^2 B(x)$ be continuous and satisfy conditions (3.61) and (3.63) (with t omitted). If conditions (A1), (A3), and (A4) hold, then for $\varphi \in C_b^2(H)$, there exists a unique solution u to Kolmogorov's backward equation, satisfying (3.74) on $[0, T[^3$ such that $u(t, \cdot) \in C_b^2(H)$ and $u(\cdot, x) \in C_b^1([0, T[)$. The solution is given by*

$$
u(t,x) = P_t \varphi(x) = E\big(\varphi\big(X^x(t)\big)\big), \quad 0 \le t \le T, x \in H, \tag{3.75}
$$

where $X^x(t)$ is the solution to (3.1) with deterministic initial condition $\xi_0 = x \in H$.

Proof We first show that $u(t, x)$ defined by (3.75) satisfies (3.74). Since the operator A is bounded, the proof follows from the Itô formula applied to the function $\varphi(x)$ and the strong solution $X^x(t)$ of the SSDE (3.1),

$$
\begin{aligned}
d\varphi\big(X^x(t)\big) = \bigg\{ &\left\langle \dfrac{d\varphi(X^x(s))}{dx}, AX^x(t) + F\big(X^x(t)\big) \right\rangle_H \\
&+ \dfrac{1}{2} \operatorname{tr}\left(\dfrac{d^2\varphi(X^x(t))}{dx^2} \big(B\big(X^x(t)\big)Q^{1/2}\big)\big(B\big(X^x(t)\big)Q^{1/2}\big)^* \right) \bigg\} dt \\
&+ \left\langle \dfrac{d\varphi(X^x(s))}{dx}, B\big(X^x(t)\big)dW_t \right\rangle_H .
\end{aligned}
$$

Let $u(t, x) = E\varphi(X^x(t))$. Then by the Lebesgue DCT,

$$
\begin{aligned}
\dfrac{\partial^+ u(0,x)}{\partial t} &= \lim_{t \to 0^+} \dfrac{u(t,x) - \varphi(x)}{t} \\
&= E \lim_{t \to 0^+} \dfrac{1}{t} \int_0^t \left\langle AX^x(s) + F\big(X^x(s)\big), \dfrac{d\varphi\big(X^x(s)\big)}{dx} \right\rangle_H ds
\end{aligned}
$$

---

[3] Such solution is called a *strict solution*.

$$+ \frac{1}{2} E \lim_{t\to 0^+} \frac{1}{t} \int_0^t \mathrm{tr}\left(\frac{d^2\varphi(X^x(s))}{dx^2} \left(B(X^x(s))Q^{1/2}\right)\left(B(X^x(s))Q^{1/2}\right)^* \right) ds$$

$$= \left\langle Ax + F(x), \frac{d\varphi(x)}{dx} \right\rangle_H + \frac{1}{2} \mathrm{tr}\left(\frac{d^2\varphi(x)}{dx^2} \left(B(x)Q^{1/2}\right)\left(B(x)Q^{1/2}\right)^* \right). \quad (3.76)$$

Now Theorem 3.9 and the fact that $\varphi \in C_b^2(H)$ imply that $u(t, x)$ is twice Fréchet differentiable in x for $0 \leq t \leq T$, and for $y, z \in H$, we have

$$\left\langle \frac{\partial u(t, x)}{\partial x}, y \right\rangle_H = E\left\langle \frac{\partial \varphi(X^x(t))}{\partial x}, DX^x(t)y \right\rangle_H,$$

$$\left\langle \frac{\partial^2 u(t, x)}{\partial x^2} y, z \right\rangle_H = E\left\langle \frac{\partial^2 \varphi(X^x(t))}{\partial x^2} DX^x(t)y, DX^x(t)z \right\rangle_H \quad (3.77)$$

$$+ E\left\langle \frac{\partial \varphi(X^x(t))}{\partial x}, D^2 X^x(t)(y, z) \right\rangle_H.$$

But $DX^x(0) = I$ and $D^2 X^x(0) = 0$, so that

$$\left\langle \frac{\partial u(0, x)}{\partial x}, y \right\rangle_H = \left\langle \frac{\partial \varphi(x)}{\partial x}, y \right\rangle_H,$$

$$\left\langle \frac{\partial^2 u(0, x)}{\partial x^2} y, z \right\rangle_H = \left\langle \frac{\partial^2 \varphi(x)}{\partial x^2} y, z \right\rangle_H,$$

giving for $x \in H$,

$$\frac{\partial^+ u(0, x)}{\partial t} = \left\langle Ax + F(x), \frac{\partial u(0, x)}{\partial x} \right\rangle_H$$

$$+ \frac{1}{2} \mathrm{tr}\left(\frac{\partial^2 u(0, x)}{\partial x^2} \left(B(x)Q^{1/2}\right)\left(B(x)Q^{1/2}\right)^* \right).$$

By (3.58), $u(t + s, x) = u(t, u(s, x))$. Hence,

$$\frac{\partial^+ u(s, x)}{\partial t} = \frac{\partial^+ u(0, u(s, x))}{\partial t} = \lim_{t\to 0^+} \frac{P_t(P_s\varphi)(x) - (P_s\varphi)(x)}{t}.$$

Note that $(P_s\varphi)(x) = u(s, x) \in C_b^2(H)$, so that we can repeat the calculations in (3.76) with $(P_s\varphi)(x)$ replacing $\varphi(x)$ to arrive at

$$\frac{\partial^+ u(s, x)}{\partial t} = \left\langle Ax + F(x), \frac{\partial u(s, x)}{\partial x} \right\rangle_H$$

$$+ \frac{1}{2} \mathrm{tr}\left(\frac{\partial^2 u(s, x)}{\partial x^2} \left(B(x)Q^{1/2}\right)\left(B(x)Q^{1/2}\right)^* \right)$$

for $x \in H$, $0 \leq s < T$. Note that the functions $\partial u(s, x)/\partial x$ and $\partial^2 u(s, x)/\partial x^2$ are continuous in t, because they depend on derivatives of φ and the derivative processes

$DX^x(t)y$ and $D^2X^x(t)(y,z)$ that are mild solutions of (3.70) and (3.71). Hence, $\partial^+u(s,x)/\partial t$ is continuous on $[0,T[$, which implies ([63], Chap. 2, Corollary 1.2) that $u(\cdot,x)$ is continuously differentiable on $[0,T[$. Thus, $u(t,x)$ satisfies (3.74) on $[0,T[$.

To prove the uniqueness, assume that $\tilde{u}(t,\cdot) \in C_b^2(H)$, $\tilde{u}(\cdot,x) \in C_b^1([0,T[)$, and $\tilde{u}(t,x)$ satisfies (3.74) on $[0,T[$. For a fixed $0 < t < T$, we use Itô's formula and, for $0 < s < t$, consider the stochastic differential

$$
d\tilde{u}\big(t-s,X^x(s)\big)
$$
$$
= \left\{ -\frac{\partial \tilde{u}(t-s,X^x(s))}{\partial t} + \left\langle \frac{\partial \tilde{u}(t-s,X^x(s))}{\partial x}, AX^x(s) + F\big(X^x(s)\big) \right\rangle_H \right.
$$
$$
\left. + \frac{1}{2}\,\mathrm{tr}\left(\frac{\partial^2 \tilde{u}(t-s,X^x(s))}{\partial x^2} \big(B(X^x(s))Q^{1/2}\big)\big(B(X^x(s))Q^{1/2}\big)^* \right) \right\} ds
$$
$$
+ \left\langle \frac{\partial \tilde{u}(t-s,X^x(s))}{\partial x}, B\big(X^x(s)\big)dW_s \right\rangle_H.
$$

Since $\tilde{u}(t,x)$ satisfies (3.74), we get

$$
\tilde{u}\big(0,X(t)\big) = \tilde{u}\big(t,X(0)\big) + \int_0^t \left\langle \frac{\partial \tilde{u}(t-s,X^x(s))}{\partial x}, B\big(X^x(s)\big)dW_s \right\rangle_H.
$$

Therefore, applying expectation to both sides and using the initial condition $\tilde{u}(0,x) = \varphi(x)$ yields

$$
\tilde{u}(t,x) = E\varphi\big(X(t)\big). \qquad \square
$$

Using Theorem 3.10 and the Yosida approximation, Da Prato and Zabczyk stated a more general result when the operator A is unbounded.

Theorem 3.11 (Kolmogorov's Backward Equation II) *Assume that F and B do not depend on t, $F : H \to H$, and $B : H \to \mathscr{L}_2(K_Q, H)$. Let the Fréchet derivatives $DF(x)$, $DB(x)$, $D^2F(x)$, and $D^2B(x)$ be continuous and satisfy conditions (3.61) and (3.63) (with t omitted). If conditions (A1), (A3), and (A4) hold, then for $\varphi \in C_b^2(H)$, there exists a unique solution u of Kolmogorov's backward equation (3.74) satisfying (3.74) on $[0,T[$ and such that*

(1) *$u(t,x)$ is jointly continuous and bounded on $[0,T] \times H$,*
(2) *$u(t,\cdot) \in C_b^2(H)$, $0 \le t < T$,*
(3) *$u(\cdot,x) \in C_b^1([0,T[)$ for any $x \in \mathscr{D}(A)$.*

Moreover, u is given by formula (3.75), where $X^x(t)$ is the solution to (3.1) with deterministic initial condition $\xi_0 = x \in H$.

Proof To prove that $u(x,t) = E\varphi(X^x(t))$ is a solution, we approximate it with the sequence $u_n(t,x) = E\varphi(X_n^x(t))$, where $X_n(t)$ are strong solutions to (3.40) with the linear terms A_n being the Yosida approximations of A. By Corollary 3.4 with $p = 1$, we know that the mild solution $X(t)$ of (3.1) is approximated in $\tilde{\mathscr{H}}_{2p}$ by the sequence $X_n^x(t)$, i.e.,

$$\lim_{n \to \infty} \sup_{0 \le t \le T} E \| X_n^x(t) - X^x(t) \|_H^2 = 0.$$

This implies, choosing subsequence if necessary, that $X_n^x(t) \to X^x(t)$ a.s., so that, by the boundedness of φ,

$$u_n(t, x) = E\varphi\big(X_n^x(t)\big) \to E\varphi\big(X^x(t)\big) = u(t, x). \tag{3.78}$$

By Theorem 3.10 we have

$$\begin{cases} \dfrac{\partial u_n(t, x)}{\partial t} = \left\langle A_n x + F(x), \dfrac{\partial u_n(t, x)}{\partial x} \right\rangle_H \\[2mm] \qquad + \dfrac{1}{2} \operatorname{tr}\left(\dfrac{\partial^2 u_n(t, x)}{\partial x^2} \big(B(x)Q^{1/2}\big)\big(B(x)Q^{1/2}\big)^* \right), \\[2mm] 0 < t < T, \ x \in H, \\[2mm] u_n(0, x) = \varphi(x). \end{cases}$$

By (3.72) we have, choosing a subsequence if necessary, that for $y, z \in H$,

$$\lim_{n \to \infty} \| DX_n^x(t)y - DX^x(t)y \|_H = 0,$$

$$\lim_{n \to \infty} \| D^2 X_n^x(t)(y, z) - D^2 X^x(t)(y, z) \|_H = 0,$$

uniformly in $[0, T]$. Consequently, using the boundedness of $d\varphi(x)/dx$,

$$\left\langle \dfrac{\partial u_n(t, x)}{\partial x}, y \right\rangle_H = E\left\langle \dfrac{d\varphi(X_n^x(t, x)}{dx}, DX_n^x(t)y \right\rangle_H$$

$$\to E\left\langle \dfrac{d\varphi(X^x(t, x)}{dx}, DX^x(t)y \right\rangle_H = \left\langle \dfrac{\partial u(t, x)}{\partial x}, y \right\rangle_H,$$

with the last equality following by direct differentiation of $E\varphi(X^x(t))$ under the expectation. Hence,

$$\left\langle \dfrac{\partial u_n(t, x)}{\partial x}, A_n x + F(x) \right\rangle_H \to \left\langle \dfrac{\partial u(t, x)}{\partial x}, Ax + F(x) \right\rangle_H. \tag{3.79}$$

Next consider

$$\operatorname{tr}\left(\dfrac{\partial^2 u_n(t, x)}{\partial x^2} \big(B(x)Q^{1/2}\big)\big(B(x)Q^{1/2}\big)^* \right)$$

$$= \sum_{k=1}^{\infty} E\left\langle \dfrac{d^2\varphi(X_n^x(t))}{dx^2} DX_n^x(t)\big(B(x)Q^{1/2}\big)\big(B(x)Q^{1/2}\big)^* e_k, DX_n^x(t)e_k \right\rangle_H$$

$$+ \sum_{k=1}^{\infty} E\left\langle \frac{d\varphi(X_n^x(t))}{dx}, D^2 X_n^x(t)\left(B(x)Q^{1/2}\right)\left(B(x)Q^{1/2}\right)^* e_k, e_k \right\rangle_H. \quad (3.80)$$

Assume that $e_k(x)$ are the eigenvectors of $(B(x)Q^{1/2})(B(x)Q^{1/2})^*$. Let us discuss the convergence of the first term. We have

$$\sum_{k=1}^{\infty} E\left| \left\langle \frac{d^2\varphi(X_n^x(t))}{dx^2} DX_n^x(t)\left(B(x)Q^{1/2}\right)\left(B(x)Q^{1/2}\right)^* e_k, DX_n^x(t)e_k \right\rangle_H \right.$$

$$\left. - \left\langle \frac{d^2\varphi(X^x(t))}{dx^2} DX^x(t)\left(B(x)Q^{1/2}\right)\left(B(x)Q^{1/2}\right)^* e_k, DX^x(t)e_k \right\rangle_H \right|$$

$$\leq \sum_{k=1}^{\infty} E\left| \left\langle \left(\frac{d^2\varphi(X_n^x(t))}{dx^2} - \frac{d^2\varphi(X^x(t))}{dx^2}\right) DX_n^x(t)\left(B(x)Q^{1/2}\right)\right.\right.$$

$$\left.\left. \left(B(x)Q^{1/2}\right)^* e_k, DX_n^x(t)e_k \right\rangle_H \right|$$

$$+ \sum_{k=1}^{\infty} E\left| \left\langle \frac{d^2\varphi(X^x(t))}{dx^2} DX_n^x(t)\left(B(x)Q^{1/2}\right)\left(B(x)Q^{1/2}\right)^* e_k, DX_n^x(t)e_k \right\rangle_H \right.$$

$$\left. - \left\langle \frac{d^2\varphi(X^x(t))}{dx^2} DX^x(t)\left(B(x)Q^{1/2}\right)\left(B(x)Q^{1/2}\right)^* e_k, DX^x(t)e_k \right\rangle_H \right|$$

$$= S_1 + S_2.$$

Now note that

$$\sup_n \sup_{\|y\|_H \leq 1} \|DX_n^x y\|_{\tilde{\mathscr{H}}_{2p}} < \infty, \quad p \geq 1. \quad (3.81)$$

Indeed, $DX_n^x(t)y$ are mild solutions to (3.70) whose coefficients satisfy the assumptions of Theorem 3.5, since we have assumed that the Fréchet derivatives of F and B satisfy conditions (3.61). By Theorem 3.5 each solution can be obtained in $\tilde{\mathscr{H}}_{2p}$, $p \geq 1$ (the initial condition is deterministic), using the iterative procedure that employs the Banach contraction principle and starting from the unit ball centered at 0. In addition, the sequence of contraction constants can be bounded by a constant strictly less than one, since $\|e^{A_n t}\|_{\mathscr{L}(H)} \leq Me^{n\alpha t/(n-\alpha)}$ by estimate (1.29) in Chap. 2.

Consider the series S_1. For each k, the sequence

$$E\left| \left\langle \left(\frac{d^2\varphi(X_n^x(t))}{dx^2} - \frac{d^2\varphi(X^x(t))}{dx^2}\right) DX_n^x(t)\left(B(x)Q^{1/2}\right)\right.\right.$$

$$\left.\left. \left(B(x)Q^{1/2}\right)^* e_k, DX_n^x(t)e_k \right\rangle_H \right|$$

$$\le E \left\| \frac{d^2\varphi(X_n^x(t))}{dx^2} - \frac{d^2\varphi(X^x(t))}{dx^2} \right\|_H \lambda_k \left\| DX_n^x(t)e_k \right\|_H^2 \to 0,$$

where $\lambda_k(x)$ is the eigenvalue corresponding to the eigenvector $e_k(x)$. The sequence converges to zero since, by (3.81) with $p = 1$, we take a scalar product in $L^2(\Omega)$ of two sequences, one converging to zero and one bounded. As functions of k, the expectations are bounded by $C\lambda_k$. We conclude that $S_1 \to 0$ as $n \to \infty$.

Consider S_2:

$$\sum_{k=1}^{\infty} E \left| \left\langle \frac{d^2\varphi(X^x(t))}{dx^2} DX_n^x(t) \big(B(x)Q^{1/2}\big)\big(B(x)Q^{1/2}\big)^* e_k, DX_n^x(t)e_k \right\rangle_H \right.$$

$$\left. - \left\langle \frac{d^2\varphi(X^x(t))}{dx^2} DX^x(t) \big(B(x)Q^{1/2}\big)\big(B(x)Q^{1/2}\big)^* e_k, DX^x(t)e_k \right\rangle_H \right|$$

$$\le \sum_{k=1}^{\infty} E \left| \left\langle \frac{d^2\varphi(X^x(t))}{dx^2} DX_n^x(t) \big(B(x)Q^{1/2}\big)\big(B(x)Q^{1/2}\big)^* e_k \right. \right.$$

$$\left. \left. - DX^x(t)\big(B(x)Q^{1/2}\big)\big(B(x)Q^{1/2}\big)^* e_k, DX^x(t)e_k \right\rangle_H \right|$$

$$+ \sum_{k=1}^{\infty} E \left| \left\langle \frac{d^2\varphi(X^x(t))}{dx^2} DX_n^x(t)\big(B(x)Q^{1/2}\big) \right. \right.$$

$$\left. \left. \big(B(x)Q^{1/2}\big)^* e_k, DX^x(t)e_k - DX_n^x(t)e_k \right\rangle_H \right|,$$

so that the convergence of S_2 to zero follows by similar arguments as above.

Finally, the second term in (3.80) is bounded by

$$\sup_{x \in H} \left\| \frac{d\varphi(x)}{dx} \right\|_H \sum_{k=1}^{\infty} E \left\| \big(D^2 X_n^x(t) - D^2 X^x(t)\big)\langle \big(B(x)Q^{1/2}\big)\big(B(x)Q^{1/2}\big)^* e_k, e_k\rangle_H \right\|_H$$

$$+ E \left\| \frac{d\varphi(X_n^x(t))}{dx} - \frac{d\varphi(X^x(t))}{dx} \right\|_H$$

$$\times \sum_{k=1}^{\infty} \left\| D^2 X^x(t)\langle \big(B(x)Q^{1/2}\big)\big(B(x)Q^{1/2}\big)^* e_k, e_k\rangle_H \right\|_H,$$

which converges to zero by similar arguments as that used for the first term, but now we need to employ the fact that an analogue to the bound (3.81) holds for $D^2 X_n^x(e_k, e_k)$,

$$\sup_{n} \sup_{\|y\|_H, \|z\|_H \le 1} \left\| D^2 X_n^x(y, z) \right\|_{\tilde{\mathcal{H}}_{2p}} < \infty, \quad p \ge 1. \tag{3.82}$$

Thus we have established that

$$\mathrm{tr}\left(\frac{\partial^2 u_n(t,x)}{\partial x^2}\left(B(x)Q^{1/2}\right)\left(B(x)Q^{1/2}\right)^*\right)$$

$$\rightarrow \mathrm{tr}\left(\frac{\partial^2 u(t,x)}{\partial x^2}\left(B(x)Q^{1/2}\right)\left(B(x)Q^{1/2}\right)^*\right). \qquad (3.83)$$

Putting together (3.79) and (3.83), we have (at least for a subsequence)

$$\lim_{n\to\infty}\frac{\partial u_n(t,x)}{\partial t} = \left\langle\frac{\partial u(t,x)}{\partial x}, Ax + F(x)\right\rangle_H$$

$$+ \mathrm{tr}\left(\frac{\partial^2 u(t,x)}{\partial x^2}\left(B(x)Q^{1/2}\right)\left(B(x)Q^{1/2}\right)^*\right). \qquad (3.84)$$

Using bounds (3.81) and (3.82) and the assumptions on φ, we conclude that the left-hand side in (3.84) is bounded as a function of t for any fixed $x \in H$, so that, by integrating both sides of (3.84) on $[0,t]$, we get that

$$\lim_{n\to\infty} u_n(t,x) = \int_0^t \left(\left\langle\frac{\partial u(s,x)}{\partial x}, Ax + F(x)\right\rangle_H\right.$$

$$\left. + \mathrm{tr}\left(\frac{\partial^2 u(s,x)}{\partial x^2}\left(B(x)Q^{1/2}\right)\left(B(x)Q^{1/2}\right)^*\right)\right)dt.$$

This, together with (3.78), proves that

$$\frac{\partial u(t,x)}{\partial t} = \left\langle\frac{\partial u(t,x)}{\partial x}, Ax + F(x)\right\rangle_H + \mathrm{tr}\left(\frac{\partial^2 u(t,x)}{\partial x^2}\left(B(x)Q^{1/2}\right)\left(B(x)Q^{1/2}\right)^*\right).$$

To prove the uniqueness, assume that $\tilde{u}(t,x)$ satisfies (3.74) on $[0,T[$ and fulfills conditions (1)–(3) of the theorem.

Let $X_n^x(t)$ be a mild solution of equation

$$\begin{cases} dX(t) = AX(t) + R_n F(X(t))dt + R_n B(X(t))\,dW_t \\ X(0) = x \in \mathscr{D}(A) \end{cases}$$

that is,

$$X_n^x(t) = S(t)x + \int_0^t S(t-s)R_n F\left(X_n^x(s)\right)ds + \int_0^t S(t-s)R_n B\left(X_n^x(s)\right)dW_s,$$

where R_n is defined in (1.21).

Since $x \in \mathscr{D}(A)$ and $R_n : H \to \mathscr{D}(A)$, the solution $X_n^x(t) \in \mathscr{D}(A)$. Hence, by Theorems 3.5 and 3.2 $X_n^x(t)$ is a strong solution. Then, using Itô's formula, we can consider the differential

$$d\tilde{u}\left(t-s, X_n^x\right) = -\frac{\partial\tilde{u}(t-s, X_n^x(s))}{\partial t}\,ds$$

$$+ \left\langle \frac{\partial \tilde{u}(t - s, X_n^x(s))}{\partial x}, AX_n^x(s) + R_n F\left(X_n^x(s)\right) \right\rangle_H ds$$

$$+ \frac{1}{2} \operatorname{tr}\left(\frac{\partial^2 \tilde{u}(t - s, X_n^x(s))}{\partial x^2} \left(R_n B\left(X_n^x(s)\right) Q^{1/2}\right)\left(R_n B\left(X_n^x(s)\right) Q^{1/2}\right)^* \right) ds$$

$$+ \left\langle \frac{\partial \tilde{u}(t - s, X_n^x(s))}{\partial x}, R_n B\left(X_n^x(s)\right) dW_s \right\rangle_H$$

$$= \left\langle \frac{\partial \tilde{u}(t - s, X_n^x(s))}{\partial x}, R_n F\left(X_n^x(s)\right) - F\left(X_n^x(s)\right) \right\rangle_H ds$$

$$+ \frac{1}{2} \operatorname{tr}\left(\frac{\partial^2 \tilde{u}(t - s, X_n^x(s))}{\partial x^2} \left(R_n B\left(X_n^x(s)\right) Q^{1/2}\right)\left(R_n B\left(X_n^x(s)\right) Q^{1/2}\right)^* \right) ds$$

$$- \frac{1}{2} \operatorname{tr}\left(\frac{\partial^2 \tilde{u}(t - s, X_n^x(s))}{\partial x^2} \left(B\left(X_n^x(s)\right) Q^{1/2}\right)\left(B\left(X_n^x(s)\right) Q^{1/2}\right)^* \right) ds$$

$$+ \left\langle \frac{\partial \tilde{u}(t - s, X_n^x(s))}{\partial x}, R_n B\left(X_n^x(s)\right) dW_s \right\rangle_H .$$

The second equality holds since $\tilde{u}(t, x)$ is a solution of (3.74) and since the terms containing A cancel.

Now, we integrate over the interval $[0, t]$ and take expectation. Then we pass to the limit as $n \to \infty$ (note that the operators R_n are uniformly bounded (see (1.20)) and follow the argument provided in (6.20)). Finally, use the initial condition to obtain that

$$\tilde{u}(t, x) = E\varphi\left(X^x(t)\right).$$

This concludes the proof.   □

3.7 Lipschitz-Type Approximation of Continuous Coefficients

We now construct sequences of Lipschitz-type coefficients F_n and B_n, taking values in H, which approximate continuous coefficients F and B uniformly on compact subsets of $C([0, T], H)$. The values of F_n and B_n are not restricted to any finite-dimensional subspaces of H.

Lemma 3.9 *Let $F : [0, T] \times H \to H$, $B : [0, T] \times H \to \mathcal{L}_2(K_Q, H)$ satisfy conditions (A1)–(A3). There exist sequences $F_n : [0, T] \times H \to H$ and $B_n : [0, T] \times H \to \mathcal{L}_2(K_Q, H)$ of functions satisfying conditions (A1)–(A4), with a universal constant in the condition (A3), such that*

$$\sup_{0 \le t \le T} \left\| F(t, x) - F_n(t, x) \right\|_H + \sup_{0 \le t \le T} \left\| B(t, x) - B_n(t, x) \right\|_{\mathcal{L}_2(K_Q, H)} \to 0$$

uniformly on any compact set in $C([0, T], H)$.

Proof The sequences F_n and B_n can be constructed as follows. Let $\{e_n\}_{n=1}^{\infty}$ be an ONB in H. Denote

$$f_n(t) = \left(\langle x(t), e_1\rangle_H, \langle x(t), e_2\rangle_H, \ldots, \langle x(t), e_n\rangle_H\right) \in R^n,$$

$$\Gamma_n(t) = f_n(kT/n) \quad \text{at } t = kT/n \text{ and linear otherwise,}$$

$$\gamma_n(t, x_0, \ldots, x_n) = x_k \quad \text{at } t = \frac{kT}{n} \text{ and linear otherwise, with } x_k \in R^n,$$

$$k = 0, 1, \ldots, n.$$

Let $g : R^n \to R$ be nonnegative, vanishing for $|x| > 1$, possessing bounded derivative, and such that $\int_{R^n} g(x)\,dx = 1$. Let $\varepsilon_n \to 0$. We define

$$F_n(t, x) = \int \cdots \int F\left(t, (\gamma_n(\cdot, x_0, \ldots, x_n), \underline{e})\right)$$

$$\times \exp\left\{-\frac{\varepsilon_n}{n}\sum_{k=0}^{n} x_k^2\right\} \prod_{k=0}^{n}\left(g\left(\frac{f_n\left(\frac{kT}{n} \wedge t\right) - x_k}{\varepsilon_n}\right)\frac{dx_k}{\varepsilon_n}\right). \quad (3.85)$$

Above, $(\gamma_n(\cdot, x_0, \ldots, x_n), \underline{e}) = \gamma_n^1 e_1 + \cdots + \gamma_n^n e_n$, where $\gamma_n^1, \ldots, \gamma_n^n$ are the coordinates of the vector γ_n in R^n, and $x_k^2 = \sum_{i=1}^{n}(x_k^i)^2$, $dx_k = dx_k^1 \ldots dx_k^n$.

The coefficients $B_n(t, x)$ are defined analogously using the ONB in K_Q. We note that conditions (A1)–(A4) are satisfied. To see that, note that the functions F_n and B_n depend on a finite collection of variables $f_n(kT/n)$, and hence the arguments of Gikhman and Skorokhod in [25] can be applied. We only need to verify the uniform convergence on compact sets of $C([0, T], H)$. We have

$$\sup_{0 \le t \le T} \left\| F_n(t, x) - F(t, x) \right\|_H$$

$$\le \int \cdots \int \sup_{0 \le t \le T} \left\| F\left(t, (\gamma_n(\cdot, f_n(0) + \overline{x}_0, \ldots, f_n(T \wedge t) + \overline{x}_n), \underline{e})\right)\right.$$

$$\left. - F\left(t, (\gamma_n(\cdot, f_n(0), \ldots, f_n(T \wedge t)), \underline{e})\right)\right\|_H \prod_{k=0}^{n} g\left(\frac{\overline{x}_k}{\varepsilon_n}\right)\frac{d\overline{x}_k}{\varepsilon_n}$$

$$+ \sup_{0 \le t \le T} \left\| F(t, x) - F\left(t, (\gamma_n(\cdot, f_n(0), \ldots, f_n(T \wedge t)), \underline{e})\right)\right\|_H$$

$$+ \int \cdots \int \sup_{0 \le t \le T} \left\| F\left(t, (\gamma_n(\cdot, x_0, \ldots, x_n), \underline{e})\right)\right\|_H \left(1 - \exp\left\{-\frac{\varepsilon_n}{n}\sum_{k=0}^{n} x_k^2\right\}\right)$$

$$\times \prod_{k=0}^{n}\left(g\left(\frac{f_n\left(\frac{kT}{n} \wedge t\right) - x_k}{\varepsilon_n}\right)\frac{dx_k}{\varepsilon_n}\right).$$

We will now verify convergence for each of the three components of the sum above.

Consider the first summand. If $K \subset C([0,T], H)$ is a compact set, then the collection of functions $\{\gamma_n(\cdot, f_n(0), \ldots, f_n(T)), \underline{e}), n \geq 1\} \subset C([0,T], H)$, $f_n(t) = (\langle x(t), e_1 \rangle_H, \ldots, \langle x(t), e_n \rangle_H)$ with $x \in K$, is a subset of some compact set K_1. This follows from a characterization of compacts in $C([0,T], H)$ (see Lemma 3.14 in the Appendix) and from Mazur's theorem, which states that a closed convex hull of a compact set in a Banach space is a compact set. Moreover,

$$\sup_{x_0, \ldots, x_n} \sup_{0 \leq t \leq T} \left\| \gamma_n(t, x_0 + z_0, \ldots, x_n + z_n) - \gamma_n(t, x_0, \ldots, x_n) \right\|_H < \varepsilon_n$$

if $|z_k| \leq \varepsilon_n$, $k = 0, \ldots, n$, $z_k \in R^n$.

The set $K_1 + B(0, \varepsilon_n)$ is not compact ($B(0, \varepsilon_n)$ denotes a ball of radius ε_n centered at 0), but $\sup_{0 \leq t \leq T} \|F(t,u) - F(t,v)\|_H$ can still be made arbitrarily small if v is sufficiently close to $u \in K_1$. Indeed, given $\varepsilon > 0$, for every $u \in K_1$ and $t \in [0,T]$, there exists δ_u^t such that if $\sup_{0 \leq t \leq T} \|v(t) - u(t)\|_H < \delta_u^t$, then $\|F(t,v) - F(t,u)\|_H < \varepsilon/2$. Because t is in a compact set and $F(t,u)$ is continuous in both variables, $\delta_u = \inf_t \delta_u^t > 0$. Therefore, for $u \in K_1$, $\sup_{0 \leq t \leq T} \|F(t,v) - F(t,u)\|_H < \varepsilon/2$ whenever $\sup_{0 \leq t \leq T} \|v(t) - u(t)\|_H < \delta_u$.

We take a finite covering $B(u_k, \delta_{u_k}/2)$ of K_1 and let $\delta = \min\{\delta_{u_k}/2\}$. If $u \in K_1$ and $\sup_{0 \leq t \leq T} \|v(t) - u(t)\|_H < \delta$, then for some k, $\sup_{0 \leq t \leq T} \|u(t) - u_k(t)\|_H < \delta_{u_k}/2$, and $\sup_{0 \leq t \leq T} \|v(t) - u_k(t)\|_H < \delta + \delta_{u_k}/2 \leq \delta_{u_k}$. Therefore, $\sup_{0 \leq t \leq T} \|F(t,v) - F(t,u)\|_H \leq \varepsilon$.

Thus taking n sufficiently large and noticing that $g(\overline{x}_k/\varepsilon_n)$ vanishes if $|\overline{x}_k| \geq \varepsilon_n$, we get

$$\sup_{0 \leq t \leq T} \left\| F\left(t, \left(\gamma_n(\cdot, f_n(0) + \overline{x}_0, \ldots, f_n(T \wedge t) + \overline{x}_n), \overline{e}\right)\right) \right.$$

$$\left. - F\left(t, \left(\gamma_n(\cdot, f_n(0), \ldots, f_n(T \wedge t)), \underline{e}\right)\right) \right\|_H < \varepsilon$$

for any ε, independently of f_n associated with $x \in K$. This gives the uniform convergence to zero on K of the first summand.

Now we consider the second summand. Let P_n be the orthogonal projection onto the linear subspace spanned by $\{e_1, \ldots, e_n\}$, and let $P_n^\perp$ denote the orthogonal projection onto the orthogonal complement of this space. We note that $\sup_{0 \leq t \leq T} \|P_n^\perp x(t)\|_H \to 0$ as $n \to \infty$; otherwise, there would be a sequence $t_n \to t_0$ with $\|P_n^\perp x(t_n)\|_H > c > 0$, and $\|P_n^\perp x(t_n)\|_H \leq \|x(t_n) - x(t_0)\|_H + \|P_n^\perp x(t_0)\|_H \to 0$.

Let $N = N(x)$ be chosen such that if $m \geq N$, then $\sup_{0 \leq t \leq T} \|P_m^\perp x(t)\|_H < \varepsilon/3$. Thus also $\sup_{0 \leq t \leq T} \|P_N^\perp (\Gamma_m(x)(t), \underline{e})\|_H < \varepsilon/3$.

There exists $M = M(x) \geq N(x)$ such that for $m \geq M$,

$$\sup_{0 \leq t \leq T} \left\| x(t) - \left(\Gamma_m(x)(t), \underline{e}\right) \right\|_H \leq \sup_{0 \leq t \leq T} \left\| P_N^\perp x(t) \right\|_H$$

$$+ \sup_{0 \leq t \leq T} \left\| P_N x(t) - P_N \left(\Gamma_m(x)(t), \underline{e}\right) \right\|_H + \sup_{0 \leq t \leq T} \left\| P_N^\perp \left(\Gamma_m(x)(t), \underline{e}\right) \right\|_H < \varepsilon,$$

because the middle term is a finite-dimensional approximation.

Actually, $\sup_{x \in K} \sup_{0 \le t \le T} \|x(t) - (\Gamma_m(x)(t), \underline{e})\|_H \to 0$ if K is compact. If $\{x_1, \ldots, x_n\}$, is an $\varepsilon/3$ net in K, then

$$\sup_{0 \le t \le T} \|x(t) - (\Gamma_m(x)(t), \underline{e})\|_H \le \sup_{0 \le t \le T} \|x(t) - x_k(t)\|_H$$

$$+ \sup_{0 \le t \le T} \|x_k(t) - (\Gamma_m x_k(t), e)\|_H + \sup_{0 \le t \le T} \|(\Gamma_m x_k(t), e) - (\Gamma_m x(t), e)\|_H < \varepsilon$$

if $m \ge \max\{M(x_k)\}$.

The continuity of F and the fact that $\{(\Gamma_m(x), \underline{e}), x \in K\} \cup K$ is a subset of a compact set, guarantees the uniform convergence to zero of the second summand. The third summand converges uniformly to zero on compact sets since it is bounded by

$$\ell\left(1 + \varepsilon_n + \sup_{0 \le t \le T} \|x(t)\|_H\right) \varepsilon_n \sup_{0 \le t \le T} \left(\|x(t)\|_H + \varepsilon_n\right)^2. \qquad \square$$

Exercise 3.14 Prove that F_n and B_n defined in (3.85) satisfy conditions (A1)–(A4).

We will now consider methods for proving the existence of solutions to SDEs and SSDEs that are based on some compactness assumptions. In the case of SDEs, we will require that the Hilbert space H be embedded compactly into some larger Hilbert space. In the case of SSDEs, the compactness of the semigroup $S(t)$ will be imposed to guarantee the tightness of the laws of the approximate solutions.

These cases will be studied separately.

3.8 Existence of Weak Solutions Under Continuity Assumption

To obtain weak convergence results, we need an estimate on moments for increments of solutions to SDEs.

Lemma 3.10 *Let* $\xi(t) = x + \int_0^t F(s, \xi) \, ds + \int_0^t B(s, \xi) \, dW_s$ *with* $x \in H$ *and the coefficients* F *and* B *satisfying conditions* (A1) *and* (A3). *Then*

$$E\|\xi(t+h) - \xi(t)\|_H^4 \le C(T, \ell)h^2$$

Proof Using Ito's formula, we obtain

$$\|\xi(t+h) - \xi(t)\|_H^4 = 4 \int_t^{t+h} \|\xi(u) - \xi(t)\|_H^2 \langle \xi(u) - \xi(t), F(u, \theta_u \xi)\rangle_H \, du$$

$$+ 2 \int_t^{t+h} \|\xi(u) - \xi(t)\|_H^2 \operatorname{tr}\left((B(u, \theta_u \xi)Q^{1/2})(B(u, \theta_u \xi)Q^{1/2})^*\right) du$$

$$+ 2 \int_t^{t+h} \left((B(u, \theta_u \xi)Q^{1/2})(B(u, \theta_u \xi)Q^{1/2})^*\right)[\xi(u) - \xi(t)]^{\otimes 2} \, du$$

By Lemma 3.6, this yields the following estimate for the fourth moment:

$$E\|\xi(t+h) - \xi(t)\|_H^4 \leq C_1 \int_t^{t+h} E\|\xi(u) - \xi(t)\|_H^3 \left(1 + \sup_{v \leq u}\|\xi(v)\|_H\right) du$$

$$+ C_2 \int_t^{t+h} E\|\xi(u) - \xi(t)\|_H^2 \left(1 + \sup_{v \leq u}\|\xi(v)\|_H^2\right) du$$

$$\leq C_3 \left(\left[\int_t^{t+h} E\|\xi(u) - \xi(t)\|_H^4 du\right]^{3/4} h^{1/4}\right.$$

$$\left. + \left[\int_t^{t+h} E\|\xi(u) - \xi(t)\|_H^4 du\right]^{1/2} h^{1/2}\right) \leq Ch.$$

Substituting repeatedly, starting with $C(u - t)$ for $E\|\xi_n(u) - \xi_n(t)\|_{\mathbb{R}^n}^4$, into the above inequality, we arrive at the desired result. $\qquad\square$

The next lemma is proved by Gikhman and Skorokhod in [25], Vol. I, Chap. III, Sect. 4.

Lemma 3.11 *The condition* $\sup_n E(\|\xi_n(t+h) - \xi_n(t)\|_H^4) \leq Ch^2$ *implies that for any* $\varepsilon > 0$,

$$\lim_{\delta \to 0} \sup_n P\left(\sup_{|t-s|<\delta} \|\xi_n(t) - \xi_n(s)\|_H > \varepsilon\right) = 0.$$

Corollary 3.6 *Let* F_n *and* B_n *satisfy conditions* (A1) *and* (A3) *with a common constant in the growth condition* (A3) *(in particular* F_n *and* B_n *can be the approximating sequences from Lemma 3.9). Let* X_n *be a sequence of solutions to the following SDEs:*

$$\begin{cases} dX_n(t) = F_n(t, X_n)\,dt + B_n(t, X_n)\,dW_t, \\ X_n(0) = x \in H. \end{cases}$$

Then

(1) *the sequence* X_n *is stochastically bounded, i.e., for every* $\varepsilon > 0$, *there exists* M_ε *satisfying*

$$\sup_n P\left(\sup_{0 \leq t \leq T} \|X_n(t)\|_H > M_\varepsilon\right) \leq \varepsilon, \qquad (3.86)$$

(2) *for any* $\varepsilon > 0$,

$$\lim_{\delta \to 0} \sup_n P\left(\sup_{|t-s|<\delta} \|X_n(t) - X_n(s)\|_H > \varepsilon\right) = 0. \qquad (3.87)$$

It is known that even a weak solution $X_t(\cdot) \in C([0, T], H)$ to the SDE (3.8) may not exist. Therefore our next step is to find a weak solution on a larger Hilbert space.

Let H_{-1} be a real separable Hilbert space such that the embedding $J : H \hookrightarrow H_{-1}$ is a compact operator with representation

$$(J) \qquad Jx = \sum_{n=1}^{\infty} \lambda_n \langle x, e_n \rangle_H h_n, \quad \lambda_n > 0, \; n = 1, 2, \ldots.$$

In general, J has always the above representation; we are only assuming that $\lambda_n \neq 0$. Here, $\{e_n\}_{n=1}^{\infty} \subset H$ and $\{h_n\}_{n=1}^{\infty} \subset H_{-1}$ are orthonormal bases. We will identify H with $J(H)$ and $P \circ X^{-1}$ with $P \circ X^{-1} \circ J^{-1}$. Thus, if $x \in H$, we will also write $x \in H_{-1}$. In particular, $e_n = \lambda_n h_n \in H_{-1}$ and $\langle x, h_n \rangle_{H_{-1}} = \lambda_n \langle x, e_n \rangle_H$. Note that $\|x\|_{H_{-1}} \leq \|J\| \|x\|_H$ and $C([0, T], H) \subset C([0, T], H_{-1})$.

Corollary 3.7 *Let F_n, B_n, X_n be as in Corollary 3.6, and $J : H \hookrightarrow H_{-1}$ be a compact embedding into a real separable Hilbert space H_{-1}. Then the sequence of measures $\{P \circ X_n^{-1}\}_{n=1}^{\infty}$ on $C([0, T], H_{-1})$ is relatively weakly compact.*

Proof For any $\varepsilon > 0$, let M_ε be the constant in condition (3.86) of Corollary 3.6. The ball $B(0, M_\varepsilon) \subset H$ of radius M_ε, centered at 0, is relatively compact in H_{-1}. Denote by $\overline{B}$ its closure in H_{-1}; then

$$P\big(J\big(X_n(t)\big) \notin \overline{B}\big) = P\big(\|X_n(t)\|_H > M_\varepsilon\big) < \varepsilon.$$

Condition (3.87) of Corollary 3.6 is also satisfied in H_{-1}. The relative compactness now follows from the tightness criterion given in Theorem 3.17 in the Appendix, and from Prokhorov's theorem.   □

In order to construct a weak solution to the SDE (3.8) on $C([0, T], H_{-1})$, we impose some regularity assumptions on the coefficients F and B with respect to the Hilbert space H_{-1}.

Assume that $F : [0, T] \times C([0, T], H_{-1}) \to H_{-1}$ and $B : [0, T] \times C([0, T], H_{-1}) \to \mathscr{L}_2(K_Q, H_{-1})$ satisfy the following conditions:

(B1) F and B are jointly measurable, and for every $0 \leq t \leq T$, they are measurable with respect to the σ-field $\tilde{\mathscr{C}}_t$ on $C([0, T], H_{-1})$ generated by cylinders with bases over $[0, t]$.

(B2) F and B are jointly continuous.

(B3) There exists a constant ℓ_{-1} such that $\forall x \in C([0, T], H_{-1})$,

$$\big\|F(t, x)\big\|_{H_{-1}} + \big\|B(t, x)\big\|_{\mathscr{L}_2(K_Q, H_{-1})} \leq \ell_{-1}\Big(1 + \sup_{0 \leq t \leq T} \|x(t)\|_{H_{-1}}\Big),$$

for $\omega \in \Omega$ and $0 \leq t \leq T$.

Equation (3.8) is now considered in H_{-1}, and in the circumstances described above, we can prove the existence of a weak solution to the SDE (3.8) on $C([0, T], H_{-1})$. However we do not need all conditions (A1)–(A3) and (B1)–(B3) to hold simultaneously. We state the existence result as follows.

Theorem 3.12 *Let H_{-1} be a real separable Hilbert space. Let the coefficients F and B of the SDE (3.8) satisfy conditions (B1)–(B3). Assume that there exists a Hilbert space H such that the embedding $J : H \hookrightarrow H_{-1}$ is a compact operator with representation (J) and that F and B restricted to H satisfy*

$$F : [0, T] \times C([0, T], H) \to H,$$

$$B : [0, T] \times C([0, T], H) \to \mathscr{L}_2(K_Q, H),$$

and the linear growth condition (A3). Then the SDE (3.8) has a weak solution $X(\cdot) \in C([0, T], H_{-1})$.

Proof Since the coefficients F and B satisfy assumptions (B1)–(B3), we can construct approximating sequences $F_n : [0, T] \times C([0, T], H_{-1}) \to H_{-1}$ and $B_n : [0, T] \times C([0, T], H_{-1}) \to \mathscr{L}_2(K_Q, H_{-1})$ as in Lemma 3.9. The sequences $F_n \to F$ and $B_n \to B$ uniformly on compact subsets of $C([0, T], H_{-1})$.

Now we consider restrictions of the functions F_n, B_n to $[0, T] \times C([0, T], H)$, and we claim that they satisfy conditions (A1), (A3), and (A4). Let us consider the sequence F_n only; similar arguments work for the sequence B_n. We adopt the notation developed in Lemma 3.9.

If $x \in C([0, T], H)$, then

$$F_n(t, x) = \int \cdots \int F\left(t, \left(\gamma_n(\cdot, x_1, \ldots, x_n), \underline{h}\right)\right)$$

$$\times \exp\left\{-\frac{\varepsilon_n}{n} \sum_{k=0}^{n} x_k^2\right\} \prod_{k=0}^{n} \left(g\left(\frac{\tilde{f}_n\left(\frac{kT}{n} \wedge t\right) - x_k}{\varepsilon_n}\right) \frac{dx_k}{\varepsilon_n}\right) \in H,$$

where

$$\tilde{f}_n(t) = \left(\langle x(t), h_1\rangle_{H_{-1}}, \ldots, \langle x(t), h_n\rangle_{H_{-1}}\right)$$

$$= \left(\lambda_1 \langle x(t), h_1\rangle_H, \ldots, \lambda_n \langle x(t), h_n\rangle_H\right) := \underline{\lambda} f_n(t),$$

$$\left(\gamma_n(\cdot, x_0, \ldots, x_n), \underline{h}\right) = \left(\gamma_n(\cdot, x_0, \ldots, x_n), \left(\frac{e_1}{\lambda_1}, \ldots, \frac{e_n}{\lambda_n}\right)\right)$$

$$= \left(\gamma_n\left(\cdot, \frac{x_0}{\underline{\lambda}}, \ldots, \frac{x_n}{\underline{\lambda}}\right), \underline{e}\right) \in C([0, T], R^n),$$

and $\frac{x_k}{\underline{\lambda}} = \left(\frac{x_k^1}{\lambda_1}, \ldots, \frac{x_k^n}{\lambda_n}\right) \in R^n$. Let $\varepsilon_n/\lambda_k < 1$, $k = 1, \ldots, n$. Then

$$F_n(t, x) = \int \cdots \int F\left(t, \left(\gamma_n\left(\cdot, \frac{x_0}{\underline{\lambda}}, \ldots, \frac{x_n}{\underline{\lambda}}\right), \underline{e}\right)\right)$$

$$\times \exp\left\{-\frac{\varepsilon_n}{n} \sum_{k=0}^{n} x_k^2\right\} \prod_{k=0}^{n} \left(g\left(\frac{\underline{\lambda}\left(f_n\left(\frac{kT}{n} \wedge t\right) - \frac{x_k}{\underline{\lambda}}\right)}{\varepsilon_n}\right) \frac{dx_k}{\varepsilon_n}\right)$$

$$= \int \cdots \int F\left(t, \left(\gamma_n(\cdot, y_0, \ldots, y_n), \underline{e}\right)\right)$$

$$\times \exp\left\{-\frac{\varepsilon_n}{n}\sum_{k=0}^{n}(y_k\underline{\lambda})^2\right\} \prod_{k=0}^{n}\left(g\left(\frac{\underline{\lambda}\left(f_n\left(\frac{kT}{n}\wedge t\right) - y_k\right)}{\varepsilon_n}\right)\frac{dy_k}{\varepsilon_n/\prod_{i=1}^{n}\lambda_i}\right).$$

First, we observe that F_n are measurable with respect to the product σ-field on $[0, T] \times C([0, T], H)$ because F_n satisfy condition (B1), depend only on finitely many variables, and $J^{-1}y = \sum_{k=1}^{\infty} 1/\lambda_k\langle y, h_k\rangle_{H_{-1}}e_k$ on $J(H)$ is measurable from $J(H)$ to H as a limit of measurable functions. The same argument justifies that F_n are adapted to the family $\{\mathscr{C}_t\}_{t \leq T}$.

The linear growth condition (A3) is satisfied with a universal constant. Indeed,

$$\|F_n(t, x)\|_H \leq \int \cdots \int \|F\left(t, \left(\gamma_n(\cdot, y_0, \ldots, y_n), \underline{e}\right)\right)\|_H$$

$$\times \prod_{k=0}^{n}\left(g\left(\frac{\underline{\lambda}\left(f_n\left(\frac{kT}{n}\wedge t\right) - y_k\right)}{\varepsilon_n}\right)\frac{dy_k}{\varepsilon_n/\prod_{i=1}^{n}\lambda_i}\right)$$

$$\leq \ell\left(1 + \sup_{0\leq t\leq T}\|x(t)\|_H + \max_{1\leq k\leq n}(\varepsilon_n/\lambda_k)\right)\int \cdots \int \prod_{k=0}^{n}g(z_k)\,dz_k$$

$$\leq \ell'\left(1 + \sup_{0\leq t\leq T}\|x(t)\|_H\right),$$

because of the choice of ε_n.

Further, the function F_n depends only on finitely many variables $f\left(\frac{kT}{n}\wedge t\right) = \left((x\left(\frac{kT}{n}\wedge t\right), e_1)_H, \ldots, (x\left(\frac{kT}{n}\wedge t\right), e_n)_H\right) \in R^n, k = 1, \ldots, n.$

By differentiating under the integral sign, we get that, with $x \in C([0, T], H)$,

$$\frac{\partial F_n(t, x)}{\partial f_l\left(\frac{kT}{n}\wedge t\right)} \leq \int \cdots \int \ell\left(1 + \max_{0\leq k\leq n}|y_k|\right)$$

$$\times \exp\left\{-\frac{\varepsilon_n}{n}\sum_{k=0}^{n}(y_k\underline{\lambda})^2\right\}\sup_z|g'(z)|\left(\frac{\lambda_l}{\varepsilon_n}\right)\frac{dy_l}{\varepsilon_n/\prod_{i=1}^{n}\lambda_i}$$

$$\times \prod_{k=0,k\neq l}^{n}\left(g\left(\frac{\underline{\lambda}\left(f_n\left(\frac{kT}{n}\wedge t\right) - y_k\right)}{\varepsilon_n}\right)\frac{dy_k}{\varepsilon_n/\prod_{i=1}^{n}\lambda_i}\right)$$

$$\leq C\left(\frac{\prod_{i=1}^{n}\lambda_i}{\varepsilon_n}\right)^n\left(\frac{\lambda_l}{\varepsilon_n}\right)\int \cdots \int \ell\left(1 + \max_{0\leq k\leq n}|y_k|\right)\exp\left\{-\frac{\varepsilon_n}{n}\sum_{k=0}^{n}(y_k\underline{\lambda})^2\right\}\prod_{k=0}^{n}dy_k$$

$$< \infty.$$

Thus, $F_n(t, x)$ has bounded partial derivatives, and hence it is a Lipschitz function with respect to the variable x. Let X_n be a sequence of strong solutions to equations

$$X_n(t) = x + \int_0^t F_n(s, X_n) \, ds + \int_0^t B_n(s, X_n) \, dW_s$$

considered on $C([0, T], H)$. By Corollary 3.12, the sequence of measures $\mu_n = P \circ X_n^{-1} = P \circ X_n^{-1} \circ J^{-1}$ on $C([0, T], H_{-1})$ (with H identified with $J(H)$) is relatively weakly compact, and therefore, by choosing a subsequence if necessary, we can assume that μ_n converges weakly to a measure μ on $C([0, T], H_{-1})$.

We now follow the ideas of Gikhman and Skorokhod in [25]. Let g_t be a bounded continuous function on $C([0, T], H_{-1})$ measurable with respect to the cylindrical σ-field generated by cylinders with bases over $[0, t]$. Then for any $u \in H_{-1}$,

$$\int \left[\langle x(t+h) - x(t), u \rangle_{H_{-1}} - \int_t^{t+h} \langle F_n(s, x), u \rangle_{H_{-1}} \, ds \right] g_t(x) \mu_n(dx)$$

$$= \left\langle E\left(\left(X_n(t+h) - X_n(t) - \int_t^{t+h} F_n(s, X_n) \, ds \right) g_t(X_n) \right), u \right\rangle_{H_{-1}} = 0. \quad (3.88)$$

Let $\nu_n(dx) = (1 + \sup_{0 \le s \le T} \|x(s)\|_{H_{-1}}) \mu_n(dx)$. As in the finite-dimensional case, the measures ν_n are uniformly bounded and are weakly convergent.

Indeed, using the tightness of the sequence μ_n, for any $\varepsilon > 0$, we find a compact set $K_\varepsilon \subset C([0, T], H_{-1})$ with $\limsup_n \mu_n(K_\varepsilon^c) < \varepsilon$, and therefore if $B_N \subset C([0, T], H_{-1})$ is a ball of radius N centered at 0 and if $K_\varepsilon \subset B_N$, we obtain

$$\int_{B_N^c} \sup_{0 \le s \le T} \|x(s)\|_{H_{-1}} \, d\mu_n \le \mu_n^{1/2}(B_N^c) \left(\int_{B_N^c} \sup_{0 \le s \le T} \|x(s)\|_{H_{-1}}^2 \, d\mu_n \right)^{1/2}$$

$$\le C \mu_n^{1/2}(K_\varepsilon^c) = C \sqrt{\varepsilon},$$

and for a real-valued, bounded, continuous function f,

$$\left| \int f(x) \left(1 + \sup_{0 \le s \le T} \|x(s)\|_{H_{-1}} \right) d\mu_n - \int f(x) \left(1 + \sup_{0 \le s \le T} \|x(s)\|_{H_{-1}} \right) d\mu \right|$$

$$\le \left| \int f(x) \left(1 + \sup_{0 \le s \le T} \|x(s)\|_{H_{-1}} \wedge N \right) d\mu_n \right.$$

$$\left. - \int f(x) \left(1 + \sup_{0 \le s \le T} \|x(s)\|_{H_{-1}} \wedge N \right) d\mu \right|$$

$$+ \|f\|_\infty (C \sqrt{\varepsilon} + \varepsilon) = O(\sqrt{\varepsilon}).$$

Thus, there exists a compact set $\tilde{K}_\varepsilon \subset C([0, T], H_{-1})$ such that $\limsup_{n \to \infty} \nu_n(\tilde{K}_\varepsilon^c) \le \varepsilon$. Thus, because of the uniform convergence $\sup_{0 \le s \le T} \|F(s, x) - F_n(s, x)\|_{H_{-1}}$

$\to 0$ on $\tilde{K}_\varepsilon$ and the growth condition,

$$\limsup_{n\to\infty} \iint_t^{t+h} \left|\langle F(s,x) - F_n(s,x), u\rangle_{H_{-1}} g_t(x)\right| ds \, \mu_n(dx)$$

$$\leq C \limsup_{n\to\infty} \iint_t^{t+h} \frac{\|F(s,x) - F_n(s,x)\|_{H_{-1}} \|u\|_{H_{-1}}}{(1 + \sup_{0\leq s\leq T} \|x(s)\|_{H_{-1}})} |g_t(x)| \, ds \, v_n(dx) = O(\varepsilon).$$

It follows that

$$\lim_{n\to\infty} \iint_t^{t+h} \left|\langle F(s,x) - F_n(s,x), u\rangle_{H_{-1}} g_t(x)\right| ds \, \mu_n(dx) = 0,$$

and the weak convergence of the measures μ_n, together with the uniform integrability, implies that, as $n \to \infty$,

$$\int \langle x(t+h) - x(t), u\rangle_{H_{-1}} g_t(x) \, d\mu_n \to \int \langle x(t+h) - x(t), u\rangle_{H_{-1}} g_t(x) \, d\mu \quad (3.89)$$

and

$$\iint_t^{t+h} \langle F(s,x), u\rangle_{H_{-1}} g_t(x) \, ds \, d\mu_n \to \iint_t^{t+h} \langle F(s,x), u\rangle_{H_{-1}} g_t(x) \, ds \, d\mu. \quad (3.90)$$

Now, taking the limit in (3.88), we obtain

$$\int \left[\langle x(t+h) - x(t), u\rangle_{H_{-1}} - \int_t^{t+h} \langle F(s,x), u\rangle_{H_{-1}} ds \right] g_t(x)\mu(dx) = 0,$$

proving that the process $y(t) = x(t) - \int_0^t F(s,x) \, ds$ is a martingale on $C([0,T], H_{-1})$ endowed with the canonical filtration. Using the equality

$$\int \left(\langle x(t+h) - x(t) - \int_t^{t+h} F_n(s,x) \, ds, u\rangle_{H_{-1}} \right)^2 g_t(x)\mu_n(dx)$$

$$= \int_t^{t+h} \left((B_n(s,x)Q^{1/2})(B_n(s,x)Q^{1/2})^* \right)[u^{\otimes 2}] g_t(x)\mu_n(dx)$$

one can prove in a similar way that the process

$$\langle y(t), u\rangle_{H_{-1}}^2 - \int_0^t \left((B(s,x)Q^{1/2})(B(s,x)Q^{1/2})^* \right)[u^{\otimes 2}] \, ds$$

is a martingale on $C([0,T], H_{-1})$ and obtain the increasing process

$$\langle\langle y(\cdot), u\rangle_{H_{-1}}\rangle_t = \int_0^t \left((B(s,x)Q^{1/2})(B(s,x)Q^{1/2})^* \right)[u^{\otimes 2}] \, ds.$$

In view of Theorem 2.7, the measure μ is a law of a $C([0, T], H_{-1})$-valued process $X(t)$, which is a weak solution to the SDE (3.8) (we can let $\Phi(s) = B(s, X)$ in Theorem 2.7). □

The following result was communicated to us by A.V. Skorokhod [69].

Theorem 3.13 *Consider the equation*

$$dX_t = a\big(X(t)\big)dt + B\big(X(t)\big)dW_t, \tag{3.91}$$

where $a : H \to H$, $B : H \to \mathcal{L}(K, H)$, *and* W_t *is a* K-*valued Wiener process with covariance* Q. *Assume the following conditions:*

(1) a *and* B *are jointly continuous and locally bounded,*
(2) $2\langle a(x), x \rangle_H + \mathrm{tr}(B(x)QB^*(x)) \le \ell(1 + \|x\|_H^2)$,
(3) *there exists a positive symmetric compact linear operator* S *on* H *such that*

$$2\langle S^{-1}a(Sx), x \rangle_H + \mathrm{tr}\big(S^{-1}B(Sx)QB^*(Sx)S^{-1}\big) \le \ell_1\big(1 + \|x\|_H^2\big).$$

Then there exists a weak solution to (3.91) with $X(0) = Sx_0$, $x_0 \in H$.

Exercise 3.15 Let us substitute conditions (1)–(3) with

(1') a *and* B *are jointly continuous,*
(2') $\|a(x)\|_H + \|B(x)\|_{\mathcal{L}}(K, H) \le \ell(1 + \|x\|_H)$,
(3') *there exists a positive symmetric compact linear operator* S *on* H *such that*

$$\big\|S^{-1}a(Sx)\big\|_H + \big\|S^{-1}B\big(S(x)\big)\big\|_{\mathcal{L}}(K, H) \le \ell_1(1 + \|x\|_H).$$

Prove the assertion of Theorem 3.13.

   Hint: define a norm $\|x\|_0 = \|S^{-1}x\|_H$ *on* $H_0 = S(H)$. *Prove that the assumptions of Theorem 3.12 are satisfied with* $J : H_0 \to H$.

3.9 Compact Semigroups and Existence of Martingale Solutions

In this section we present an existence theorem for martingale solutions in case where the operator A generates a compact semigroup. This extends Theorem 8.1 in [11], since we include coefficients F and B of (3.1) that may depend on the entire past of a solution.

   The following technical lemma will allow us to prove the tightness of a family of probability measures. We will use it in the proof of Theorem 3.14 and again in Sect. 7.4.

Lemma 3.12 *Let* $p > 1$ *and* $1/p < \alpha \le 1$. *Consider the operator* G_α *defined in (3.24),*

$$G_\alpha f(t) = \int_0^t (t - s)^{\alpha - 1} S(t - s) f(s) \, ds, \quad f \in L^p\big([0, T], H\big).$$

Assume that $\{S(t),\ t \geq 0\}$ *is a compact* C_0-*semigroup on* H. *Then* G_α *is compact from* $L^p([0, T], H)$ *into* $C([0, T], H)$.

Proof It is enough to show that

$$G_\alpha(\{f \in L^p([0, T], H) : \|f\|_{L^p} \leq 1\})$$

is relatively compact in $C([0, T], H)$. We will show that conditions (1) and (2) of Lemma 3.14, in the Appendix, hold, i.e., that for any fixed $0 \leq t \leq T$, the set

$$\{G_\alpha f(t) : \|f\|_{L^p} \leq 1\} \tag{3.92}$$

is relatively compact in H and that

$$\lim_{\delta \to 0^+} \sup_{|t-s|<\delta, t,s \in [0,T]} \|G_\alpha f(t) - G_\alpha f(s)\|_H = 0. \tag{3.93}$$

For $\varepsilon \in (0, t]$, $f \in L^p([0, T], H)$, define

$$(G_\alpha^\varepsilon f)(t) = \int_0^{t-\varepsilon} (t-s)^{\alpha-1} S(t-s) f(s)\, ds.$$

Then

$$G_\alpha^\varepsilon f = S(\varepsilon) \int_0^{t-\varepsilon} (t-s)^{\alpha-1} S(t-\varepsilon-s) f(s)\, ds.$$

Since $S(\varepsilon)$ is compact, then so is G_α^ε. Let $q = p/(p-1)$. Now, using Hölder's inequality, we have

$$\left\| (G_\alpha f)(t) - (G_\alpha^\varepsilon f)(t) \right\|_H$$

$$= \left\| \int_{t-\varepsilon}^t (t-s)^{\alpha-1} S(t-s) f(s)\, ds \right\|_H$$

$$\leq \left(\int_{t-\varepsilon}^t (t-s)^{(\alpha-1)q} \|S(t-s)\|^q\, ds \right)^{\frac{1}{q}} \left(\int_{t-\varepsilon}^t \|f(s)\|_{L^p}^p\, ds \right)^{\frac{1}{p}}$$

$$\leq M \left(\frac{\varepsilon^{(\alpha-1)q+1}}{(\alpha-1)q+1} \right)^{\frac{1}{q}} \|f\|_{L^p}$$

with $M = \sup_{s \in [0,T]} \|S(s)\|_{\mathscr{L}(H)}$. Since $(\alpha-1)q + 1 = \alpha - 1/p > 0$, $G_\alpha^\varepsilon \to G_\alpha$ in the operator norm in $L(L^p([0, T], H), H)$; thus the limit G_α is a compact operator from $L^p([0, T], H)$ to H, and the relative compactness of the set (3.92) follows.

Consider now the difference

$$\|G_\alpha f(t) - G_\alpha f(s)\|_H$$

$$\leq \int_0^s \left\| (t-u)^{\alpha-1} S(t-u) - (s-u)^{\alpha-1} S(s-u) \right\|_{\mathscr{L}(H)} \|f(u)\|_H\, du$$

$$+ \int_s^t \left\| (t-u)^{\alpha-1} S(t-u) f(u) \right\|_H du$$

$$\leq \left(\int_0^T \left\| (v+u)^{\alpha-1} S(v+u) - v^{\alpha-1} S(v) \right\|_{\mathscr{L}(H)}^q dv \right)^{\frac{1}{q}} \| f \|_{L^p}$$

$$+ M \left(\int_0^{t-s} v^{(\alpha-1)q} dv \right)^{\frac{1}{q}} \| f \|_{L^p}$$

$$= I_1 + I_2,$$

where $u = t - s > 0$. Since, as $u \to 0$, the expression

$$\left\| (v+u)^{\alpha-1} S(v+u) - v^{\alpha-1} S(v) \right\|_{\mathscr{L}(H)}^q \to 0$$

by the compactness of the semigroup, see (1.10), and it is bounded by $(2M)^q v^{(\alpha-1)q}$, we conclude by the Lebesgue DCT that $I_1 \to 0$ as $u = t - s \to 0$. Also, the second term

$$I_2 \leq M \frac{(t-s)^{\alpha-1/p}}{((\alpha-1)q+1)^{1/q}} \| f \|_{L^p} \to 0$$

as $t - s \to 0$. This concludes the proof. $\qquad\square$

Theorem 3.14 *Assume that A is the infinitesimal generator of a compact C_0-semigroup $S(t)$ on a real separable Hilbert space H. Let the coefficients of the SSDE (3.1) satisfy conditions (A1)–(A3). Then (3.1) has a martingale solution.*

Proof As in the proof of Theorem 3.12, we begin with a sequence of mild solutions X_n to equations

$$X_n(t) = x + \int_0^t \left(A X_n(s) + F_n(s, X_n) \right) ds + \int_0^t B_n(s, X_n) dW_s.$$

The coefficients F_n and B_n are the Lipschitz approximations of F and B as in Lemma 3.9. Lemma 3.6 guarantees that

$$\sup_n E \sup_{0 \leq t \leq T} \| X_n(t) \|_H^{2p} < C, \quad p > 1. \tag{3.94}$$

For $p > 1$ fixed, define

$$Y_n(t) = \int_0^t (t-s)^{-\alpha} S(t-s) B_n(s, X_n) dW_s$$

with $1/(2p) < \alpha < 1/2$, then condition (A3) and inequality (3.94) imply that

$$E \int_0^T \| Y_n(s) \|_H^{2p} ds < C'. \tag{3.95}$$

Using the factorization technique, as in Lemma 3.3, we can express $X_n(t)$ as follows:

$$X_n(t) = S(t)x + \int_0^t S(t-s)F_n(s, X_n)\,ds + \int_0^t S(t-s)B_n(s, X_n)\,dW_s$$

$$= S(t)x + G_1 F_n(\cdot, X_n)(t) + \frac{\sin \pi \alpha}{\pi} G_\alpha Y_n(t),$$

where $G_\alpha : L^{2p}([0, T], H) \to C([0, T], H)$, defined in (3.24), is a compact operator for $1/2p < \alpha \le 1$ by Lemma 3.12.

Inequalities (3.94) and (3.95) and the growth condition (A3) imply that for any $\varepsilon > 0$, there exists $\nu > 0$ such that for all $n \ge 1$,

$$P\left(\left\{\left(\int_0^T \|Y_n(s)\|_H^{2p}\,ds\right)^{\frac{1}{2p}} \le \frac{\pi}{\sin \pi \alpha}\nu\right\} \cap \left\{\left(\int_0^T \|F_n(s, X_n)\|_H^{2p}\,ds\right)^{\frac{1}{2p}} \le \nu\right\}\right)$$
$$\ge 1 - \varepsilon.$$

By the compactness of G_α and $S(t)$ and the continuity of the mapping $t \to S(t)x$, we conclude that the set

$$K = \left\{S(\cdot)x + G_\alpha f(\cdot) + G_1 g(\cdot) : \|f\|_{L^{2p}} \le \nu,\ \|g\|_{L^{2p}} \le \nu\right\}$$

is compact in $C([0, T], H)$ and obtain the tightness of the measures $\mu_n = \mathcal{L}(X_n)$ on $C([0, T], H)$.

We will now prove that

$$X_n(t) = x + A\left(\int_0^t X_n(s)\,ds\right) + \int_0^t F(s, X_n)\,ds + \int_0^t B(s, X_n)\,dW_s. \quad (3.96)$$

Let $A_m = m R_m = m A(m - A)^{-1}$, $m > \alpha$, be the Yosida approximations of A (with α determined by the Hille-Yosida Theorem 1.4). Consider the equations

$$X_{n,m} = x + \int_0^t A_m X_{n,m}(s)\,ds + \int_0^t F_n(s, X_{n,m})\,ds + \int_0^t B_n(s, X_{n,m})\,dW_s,$$

whose strong and mild solutions coincide. Moreover, by Proposition 3.2, as $m \to \infty$,

$$E \sup_{0 \le t \le T} \|X_{n,m}(t) - X_n(t)\|_H^{2p}\,ds \to 0,$$

which implies, by selecting a subsequence if necessary, that a.s., as $m \to \infty$,

$$X_{n,m} \to X_n \quad \text{in } C([0, T], H).$$

Using the Lipschitz-type condition (A4), we obtain that

$$\int_0^t F_n(s, X_{n,m})\,ds \to \int_0^t F_n(s, X_n)\,ds,$$

$$\int_0^t B_n(s, X_{n,m}) \, dW_s \to \int_0^t B_n(s, X_n) \, dW_s$$

a.s. in $C([0, T], H)$, which implies that $A_m \int_0^t X_{n,m}(s) \, ds$ is an a.s. convergent sequence in $C([0, T], H)$. Because

$$\int_0^t X_{n,m}(s) \, ds \to \int_0^t X_n(s) \, ds$$

a.s. in $C([0, T], H)$, $A_m = A(m(m - A)^{-1}) = AR_m$, with the operators R_m uniformly bounded, and $R_m(u) \to u$ for $u \in H$, we obtain that, a.s.,

$$R_m\left(\int_0^t X_{n,m}(s) \, ds\right) \to \int_0^t X_n(s) \, ds \quad \text{for each } t.$$

The operator A is closed, and $A(R_m(\int_0^t X_{n,m}(s) \, ds)) = A_m(\int_0^t X_{n,m}(s) \, ds)$ converges; therefore, $\int_0^t X_n(s) \, ds \in \mathscr{D}(A)$ and $A_m(\int_0^t X_{n,m}(s) \, ds) \to A \int_0^t X_n(s) \, ds$, and, as we showed above, the convergence actually holds a.s. in $C([0, T], H)$. This proves that the representation (3.96) is valid.

Using the tightness of the measures $\mu_n = \mathscr{L}(X_n)$, we can assume, passing to a subsequence if necessary, that $X_n \to X$ weakly for some process $X(\cdot) \in C([0, T], H)$. Using the Skorokhod theorem and changing the underlying probability space, we can assume that $X_n \to X$ a.s. as $C([0, T], H)$-valued random variables.

The process

$$M_n(t) = X_n(t) - x - A \int_0^t X_n(s) \, ds - \int_0^t F_n(s, X_n) \, ds$$

$$= \int_0^t B_n(s, X_n) \, dW_s \tag{3.97}$$

is a martingale with respect to the family of σ-fields $\mathscr{F}_n(t) = \sigma(X_n(s), s \le t)$. Because the operator A is unbounded, we cannot justify direct passage to the limit with $n \to \infty$ in (3.97), as we did in Theorem 3.12, but we follow an idea outlined in [11]. Consider the processes

$$N_{n,\lambda}(t) = (A - \lambda I)^{-1} M_n(t)$$

$$= (A - \lambda I)^{-1} X_n(t) - (A - \lambda I)^{-1} x - (A - \lambda I)^{-1} A \int_0^t X_n(s) \, ds$$

$$- \int_0^t (A - \lambda I)^{-1} F_n(s, X_n) \, ds$$

with $\lambda > \alpha$ (thus λ is in the resolvent set of A). The martingale M_n is square integrable, and so is the martingale $N_{n,\lambda}$. The quadratic variation process of $N_{n,\lambda}$ has the form

$$\langle\!\langle N_{n,\lambda}\rangle\!\rangle_t = \int_0^t \left[(A - \lambda I)^{-1} B_n(s, X_n) Q^{1/2}\right]\left[(A - \lambda I)^{-1} B_n(s, X_n) Q^{1/2}\right]^* ds.$$

We note that the operator $(A - \lambda I)^{-1} A$ is bounded on $\mathscr{D}(A)$ and can be extended to a bounded operator on H. We denote this extension by A^λ.

Observe that we are now in a position to repeat the proof of Theorem 3.12 in the current situation (the assumption concerning a compact embedding J in Theorem 3.12 is unnecessary if the sequence of measures is weakly convergent, which now is the case, and we carry out the proof for $C([0, T], H)$-valued processes). The coefficients $F_n^\lambda(s, x) = A^\lambda x(s) + (A - \lambda I)^{-1} F_n(s, x)$ and $B_n^\lambda(s, x) = (A - \lambda I)^{-1} B_n(s, x)$ satisfy assumptions (A1)–(A4), and the coefficients $F^\lambda(s, x) = A^\lambda x(s) + (A - \lambda I)^{-1} F(s, x)$ and $B^\lambda(s, x) = (A - \lambda I)^{-1} B(s, x)$ satisfy assumptions (A1)–(A3) with F_n^λ and B_n^λ converging to F^λ and B^λ, respectively, uniformly on compact subsets of $C([0, T], H)$.

Moreover, by inequality (3.94), we have the uniform integrability,

$$E \sup_{0 \le t \le T} \left\| N_{n,\lambda}(t) \right\|_H^2 \le \frac{M^2}{(\lambda - \alpha)^2} E \sup_{0 \le t \le T} \left\| M_n(t) \right\|_H^2 < \infty.$$

Thus we conclude, as in the proof of Theorem 3.12, that the process

$$Y^\lambda(t) = (A - \lambda I)^{-1} X(t) - (A - \lambda I)^{-1} x$$
$$- A^\lambda \int_0^t X(s)\, ds - \int_0^t (A - \lambda I)^{-1} F(s, X)\, ds$$

is a square-integrable martingale on $C([0, T], H)$ with respect to the family of σ-fields $\mathscr{F}_t = \sigma(X(s), s \le t)$, and its quadratic variation process has the form

$$\langle\!\langle Y^\lambda\rangle\!\rangle_t = \int_0^t \left[(A - \lambda I)^{-1} B(s, X) Q^{1/2}\right]\left[(A - \lambda I)^{-1} B(s, X) Q^{1/2}\right]^* ds.$$

The representation theorem, Theorem 2.7, implies the existence of a Q-Wiener process W_t on a filtered probability space $(\Omega \times \tilde{\Omega}, \mathscr{F} \times \tilde{\mathscr{F}}, \{\mathscr{F}_t \times \tilde{\mathscr{F}}_t\}, P \times \tilde{P})$ such that

$$(A - \lambda I)^{-1} X(t) - (A - \lambda I)^{-1} x - A^\lambda \int_0^t X(s)\, ds - \int_0^t (A - \lambda I)^{-1} F(s, X)\, ds$$
$$= \int_0^t (A - \lambda I)^{-1} B(s, X)\, dW_s.$$

Consequently,

$$X(t) = x + (A - \lambda I) A^\lambda \int_0^t X(s)\, ds + \int_0^t F(s, X)\, ds + \int_0^t B(s, X)\, dW_s,$$

and because for $u \in \mathscr{D}(A^*)$, $((A - \lambda I)A^{\lambda})^*(u) = A^*(u)$, we obtain

$$\langle X(t), u \rangle_H = \langle x, u \rangle_H + \int_0^t \langle X(s), A^* u \rangle_H \, ds + \int_0^t \langle F(s, X), u \rangle_H \, ds$$

$$+ \int_0^t \langle u, B(s, X) \, dW_s \rangle_H.$$

It follows by Theorem 3.2 that the process $X(t)$ is a mild solution to (3.1). $\qquad\square$

Exercise 3.16 Show that the operator $(A - \lambda I)^{-1}A$ is bounded on $\mathscr{D}(A)$.

3.10 Mild Solutions to SSDEs Driven by Cylindrical Wiener Process

We now present an existence and uniqueness result from [12], which later will be useful in discussing an innovative method for studying invariant measures in the case of a compact semigroup (see Sect. 7.4.3).

Let K and H be real separable Hilbert spaces, and $\widetilde{W}_t$ be a cylindrical Wiener process in K defined on a complete filtered probability space $(\Omega, \mathscr{F}, \{\mathscr{F}_t\}_{t \leq T}, P)$ with the filtration $\{\mathscr{F}_t\}_{t \leq T}$ satisfying the *usual conditions*. We consider the following SSDE on $[0, T]$ in H, with an $\mathscr{F}_0$-measurable initial condition ξ:

$$\begin{cases} dX(t) = (AX(t) + F(X(t))) \, dt + B(X(t)) \, d\widetilde{W}_t, \\ X(0) = \xi. \end{cases} \qquad (3.98)$$

Let the coefficients of (3.98) satisfy the following assumptions,

(DZ1) A is the infinitesimal generator of a strongly continuous semigroup $\{S(t), t \geq 0\}$ on H.

(DZ2) $F : H \to H$ is a mapping such that for some $c_0 > 0$,

$$\|F(x)\|_H \leq c_0(1 + \|x\|_H), \quad x \in H,$$

$$\|F(x) - F(y)\|_H \leq c_0 \|x - y\|, \quad x, y \in H.$$

(DZ3) $B : H \to \mathscr{L}(K, H)$ is such that for any $k \in K$, the mapping $x \to B(x)k$ is continuous from H to H and for any $t > 0$ and $x \in H$, $S(t)B(x) \in \mathscr{L}_2(K, H)$, and there exists a locally square-integrable mapping $\mathscr{K} : [0, \infty) \to [0, \infty)$ such that

$$\|S(t)B(x)\|_{\mathscr{L}_2(K,H)} \leq \mathscr{K}(t)(1 + \|x\|_H) \quad \text{for } t > 0, \ x \in H,$$

$$\|S(t)B(x) - S(t)B(y)\|_{\mathscr{L}_2(K,H)} \leq \mathscr{K}(t)\|x - y\|_H, \quad \text{for } t > 0, \ x, y \in H.$$

(DZ4) There exists $\alpha \in (0, 1/2)$ such that

$$\int_0^1 t^{-2\alpha} \mathscr{K}^2(t) \, dt < \infty.$$

We are interested here only in mild solutions and, taking into account the assumption (DZ3), we have the following definition.

Definition 3.3 A stochastic process $X(t)$ defined on a filtered probability space $(\Omega, \mathscr{F}, \{\mathscr{F}_t\}_{t \leq T}, P)$, adapted to the filtration $\{\mathscr{F}_t\}_{t \leq T}$, is a *mild* solution of (3.98) if

$$P\left(\int_0^T \|X(t)\|_H \, dt < \infty\right) = 1, \tag{3.99}$$

$$P\left(\int_0^T \left(\|F(X(t))\|_H + \|S(t-s)B(X(t))\|^2_{\mathscr{L}_2(K,H)}\right) dt < \infty\right) = 1, \tag{3.100}$$

and for all $t \leq T$, P-a.s.,

$$X(t) = S(t)\xi + \int_0^t S(t-s)F(X(s)) \, ds + \int_0^t S(t-s)B(X(s)) \, d\widetilde{W}_s. \tag{3.101}$$

Recall from Sect. 3.3 that $\widetilde{\mathscr{H}}_{2p}$ denotes a Banach space of H-valued stochastic processes X, measurable as mappings from $([0, T] \times \Omega, \mathscr{B}([0, T]) \otimes \mathscr{F})$ to $(H, \mathscr{B}(H))$, adapted to the filtration $\{\mathscr{F}_t\}_{t \leq T}$, and satisfying $\sup_{0 \leq s \leq T} E\|\xi(s)\|_H^{2p} < \infty$ with the norm

$$\|X\|_{\widetilde{\mathscr{H}}_{2p}} = \left(\sup_{0 \leq t \leq T} E\|X(t)\|_H^{2p}\right)^{\frac{1}{2p}}.$$

We will need the following lemmas.

Lemma 3.13 *Let $\Phi \in \Lambda_2(K, H)$, $p \geq 1$, then*

$$\sup_{0 \leq t \leq T} E \left\|\int_0^t \Phi(s) \, d\widetilde{W}_s\right\|_H^{2p}$$

$$\leq (p(2p-1))^p \left(\int_0^T \left(E\|\Phi(s)\|^{2p}_{\mathscr{L}(K,H)}\right)^{1/p} ds\right)^p. \tag{3.102}$$

Proof For $p = 1$, the result is just the isometric property (2.32) of the stochastic integral. For $p > 1$, let $\widetilde{M}(t) = \int_0^t \Phi(s) \, d\widetilde{W}_s$, and we apply the Itô formula (2.61) to $\|\widetilde{M}(t)\|_H^{2p}$ and, as in the proof of Lemma 3.1, obtain

$$E\|\widetilde{M}(s)\|^{2p} \leq p(2p-1)E\left(\int_0^s \|\widetilde{M}(u)\|^{2(p-1)}_H \|\Phi(u)\|^2_{\mathscr{L}_2(K,H)} \, du\right).$$

Using the Hölder inequality, we have

$$E\|\tilde{M}(s)\|^{2p} \leq p(2p-1)\int_0^s \left(E\|\tilde{M}(u)\|_H^{2p}\right)^{(p-1)/p} \left(E\|\Phi(u)\|_{\mathscr{L}_2(K,H)}^{2p}\right)^{1/p} du$$

$$\leq p(2p-1)\int_0^s \left(\sup_{0\leq v\leq u} E\|\tilde{M}(v)\|_H^{2p}\right)^{(p-1)/p} \left(E\|\Phi(u)\|_{\mathscr{L}_2(K,H)}^{2p}\right)^{1/p} du.$$

Consequently,

$$\sup_{0\leq t\leq T} E\|\tilde{M}(t)\|^{2p}$$

$$\leq p(2p-1)\int_0^T \left(\sup_{0\leq s\leq t} E\|\tilde{M}(s)\|_H^{2p}\right)^{(p-1)/p} \left(E\|\Phi(t)\|_{\mathscr{L}_2(K,H)}^{2p}\right)^{1/p} dt$$

$$\leq p(2p-1)\left(\sup_{0\leq t\leq T} E\|\tilde{M}(s)\|_H^{2p}\right)^{(p-1)/p}\int_0^T \left(E\|\Phi(t)\|_{\mathscr{L}_2(K,H)}^{2p}\right)^{1/p} dt,$$

and (3.102) follows. □

Theorem 3.15 (a) *Under conditions* (DZ1)–(DZ3), *for an arbitrary* $\mathscr{F}_0$*-measurable initial condition* ξ *such that* $E\|\xi\|_H^{2p} < \infty$, $p \geq 1$, *there exist a unique mild solution to* (3.98) *and a constant* C, *independent of the initial condition* ξ, *such that*

$$\|X\|_{\tilde{\mathscr{H}}_{2p}}^{2p} \leq C\left(1 + E\|\xi\|_H^{2p}\right). \tag{3.103}$$

(b) *If, in addition, condition* (DZ4) *holds, then the solution* $X(t)$ *is continuous* P-*a.s.*

Proof For $X \in \tilde{\mathscr{H}}_{2p}$, let

$$\tilde{I}(X)(t) = S(t)\xi + \int_0^t S(t-s)F(X(s))\,ds + \int_0^t S(t-s)B(X(s))\,d\tilde{W}_s.$$

First let us show that

$$\|\tilde{I}(X)\|_{\tilde{\mathscr{H}}_{2p}}^{2p} \leq C\left(1 + \|X\|_{\tilde{\mathscr{H}}_{2p}}^{2p}\right), \tag{3.104}$$

where the constant C may depend on ξ. We have

$$\|\tilde{I}(X)\|_{\tilde{\mathscr{H}}_{2p}}^{2p} \leq \sup_{t\leq t\leq T} 3^{2p-1}\left\{\|S(t)\|_{\mathscr{L}(H)}^{2p} E\|\xi\|_H^{2p}\right.$$

$$+ E\left(\int_0^t \|S(t-s)F(X(s))\|_H\,ds\right)^{2p}$$

$$
+ E \left\| \int_0^t S(t-s) B(X(s)) \, d\widetilde{W}_s \right\|_H^{2p} \Bigg\}
$$

$$
\leq C \Bigg\{ E \|\xi\|_H^{2p} + \left(1 + \sup_{0 \leq t \leq T} E \|X(t)\|_H^{2p} \right)
$$

$$
\sup_{0 \leq t \leq T} \left[\int_0^t \left(E \| S(t-s) B(X(s)) \|_{\mathscr{L}_2(K,H)}^{2p} \right)^{1/p} \right]^p \Bigg\}
$$

$$
\leq C \Bigg\{ E \|\xi\|_H^{2p} + 1 + \|X\|_{\widetilde{\mathscr{H}}_{2p}}^{2p}
$$

$$
+ 2^{2p-1} \left(\int_0^T \mathscr{K}^2(t-s) \, ds \right)^p \left(1 + \sup_{0 \leq t \leq T} E \|X(t)\|_H^{2p} \right) \Bigg\}
$$

$$
\leq C_\xi \left(1 + \|X\|_{\widetilde{\mathscr{H}}_{2p}}^{2p} \right),
$$

where we have used Lemma 3.13 to find a bound on the norm of the stochastic integral. Next, we compute

$$
\left\| \int_0^t S(t-s) \big(F(X(s)) - F(Y(s)) \big) \, ds \right.
$$

$$
\left. + \int_0^t S(t-s) \big(B(X(s)) - B(Y(s)) \big) \, d\widetilde{W}_s \right\|_H^{2p}
$$

$$
\leq 2^{2p-1} \left(\left\| \int_0^t S(t-s) \big(F(X(s)) - F(Y(s)) \big) \, ds \right\|_H^{2p} \right.
$$

$$
\left. + \left\| \int_0^t S(t-s) \big(B(X(s)) - B(Y(s)) \big) \, d\widetilde{W}_s \right\|_H^{2p} \right)
$$

$$
\leq 2^{2p-1} \left(C_1 \int_0^t E \|X(s) - Y(s)\|_H^{2p} \, ds \right.
$$

$$
\left. + C_2 \left(\int_0^t \mathscr{K}^2(t-s) \big(E \|X(s) - Y(s)\|_H^{2p} \big)^{1/p} \, ds \right)^p \right).
$$

Let $L_1 = 2^{2p-1} \max\{C_1, C_2\}$ and $L_2 = 2L_1(1 + p \int_0^T \mathscr{K}^2(t) \, dt)$. Let, as in the proof of Theorem 3.5, $\mathfrak{B}$ denote the Banach space obtained from $\widetilde{H}_{2p}$ by modifying its norm to an equivalent norm

$$
\|X\|_{\mathfrak{B}} = \left(\sup_{0 \leq t \leq T} e^{-L_2 t} E \|X(t)\|_H^{2p} \right)^{\frac{1}{2p}}.
$$

Then,

$$
\left\| \widetilde{I}(X) - \widetilde{I}(Y) \right\|_{\mathfrak{B}}^{2p}
$$

$$
\leq \sup_{0 \leq t \leq T} e^{-L_2 t} L_1 \left(\int_0^t E \|X(s) - Y(s)\|_H^{2p} \, ds \right.
$$

$$+ \left(\int_0^t \mathcal{K}^2(t-s) \left(E \|X(s) - Y(s)\|_H^{2p} \right)^{1/p} ds \right)^p \right)$$

$$= \sup_{0 \le t \le T} e^{-L_2 t} L_1 \left(\int_0^t E \|X(s) - Y(s)\|_H^{2p} e^{L_2 s} e^{-L_2 s} ds \right.$$

$$+ \left(\int_0^t \mathcal{K}^2(t-s) \left(e^{L_2 s} e^{-L_2 s} E \|X(s) - Y(s)\|_H^{2p} \right)^{1/p} ds \right)^p \right)$$

$$\le \sup_{0 \le t \le T} e^{-L_2 t} L_1 \left(\int_0^t \left(\sup_{0 \le s \le T} e^{-L_2 s} E \|X(s) - Y(s)\|_H^{2p} \right) e^{L_2 s} ds \right.$$

$$+ \left(\int_0^t \mathcal{K}^2(t-s) e^{L_2 s/p} \left(\sup_{0 \le s \le T} \left(E^{-L_2 s} E \|X(s) - Y(s)\|_H^{2p} \right) \right)^{1/p} ds \right)^p \right)$$

$$\le L_1 \|X - Y\|_{\mathfrak{B}}^{2p} \sup_{0 \le t \le T} e^{-L_2 t} \left(\int_0^t e^{L_2 s} ds + \left(\int_0^t \mathcal{K}^2(t-s) e^{L_2 s/p} ds \right)^p \right)$$

$$\le L_1 \|X - Y\|_{\mathfrak{B}}^{2p} \sup_{0 \le t \le T} e^{-L_2 t} \left(\frac{e^{L_2 t} - 1}{L_2} + \left(\int_0^t \mathcal{K}^2(s) ds \right)^p \left(\int_0^t e^{L_2 s/p} ds \right)^p \right)$$

$$\le L_1 \left(\frac{1 - e^{-L_2 T}}{L_2} + \left(\int_0^t \mathcal{K}^2(s) ds \right)^p p^p \left(\frac{1 - e^{-L_2 T}}{L_2} \right)^p \right) \|X - Y\|_{\mathfrak{B}}^{2p}$$

$$\le C_{\mathfrak{B}} \|X - Y\|_{\mathfrak{B}}^{2p}$$

with the constant $C_{\mathfrak{B}} < 1$.

Hence, $\tilde{I}$ is a contraction on $\mathfrak{B}$, and it has a unique fixed point, which is the solution to (3.98).

To prove (3.103), note that

$$\|\tilde{I}(0)\|_{\tilde{\mathcal{H}}_{2p}}^{2p} = \sup_{0 \le t \le T} \left\| S(t)\xi + \int_0^t S(t-s) F(0) ds + \int_0^t S(t-s) B(0) d\tilde{W}_s \right\|_H^{2p}$$

$$\le C_0 \left(1 + \|\xi\|_H^{2p} \right)$$

for a suitable constant C_0.

Since the fixed point can be obtained as a limit, $X = \lim_{n \to \infty} \tilde{I}^n(0)$ in $\mathfrak{B}$, using the equivalence of the norms in $\mathcal{H}_{2p}$ and $\mathfrak{B}$, we have

$$\|X\|_{\tilde{\mathcal{H}}_{2p}} \le C_1 \left(\sum_{n=1}^\infty \|\tilde{I}^{n+1}(0) - \tilde{I}^n(0)\|_{\mathfrak{B}} + \|\tilde{I}(0)\|_{\tilde{\mathcal{H}}_{2p}} \right)$$

$$\le C_2 \sum_{n=0}^\infty C_{\mathfrak{B}}^n \|\tilde{I}(0)\|_{\tilde{\mathcal{H}}_{2p}}$$

$$\le C \left(1 + \|\xi\|_H^{2p} \right)$$

for suitable constants C, C_1, C_2.

To prove part (b), the continuity of $X(t)$, it is enough to show that the stochastic convolution with respect to a cylindrical Wiener process,

$$S \tilde{*} B(X)(t) = \int_0^t S(t-s) B(X(s)) d\tilde{W}_s,$$

has a continuous version. Similarly as in the proof of Lemma 3.3, let $\frac{1}{2p} < \alpha < \frac{1}{2}$ and define

$$Y(s) = \int_0^s (s-\sigma)^{-\alpha} S(s-\sigma) B(X(\sigma)) d\tilde{W}_\sigma.$$

Using the stochastic Fubini Theorem 2.3 for a cylindrical Wiener process, we have

$$\int_0^t S(t-s) B(X(s)) d\tilde{W}_s = \frac{\sin \pi \alpha}{\pi} \int_0^t (t-s)^{\alpha-1} S(t-s) Y(s) ds.$$

Using the Hölder inequality, we have

$$\sup_{0 \le t \le T} \left\| \int_0^t S(t-s) B(X(s)) d\tilde{W}_s \right\|_H^{2p}$$

$$= \sup_{0 \le t \le T} \left\| \frac{\sin \pi \alpha}{\pi} \int_0^t (t-s)^{\alpha-1} S(t-s) Y(s) ds \right\|_H^{2p}$$

$$\le C \int_0^T \|Y(s)\|_H^{2p} ds.$$

However,

$$\int_0^T E \|Y(s)\|_H^{2p} ds = \int_0^T E \left\| \int_0^s (s-\sigma)^{-\alpha} S(s-\sigma) B(X(\sigma)) d\tilde{W}_\sigma \right\|_H^{2p} ds$$

$$\le C \int_0^T E \left(\int_0^s (s-\sigma)^{-2\alpha} \|S(s-\sigma) B(X(\sigma))\|_{\mathscr{L}_2(K,H)}^2 d\sigma \right)^p ds$$

$$\le C \int_0^T \left(\int_0^s (s-\sigma)^{-2\alpha} \mathscr{K}^2(s-\sigma)(1 + \|X(\sigma)\|_H^2) d\sigma \right)^p ds$$

$$\le C (1 + \|X\|_{\mathscr{H}_{2p}}^{2p}) \int_0^T \left(\int_0^s (s-\sigma)^{-2\alpha} \mathscr{K}^2(s-\sigma) d\sigma \right)^p ds < \infty.$$

Since the process $Y(t)$ has almost surely $2p$-integrable paths, Lemma 3.2 implies that

$$S \tilde{*} B(X)(t) = \frac{\sin \pi \alpha}{\pi} G_\alpha Y(t)$$

has a continuous version. □

We conclude with two important properties of the solutions to (3.98).

Proposition 3.3 *The solution $X^\xi(t)$ of (3.98) as a function of the initial condition ξ is continuous as mapping from $L^{2p}(\Omega, H)$, $p \geq 1$, into itself, and there exists a constant C such that for $\xi, \eta \in L^{2p}(\Omega, H)$,*

$$E\|X^\xi(t) - X^\eta(t)\|_H^{2p} \leq CE\|\xi - \eta\|_H^{2p}. \tag{3.105}$$

Proof Using assumptions (DZ2) and (DZ3), Lemma 3.13, and Exercise 3.7, we calculate

$$
\begin{aligned}
E\|X^\xi(t) - X^\eta(t)\|_H^{2p} &\leq C\Bigg\{ E\|\xi - \eta\|_H^{2p} + E\left(\int_0^t \|X^\xi(s) - X^\eta(s)\|_H \, ds\right)^{2p} \\
&\quad + \left(\int_0^t \left(E\big(\mathscr{K}(t-s)\|X^\xi(s) - X^\eta(s)\|_H\big)^{2p}\right)^{1/p} ds\right)^p \Bigg\} \\
&\leq C\Bigg\{ E\|\xi - \eta\|_H^{2p} + E\left(\int_0^t \|X^\xi(s) - X^\eta(s)\|_H \, ds\right)^{2p} \\
&\quad + \left(\int_0^t \mathscr{K}^2(t-s)\big(E\|X^\xi(s) - X^\eta(s)\|_H^{2p}\big)^{1/p} ds\right)^p \Bigg\} \\
&\leq C\Bigg\{ E\|\xi - \eta\|_H^{2p} + E\left(\int_0^t \|X^\xi(s) - X^\eta(s)\|_H \, ds\right)^{2p} \\
&\quad + \left(\int_0^t \mathscr{K}^2(s) \, ds \int_0^t \big(E\|X^\xi(s) - X^\eta(s)\|_H^{2p}\big)^{1/p} ds\right)^p \Bigg\} \\
&\leq C\Bigg\{ E\|\xi - \eta\|_H^{2p} + \int_0^t E\|X^\xi(s) - X^\eta(s)\|_H^{2p} \, ds \Bigg\}.
\end{aligned}
$$

Now an appeal to the Gronwall lemma concludes the proof. $\qquad\square$

Proposition 3.4 *The solution of (3.98) is a homogeneous Markov and Feller process.*

We omit the proof since it follows nearly word by word the proof of Theorem 3.6 and the discussion in the remainder of Sect. 3.5 with the σ-field $\mathscr{F}^{W,\xi}$ being replaced by

$$\mathscr{F}^{\tilde{W},\xi} = \sigma\left(\bigcup_{j=1}^\infty \sigma\big(\tilde{W}_s(f_j), \, s \leq t\big) \cup \sigma(\xi)\right).$$

Appendix: Compactness and Tightness of Measures in $C([0, T], \mathcal{M})$

In this book we are dealing with the space $\mathcal{X} = C([0, T], \mathcal{M})$, where $(\mathcal{M}, \rho_{\mathcal{M}})$ is a complete separable metric space. For this case, we obtain analytic conditions for the family of measures $\{\mu_n\} \subset \mathcal{P}(\mathcal{X})$ to be tight.

The space $\mathcal{X}$ with the usual supremum metric $\rho_\infty(x, y) = \sup_{0 \le t \le T} \rho_{\mathcal{M}}(x(t) - y(t))$, $x, y \in \mathcal{M}$, is itself a complete separable metric space. We denote by $\mathcal{B}(\mathcal{X})$ the σ-field of subsets of $\mathcal{X}$ generated by the open subsets of $\mathcal{X}$. The elements of $\mathcal{B}(\mathcal{X})$ will be referred to as Borel subsets of $\mathcal{X}$. The object is to study probability measures on the measurable space $(\mathcal{X}, \mathcal{B}(\mathcal{X}))$. We denote the class of all probability measures on $\mathcal{X}$ by $\mathcal{P}(\mathcal{X})$. We recall that a sequence of probability measures $\{\mu_n\}_{n=1}^\infty \subset \mathcal{P}(\mathcal{X})$ converges weakly to $\mu \in \mathcal{P}(\mathcal{X})$ if for every bounded continuous function $f : \mathcal{X} \to \mathbb{R}$,

$$\int_{\mathcal{X}} f(x)\,\mu_n(dx) \to \int_{\mathcal{X}} f(x)\,\mu(dx).$$

The above convergence defines a metric ρ on $\mathcal{P}(\mathcal{X})$ (Prokhorov's metric), such that $(\mathcal{P}(\mathcal{X}), \rho)$ is a complete separable metric space (see [4], Theorem 5, p. 238). The concept of tightness is essential in verifying that a family of measures $\{\mu_n\} \subset \mathcal{P}(\mathcal{X})$ possesses weak limit points.

Definition 3.4 A family of measures $\{\mu_n\}_{n=1}^\infty \subset \mathcal{P}(\mathcal{X})$ is *tight* if for each $\varepsilon > 0$, there exists a compact set $K_\varepsilon \subset \mathcal{X}$ such that

$$\mu_n\left(K_\varepsilon^c\right) < \varepsilon$$

for all n.

The following theorem is due to Prokhorov [65].

Theorem 3.16 *A family of measures* $\{\mu_n\}_{n=1}^\infty \subset (\mathcal{P}(\mathcal{X}), \rho)$ *is relatively compact if and only if it is tight.*

The modulus of continuity of a function $x \in C([0, T], \mathcal{M})$ is defined by

$$w(x, \delta) = \sup_{|t-s|<\delta} \rho_{\mathcal{M}}\big(x(t), x(s)\big). \tag{3.106}$$

Lemma 3.14 (Compactness) *A set* $A \subset C([0, T], \mathcal{M})$ *has a compact closure if and only if the two following conditions are fulfilled:*

(1) *There exists a dense subset* $T' \subset [0, T]$ *such that for every* $t \in T'$, *the closure of the set* $\{x(t) : x \in A\}$ *is a compact subset of* $\mathcal{M}$,

(2) $\lim_{\delta \to 0^+} \sup_{x \in A} w(x, \delta) = 0$.

Proof The compactness of $\overline{A}$ implies condition (1) easily. Condition (2) follows from the fact that the function $w(x, \delta)$ is a continuous function of x, monotonically decreasing to 0 as $\delta \to 0$. By Dini's theorem, $w(x, \delta)$ converges to 0 uniformly on a compact set $\overline{A}$.

If conditions (1) and (2) hold, then let $x_n(t)$ be a sequence of elements in A. We form a sequence $t_1, t_2, \ldots$ from all elements of T' and select a convergent subsequence $x_{n_{k_1}}(t_1)$. From this subsequence we select a convergent subsequence $x_{n_{k_2}}(t_2)$, etc. Using the diagonal method, we construct a subsequence x_{n_k} such that $x_{n_k}(t)$ converges for every $t \in T'$. Denote the limit by $y(t)$ and let $y_k(t) = x_{n_k}(t)$. Then $y_k(t) \to y(t)$ pointwise on T'. For any $\varepsilon > 0$, let δ be chosen, so that $\sup_{x \in A} w(x, \delta) < \varepsilon/3$. Let $t_1, \ldots, t_N$ be such that the length of each of the intervals $[0, t_1], [t_1, t_2], \ldots, [t_N, T]$ is less than δ. Then

$$\sup_{0 \le t \le T} \rho_{\mathcal{M}}\big(y_k(t), y_l(t)\big) \le \sup_{1 \le i \le N} \rho_{\mathcal{M}}\big(y_k(t_i), y_l(t_i)\big)$$

$$+ \sup_{|t - t_i| \le \delta} \big(\rho_{\mathcal{M}}\big(y_k(t_i), y_k(t)\big) + \rho_{\mathcal{M}}\big(y_l(t_i), y_l(t)\big)\big) < \varepsilon$$

for k, l sufficiently large.

Hence, y_k is a Cauchy sequence in the ρ_∞ metric. Therefore, $\rho_\infty(y_k, y) \to 0$. $\square$

Theorem 3.17 (Tightness) *A family P_n of probability measures on $C([0, T], \mathcal{M})$ is tight if and only if the following two conditions are fulfilled,*

(1) *There exists a dense subset $T' \subset [0, T]$ such that for every $t \in T'$, the family of measures $P_n \circ x(t)^{-1}$ on $\mathcal{M}$ is tight,*
(2) $\lim_{\delta \to 0} \lim \sup_n P_n(w(x, \delta) > \varepsilon) = 0$ *for all $\varepsilon > 0$.*

Proof Conditions (1) and (2) follow easily from the tightness assumption and Lemma 3.14. Conversely, let $\{t_k\}_{k \in \mathbb{Z}_+}$ be a countably dense set in $[0, T]$. For each k, we can find a compact set $C_k \subset \mathcal{M}$ and $\delta_k > 0$ such that $\sup_n P_n(x(t_k) \notin C_k) \le \varepsilon/2^{k+2}$ (by Prokhorov's theorem) and, for some n_0, $\sup_{n \ge n_0} P_n(w(x, \delta_k) > \frac{1}{k}) \le \varepsilon/2^{k+2}$. Let

$$K_\varepsilon = \left(\bigcup_{k > n_0} \big(\{x(t_k) \notin C_k\} \cup \{w(x, \delta_k) > 1/k\}\big) \right)^c$$

$$= \bigcap_{k > n_0} \{x(t_k) \in C_k\} \cap \{w(x, \delta_k) \le 1/k\}.$$

The set K_ε satisfies the assumptions of Lemma 3.14; therefore it has a compact closure. Moreover,

$$\sup_{n \ge n_0} P_n\big(K_\varepsilon^c\big) \le \sum_{k=0}^\infty 2 \cdot \varepsilon/2^{k+2} = \varepsilon \Rightarrow \inf_{n \ge n_0} P_n(K_\varepsilon) \ge 1 - \varepsilon.$$

$\square$

Chapter 4
Solutions by Variational Method

4.1 Introduction

The purpose of this chapter is to study both weak and strong solutions of nonlinear stochastic partial differential equations, or SPDEs. The first work in this direction was done by Viot [75]. Since then, Pardoux [62] and Krylov and Rozovskii [42] studied strong solutions of nonlinear SPDEs. We will utilize the recent publication by Prévôt and Röckner [64] to study strong solutions.

In all these publications, the SPDEs are recast as evolution equations in a Gelfand triplet,

$$V \hookrightarrow H \hookrightarrow V^*,$$

where H is a real separable Hilbert space identified with its dual H^*. The space V is a Banach space embedded continuously and densely in H. Then for its dual space V^*, the embedding $H \hookrightarrow V^*$ is continuous and dense, and V^* is necessarily separable. The norms are denoted by $\| \cdot \|_V$, and similarly for the spaces H and V^*. The duality on $V \times V^*$ is denoted by $\langle \cdot, \cdot \rangle$, and it agrees with the scalar product in H, i.e., $\langle v, h \rangle = \langle v, h \rangle_H$ if $h \in H$.

By using the method of compact embedding of Chap. 3, the ideas from [36], and the stochastic analogue of Lion's theorem from [42], we show the existence of a weak solution X in the space $C([0, T], H) \cap L^\infty([0, T], H) \cap L^2([0, T] \times \Omega, V)$ such that

$$E\left(\sup_{t \in [0, T]} \| X(t) \|_H^2 \right) < \infty, \tag{4.1}$$

under the assumption that the injection $V \hookrightarrow H$ is compact without using the assumption of monotonicity. In the presence of monotone coefficients, as in [36], we obtain a unique strong solution using pathwise uniqueness.

The approach in [64] is to consider monotone coefficients. Under weakened assumptions on growth and without assuming compact embedding, using again the stochastic analogue of Lion's theorem, a unique strong solution is produced in $C([0, T], H) \cap L^2([0, T] \times \Omega, V)$, which again satisfies (4.1). We will present this method in Sect. 4.3.

L. Gawarecki, V. Mandrekar, *Stochastic Differential Equations in Infinite Dimensions*, 151
Probability and Its Applications,
DOI 10.1007/978-3-642-16194-0_4, © Springer-Verlag Berlin Heidelberg 2011

Now, let K be another real separable Hilbert space, $Q \in \mathcal{L}_1(K)$ be a symmetric nonnegative definite operator, and $\{W_t, t \geq 0\}$ be a K-valued Q-Wiener process defined on a filtered probability space $(\Omega, \mathcal{F}, \{\mathcal{F}_t\}_{t \geq 0}, P)$.

Consider the variational SDE

$$dX(t) = A\big(t, X(t)\big)\,dt + B\big(t, X(t)\big)\,dW_t \tag{4.2}$$

with coefficients

$$A : [0, T] \times V \to V^* \quad \text{and} \quad B : [0, T] \times V \to \mathcal{L}_2(K_Q, H)$$

and H-valued $\mathcal{F}_0$-measurable initial condition $\xi_0 \in L^2(\Omega, H)$. Recall that

$$\big(B(t, v)Q^{1/2}\big)\big(B(t, v)Q^{1/2}\big)^* \in \mathcal{L}_1(H).$$

Let us now define different notions of a solution to (4.2). Such solutions are often called *variational solutions*.

Definition 4.1 An H-valued stochastic process $X(t)$ defined on a given filtered probability space $(\Omega, \mathcal{F}, \{\mathcal{F}_t\}_{t \leq T}, P)$ is a *strong solution* of (4.2) if

(1) $E \displaystyle\int_0^T \|X(t)\|_V^2\,dt < \infty,$

(2) $P\left(\displaystyle\int_0^T \|A\big(t, X(t)\big)\|_{V^*}\,dt < \infty\right) = 1,$

(3) $\displaystyle\int_0^t B\big(s, X(s)\big)\,dW_s$ is a square-integrable H-valued martingale,

(4) $X(t) = \xi_0 + \displaystyle\int_0^t A\big(s, X(s)\big)\,ds + \int_0^t B\big(s, X(s)\big)\,dW_s$ P-a.s.

The integrants $A(t, X(t))$ and $B(t, X(t))$ are evaluated at a V-valued $\mathcal{F}_t$-measurable version of $X(t)$ in $L^2([0, T] \times \Omega, V)$.

A *weak solution* of (4.2) is a system $((\Omega, \mathcal{F}, \{\mathcal{F}_t\}_{t \leq T}, P), W, X)$, where W_t is a K-valued Q-Wiener process with respect to the filtration $\{\mathcal{F}_t\}_{t \leq T}$, X_t is an H-valued process adapted to $\mathcal{F}_t$ and satisfies conditions (1)–(3) above,

$$X(t) = X(0) + \int_0^t A\big(s, X(s)\big)\,ds + \int_0^t B\big(s, X(s)\big)\,dW_s, \quad P\text{-a.s.},$$

and $P \circ (X(0))^{-1} = \mathcal{L}(\xi_0)$.

Unlike in the case of a strong solution, the filtered probability space and the Wiener process are part of a weak solution and are not given in advance.

4.2 Existence of Weak Solutions Under Compact Embedding

In this section we study the existence of weak solutions using mainly the techniques in [36] and [22]. We assume that in the Gelfand triplet $V \hookrightarrow H \hookrightarrow V^*$, V is a real separable Hilbert space.

Let the coefficients A and B satisfy the following joint continuity and growth conditions:

(JC) (Joint Continuity) The mappings

$$(t, v) \to A(t, v) \in V^* \quad \text{and} \quad (t, v) \to B(t, v)QB^*(t, v) \in \mathcal{L}_1(H) \quad (4.3)$$

are continuous.

For some constant $\theta \geq 0$,

(G-A)

$$\left\| A(t, v) \right\|_{V^*}^2 \leq \theta \left(1 + \|v\|_H^2 \right), \quad v \in V. \quad (4.4)$$

(G-B)

$$\left\| B(t, v) \right\|_{\mathcal{L}_2(K_Q, H)}^2 \leq \theta \left(1 + \|v\|_H^2 \right), \quad v \in V. \quad (4.5)$$

In addition, we will impose the following coercivity condition on A and B:

(C) There exist constants $\alpha > 0$, $\gamma, \lambda \in \mathbb{R}$ such that for $v \in V$,

$$2\langle A(t, v), v \rangle + \left\| B(t, v) \right\|_{\mathcal{L}_2(K_Q, H)}^2 \leq \lambda \|v\|_H^2 - \alpha \|v\|_V^2 + \gamma. \quad (4.6)$$

Finally, we will require that the initial condition of (4.2) satisfies

(IC)

$$E\left\{ \|\xi_0\|_H^2 \left(\ln \left(3 + \|\xi_0\|_H^2 \right) \right)^2 \right\} < c_0 \quad (4.7)$$

for some constant c_0. It will become clear that this property will be used to ensure the uniform integrability of the squared norm of the approximate solutions.

We will first consider a finite-dimensional SDE related to the infinite-dimensional equation (4.2). Let $\{\varphi_j\}_{j=1}^\infty \subset V$ be a complete orthonormal system in H, and $\{f_k\}_{k=1}^\infty$ be a complete orthonormal system in K. Define the map $J_n : \mathbb{R}^n \to V$ by

$$J_n(x) = \sum_{j=1}^n x_j \varphi_j = u \in V$$

and the coefficients $(a^n(t, x))_j : [0, T] \times \mathbb{R}^n \to \mathbb{R}^n$, $(b^n(t, x))_{i,j} : [0, T] \times \mathbb{R}^n \to \mathbb{R}^n \times \mathbb{R}^n$, and $(\sigma^n(t, x))_{i,j} : [0, T] \times \mathbb{R}^n \to \mathbb{R}^n \times \mathbb{R}^n$ and the initial condition ξ_0^n by

$$\begin{aligned}
\left(a^n(t, x) \right)_j &= \langle \varphi_j, A(t, J_n x) \rangle, \quad 1 \leq j \leq n, \\
\left(b^n(t, x) \right)_{i,j} &= \langle Q^{1/2} B^*(t, J_n x)\varphi_i, f_j \rangle_K, \quad 1 \leq i, j \leq n, \\
\left(\sigma^n(t, x) \right)_{i,j} &= \left(b^n(t, x) \left(b^n(t, x) \right)^T \right)_{i,j}, \\
\left(\xi_0^n \right)_j &= \langle \xi_0, \varphi_j \rangle_H.
\end{aligned} \quad (4.8)$$

Note that

$$\left(\sigma^n(t, x)\right)_{i,j} = \sum_{k=1}^{n} \left(b^n(t, x)\right)_{i,k}\left(b^n(t, x)\right)_{j,k}$$

$$= \sum_{k=1}^{n} \left\langle Q^{1/2} B^*(t, J_n x)\varphi_i, f_k\right\rangle_K \left\langle Q^{1/2} B^*(t, J_n x)\varphi_j, f_k\right\rangle_K.$$

Lemma 4.1 *The growth conditions* (4.4) *and* (4.5) *assumed for the coefficients A and B imply the following growth conditions on a^n and b^n:*

$$\left\|a^n(t, x)\right\|_{\mathbb{R}^n}^2 \le \theta_n\left(1 + \|x\|_{\mathbb{R}^n}^2\right), \tag{4.9}$$

$$\text{tr}\left(\sigma^n(t, x)\right) = \text{tr}\left(b^n(t, x)\left(b^n(t, x)\right)^T\right) \le \theta\left(1 + \|x\|_{\mathbb{R}^n}^2\right). \tag{4.10}$$

The coercivity condition (4.6) *implies that*

$$2\langle a^n(t, x), x\rangle_{\mathbb{R}^n} + \text{tr}\left(b^n(t, x)\left(b^n(t, x)\right)^T\right)$$

$$\le 2\langle J_n x, A(t, J_n x)\rangle + \text{tr}\left(\left(B(t, J_n x)Q^{1/2}\right)\left(B(t, J_n x)Q^{1/2}\right)^*\right)$$

$$\le \lambda\|J_n x\|_H^2 - \alpha\|J_n x\|_V^2 + \gamma. \tag{4.11}$$

In particular, for a large enough value of θ, the coercivity condition (4.6) *implies that*

$$2\langle a^n(t, x), x\rangle_{\mathbb{R}^n} + \text{tr}\left(b^n(t, x)\left(b^n(t, x)\right)^T\right) \le \theta\left(1 + \|x\|_{\mathbb{R}^n}^2\right). \tag{4.12}$$

The constant θ_n depends on n, but θ does not.
The distribution μ_0^n of ξ_0^n on $\mathbb{R}^n$ satisfies

$$E\left\{\left\|\xi_0^n\right\|_{\mathbb{R}^n}^2\left(\ln\left(3 + \left\|\xi_0^n\right\|_{\mathbb{R}^n}^2\right)\right)^2\right\} < c_0. \tag{4.13}$$

Exercise 4.1 Prove Lemma 4.1. In addition, show that for $k \ge n$ and $x \in \mathbb{R}^k$, the following estimate holds true:

$$\sum_{j=1}^{n} \left(\left(a^k(t, x)\right)_j\right)^2 \le \theta_n\left(1 + \|x\|_{\mathbb{R}^k}^2\right). \tag{4.14}$$

We will need the following result, Theorem V.3.10 in [17], on the existence of a weak solution. Consider the following finite-dimensional SDE,

$$dX(t) = a\left(t, X(t)\right)dt + b\left(t, X(t)\right)dB_t^n, \tag{4.15}$$

with an $\mathbb{R}^n$-valued $\mathscr{F}_0$-measurable initial condition ξ_0^n. Here B_t^n is a standard Brownian motion in $\mathbb{R}^n$.

Theorem 4.1 *There exists a weak solution to (4.15) if $a : [0, \infty] \times \mathbb{R}^n \to \mathbb{R}^n$ and $b :$ $[0, \infty] \times \mathbb{R}^n \to \mathbb{R}^n \otimes \mathbb{R}^n$ are continuous and satisfy the following growth conditions:*

$$
\begin{aligned}
\|b(t,x)\|^2_{\mathscr{L}(\mathbb{R}^n)} &\le K\left(1 + \|x\|^2_{\mathbb{R}^n}\right), \\
\langle x, a(t,x)\rangle_{\mathbb{R}^n} &\le K\left(1 + \|x\|^2_{\mathbb{R}^n}\right)
\end{aligned}
\tag{4.16}
$$

for $t \ge 0$, $x \in \mathbb{R}^n$, and some constant K.

We will use the ideas developed in [70], Sect. 1.4, for proving the compactness of probability measures on $C([0, T], \mathbb{R}^n)$. The method was adapted to the specific case involving linear growth and coercivity conditions in [36]. Our first step in proving the existence result in the variational problem will be establishing the existence and properties of finite-dimensional Galerkin approximations in the following lemma.

Lemma 4.2 *Assume that the coefficients A and B of (4.2) satisfy the assumptions of joint continuity (4.3), growth (4.4), (4.5), and coercivity (4.6) and that the initial condition ξ_0 satisfies (4.7). Let a^n, b^n, and ξ_0^n be defined as in (4.8), and B_t^n be an n-dimensional standard Brownian motion. Then the finite-dimensional equation*

$$
dX(t) = a^n\left(t, X(t)\right) dt + b^n\left(t, X(t)\right) dB_t^n
\tag{4.17}
$$

with the initial condition ξ_0^n has a weak solution $X^n(t)$ in $C([0, T], \mathbb{R}^n)$. The laws $\mu^n = P \circ (X^n)^{-1}$ have the property that for any $R > 0$,

$$
\sup_n \mu^n \left\{ x \in C\left([0, T], \mathbb{R}^n\right) : \sup_{0 \le t \le T} \|x(t)\|_{\mathbb{R}^n} > R \right\}
$$
$$
\le 2c_0 e^{C(\theta)T} / \left(1 + R^2\right)\left(\ln\left(3 + R^2\right)\right)^2
\tag{4.18}
$$

and that

$$
\int_{C([0,T],\mathbb{R}^n)} \sup_{0 \le t \le T} \left(1 + \|x(t)\|^2_{\mathbb{R}^n}\right) \ln\ln\left(3 + \|x(t)\|^2_{\mathbb{R}^n}\right) \mu^n(dx) < C
\tag{4.19}
$$

for some constant C.

Proof Since the coefficients a^n and b^n satisfy conditions (4.9) and (4.10), we can use Theorem 4.1 to construct a weak solution $X^n(t)$ to (4.17) for every n. Let

$$
f(x) = \left(1 + \|x\|^2_{\mathbb{R}^n}\right)\left(\ln\left(3 + \|x\|^2_{\mathbb{R}^n}\right)\right)^2, \quad x \in \mathbb{R}^n.
$$

Define for $g \in C_0^2(\mathbb{R}^n)$ the differential operator

$$
\left(L_t^n g\right)(x) = \sum_{i=1}^n \frac{\partial g}{\partial x_i}(x)\left(a^n(t,x)\right)_i + \frac{1}{2}\sum_{i=1}^n\sum_{j=1}^n \frac{\partial^2 g}{\partial x_i \partial x_j}(x)\left(\sigma^n(t,x)\right)_{i,j}.
$$

We leave as an exercise, see Exercise 4.2, to prove that

$$\left\| f_x(x) \right\|_{\mathbb{R}^n}^2 \leq C f(x) \left(\ln\left(3 + \|x\|_{\mathbb{R}^n}^2\right) \right)^2 \tag{4.20}$$

and that the coercivity condition (4.6) implies

$$L_t^n f(x) \leq C f(x). \tag{4.21}$$

Using Itô's formula for the function $f(x)$, we have

$$f\left(X^n(t)\right) = f\left(X^n(0)\right) + \int_0^t L_s^n f\left(X^n(s)\right) ds + M_t, \tag{4.22}$$

where M_t is a local martingale. Define a stopping time

$$\tau_R = \inf\left\{ t : \left\|X^n(t)\right\|_{\mathbb{R}^n} > R \right\} \quad \text{or} \quad T. \tag{4.23}$$

Then $M_{t \wedge \tau_R}$ is a square-integrable martingale with the increasing process

$$
\begin{aligned}
\langle M \rangle_{t \wedge \tau_R} &= \int_0^{t \wedge \tau_R} \left\| b^n\left(s, X^n(s)\right) f_x\left(X^n(s)\right) \right\|_{\mathbb{R}^n}^2 ds \\
&\leq C\theta \int_0^{t \wedge \tau_R} \left(1 + \left\|X^n(s)\right\|_{\mathbb{R}^n}^2\right) f\left(X^n(s)\right) \left(\ln\left(3 + \left\|X^n(s)\right\|_{\mathbb{R}^n}\right) \right)^2 ds \\
&= C\theta \int_0^{t \wedge \tau_R} f^2\left(X^n(s)\right) ds \\
&\leq C\theta \left(\sup_{0 \leq s \leq t \wedge \tau_R} f\left(X^n(s)\right) \right) \int_0^{t \wedge \tau_R} f\left(X^n(s)\right) ds,
\end{aligned}
$$

where we have applied (4.20).

Using Burkholder's inequality, Theorem 3.28 in [38], we calculate

$$
\begin{aligned}
E\left(\sup_{0 \leq s \leq t} \left|M_{t \wedge \tau_R}\right| \right) &\leq 4E\left(\langle M \rangle_{t \wedge \tau_R} \right)^{1/2} \\
&\leq 4(C\theta)^{1/2} E\left\{ \left(\sup_{0 \leq s \leq t} f\left(X^n(s \wedge \tau_R)\right) \right)^{1/2} \left(\int_0^{t \wedge \tau_R} f\left(X^n(s)\right) ds \right)^{1/2} \right\} \\
&\leq E\left\{ \left(\sup_{0 \leq s \leq t} f\left(X^n(s \wedge \tau_R)\right) \right)^{1/2} \left(16C\theta \int_0^t \sup_{0 \leq r \leq s} f\left(X^n(r \wedge \tau_R)\right) ds \right)^{1/2} \right\} \\
&\leq \frac{1}{2} E\left\{ \left(\sup_{0 \leq s \leq t} f\left(X^n(s \wedge \tau_R)\right) \right) + \left(16C\theta \int_0^t \sup_{0 \leq r \leq s} f\left(X^n(r \wedge \tau_R)\right) ds \right) \right\}.
\end{aligned}
$$

Then, by (4.21) and (4.22),

$$f\left(X^n(s \wedge \tau_R)\right) \leq f\left(X^n(0)\right) + C \int_0^s f\left(X^n(r \wedge \tau_R)\right) dr + M_{s \wedge \tau_R},$$

and hence,

$$E \sup_{0 \le s \le t} f\left(X^n(s \wedge \tau_R)\right) \le 2c_0 + (2C + 16C\theta)E \int_0^t \sup_{0 \le r \le s} f\left(X^n(r \wedge \tau_R)\right) ds.$$

By applying Gronwall's lemma, we obtain the bound

$$E \sup_{0 \le s \le t} f\left(X^n(s \wedge \tau_R)\right) \le 2c_0 e^{(2C+16C\theta)T}, \quad 0 \le t \le T.$$

Then

$$P\left(\sup_{0 \le s \le t} \|X^n(s)\|_{\mathbb{R}^n} > R \right) \le E\left(\sup_{0 \le s \le t} f\left(X^n(s \wedge \tau_R)\right) / f(R) \right)$$

$$\le 2c_0 e^{C(\theta)T} / \left(1 + R^2\right)\left(\ln(3 + R^2)\right)^2,$$

proving (4.18). To prove (4.19), denote $g(r) = (1+r^2)\ln\ln(3+r^2)$, $r \ge 0$. Since g is increasing, we have

$$\int_{C([0,T],\mathbb{R}^n)} \sup_{0 \le t \le T} \left(1 + \|x(t)\|_{\mathbb{R}^n}^2\right) \ln\ln\left(3 + \|x(t)\|_{\mathbb{R}^n}^2\right) \mu^n(dx)$$

$$= \int_{C([0,T],\mathbb{R}^n)} \sup_{0 \le t \le T} g\left(\|x(s)\|_{\mathbb{R}^n}\right) \mu^n(dx)$$

$$= \int_0^\infty \mu^n\left(\sup_{0 \le t \le T} g\left(\|x(s)\|_{\mathbb{R}^n}\right) > p \right) dp$$

$$= \ln\ln 3 + \int_{\ln\ln 3}^\infty \mu^n\left(\sup_{0 \le t \le T} \|x(s)\|_{\mathbb{R}^n} > g^{-1}(p) \right) dp$$

$$\le \ln\ln 3 + \int_0^\infty \mu^n\left(\sup_{0 \le t \le T} \|x(s)\|_{\mathbb{R}^n} > r \right) g'(r) dr$$

$$\le \ln\ln 3 + 2c_0 e^{C(\theta)T} \int_0^\infty \frac{g'(r)}{(1+r^2)(\ln(3+r^2))^2} dr < \infty,$$

with the very last inequality being left to prove for the reader in Exercise 4.3.   □

Exercise 4.2 Prove (4.20) and (4.21).

Exercise 4.3 With the notation of Theorem 4.2, prove that

$$\int_0^\infty \frac{g'(r)}{(1+r^2)(\ln(3+r^2))^2} dr < \infty.$$

We will need the following lemma from [36] (see also Sect. 1.4 in [70]).

Lemma 4.3 *Consider the filtered probability space* $(C([0, T], \mathbb{R}^n), \mathcal{B}, \{\mathcal{C}_t\}_{0 \le t \le T},$ $P)$, *where* $\mathcal{B}$ *is the Borel* σ*-field, and* $\mathcal{C}_t$ *is the* σ*-field generated by the cylinders with bases over* $[0, t]$. *Let the coordinate process* m_t *be a square-integrable* $\mathcal{C}_t$*-martingale with quadratic variation* $\langle\langle m \rangle\rangle_t$ *satisfying*

$$\langle m \rangle_t - \langle m \rangle_s = \text{tr}\big(\langle\langle m \rangle\rangle_t - \langle\langle m \rangle\rangle_s\big) \le \beta(t - s)$$

for some constant β *and all* $0 \le s < t \le T$. *Then for all* $\varepsilon, \eta > 0$, *there exists* $\delta > 0$, *depending possibly on* β, ε, η, *and* T, *but not on* n, *such that*

$$P\Big(\sup_{|t-s|<\delta} \|m_t - m_s\|_{\mathbb{R}^n} > \varepsilon \Big) < \eta.$$

Proof Define the sequence of stopping times

$$\tau_0 = 0, \quad \tau_j = \inf\{\tau_{j-1} < s \le T : \|m_s - m_{\tau_{j-1}}\|_{\mathbb{R}^n} > \varepsilon/4\} \quad \text{or } \tau_j = T, \ j \ge 1.$$

Let $N = \inf\{j : \tau_j = T\}$ and $\alpha = \inf\{\tau_j - \tau_{j-1} : 0 \le j \le N\}$. It is left as an exercise to show that

$$\Big\{ \sup_{|t-s|<\delta} \|m_t - m_s\|_{\mathbb{R}^n} > \varepsilon \Big\} \subset \{\alpha < \delta\}. \tag{4.24}$$

Therefore, for any positive integer k,

$$P\Big(\sup_{|t-s|<\delta} \|m_t - m_s\|_{\mathbb{R}^n} > \varepsilon \Big) \le P(\alpha \le \delta)$$

$$\le P(\tau_j - \tau_{j-1} < \delta \text{ for some } j \le N, \ j \le k)$$

$$+ P(\tau_j - \tau_{j-1} < \delta \text{ for some } j \le N, \ j > k)$$

$$\le P(\tau_j - \tau_{j-1} < \delta \text{ for some } j \le k) + P(N > k). \tag{4.25}$$

Let from now on $\varepsilon, \eta > 0$ be fixed but arbitrary.

First, for any stopping time τ, consider the sub-σ-field of $\mathcal{B}$

$$\mathcal{C}_\tau = \{A \in \mathcal{B} : A \cap \{\tau \le t\} \in \mathcal{C}_t, \text{ for all } 0 \le t \le T\} \tag{4.26}$$

and a regular conditional probability distribution of P given $\mathcal{C}_\tau$, denoted by $P^\tau(A, \omega)$. Since $C([0, T], \mathbb{R}^n)$ is a Polish space, a regular conditional probability distribution exists (see [25], Vol. I, Chap. I, Theorem 3 and [70] Sect. 1.3 for a specific construction). The important property of the measure P^τ is that it preserves the martingale property. Specifically, if m_t is a $\mathcal{C}_t$-martingale with respect to the measure P, then it also is a $\mathcal{C}_t$-martingale for $t \ge \tau(\omega)$ with respect to each conditional distribution $P^\tau(\cdot, \omega)$, except possibly for ω outside of a set E of P-measure zero.

It follows that for $t \ge 0$, the process $m_t - m_{t \wedge \tau}$ is a $\mathcal{C}_t$-martingale with respect to $P^\tau(\cdot, \omega)$, except possibly for $\omega \in E$. This is left to the reader to verify as an exercise (see Exercise 4.5), and more general results can be found in Sects. 1.2 and 1.3 of [70].

We are now going to find bounds for the probabilities in the last line of (4.25). We have, P-a.s.,

$$P(\tau_j - \tau_{j-1} < t | \mathscr{F}_{\tau_{j-1}}) = P^{\tau_j}(\tau_j - \tau_{j-1} < t)$$

$$= P^{\tau_j}\left(\sup_{0 \le s \le t} \|m_{s+\tau_{j-1}} - m_{\tau_{j-1}}\|_{\mathbb{R}^n} > \varepsilon/4\right)$$

$$= P^{\tau_j}\left(\sup_{0 \le s \le t+\tau_{j-1}} \|m_s - m_{s \wedge \tau_{j-1}}\|_{\mathbb{R}^n} > \varepsilon/4\right)$$

$$\le \frac{64}{\varepsilon^2} E\left(\|m_{t+\tau_{j-1}} - m_{\tau_{j-1}}\|_{\mathbb{R}^n}^2 | \mathscr{F}_{\tau_j}\right)$$

$$\le \frac{64}{\varepsilon^2} E\left(\langle m \rangle_{t+\tau_{j-1}} - \langle m \rangle_{\tau_{j-1}} | \mathscr{F}_{\tau_j}\right)$$

$$\le \frac{64 \beta t}{\varepsilon^2},$$

where we have used properties of the regular conditional probability distribution in the last two lines. Next, for $t > 0$, P-a.s.,

$$E\left(e^{-(\tau_j - \tau_{j-1})} | \mathscr{F}_{\tau_{j-1}}\right)$$

$$\le P(\tau_j - \tau_{j-1} < t | \mathscr{F}_{\tau_{j-1}}) + e^{-t} P(\tau_j - \tau_{j-1} \ge t | \mathscr{F}_{\tau_{j-1}})$$

$$\le e^{-t} + (1 - e^{-t}) P(\tau_j - \tau_{j-1} < t | \mathscr{F}_{\tau_{j-1}})$$

$$\le e^{-t} + (1 - e^{-t}) 64 \beta t / \varepsilon^2 = \lambda < 1$$

for t small enough. Hence,

$$E\left(e^{-\tau_j} | \mathscr{F}_{\tau_{j-1}}\right) = e^{-\tau_{j-1}} E\left(e^{-(\tau_j - \tau_{j-1})} | \mathscr{F}_{\tau_{j-1}}\right)$$

$$\le \lambda e^{-\tau_{j-1}} \le \cdots \le \lambda^j,$$

so that

$$P(N > k) = P(\tau_k < T) \le P\left(e^{-\tau_k} > e^{-T}\right) \le e^T \lambda^k < \eta/2$$

for k large enough, depending on T, λ, and η. Finally,

$$P(\tau_j - \tau_{j-1} < \delta \text{ for some } j \le k) \le \sum_{j=1}^{k} P(\tau_j - \tau_{j-1} < \delta \text{ for some } j \le k)$$

$$\le k\left(64 \beta \delta / \varepsilon^2\right) < \eta/2$$

for δ small enough, depending on k, β, ε, and η. Combining the last two inequalities proves the lemma. $\square$

Exercise 4.4 Prove (4.24).

Exercise 4.5 Let $(\Omega, \mathscr{F}, \{\mathscr{F}_t\}_{0 \le t \le T}, P)$ be a filtered probability space, with Ω a Polish space and $\mathscr{F}$ its Borel σ-field. Assume that M_t is an $\mathbb{R}^n$-valued continuous martingale and τ is a stopping time.

Show that $M_t - M_{t \wedge \tau}$ is an $\mathscr{F}_t$-martingale with respect to the conditional probability $P^\tau(\cdot, \omega)$, except possibly for ω in a set E of P-measure zero.

Hint: prove that

$$\int_{B \cap \{\tau \le s\}} \left(\int_A m_t(\omega'') \, dP^\tau(\omega'', \omega') \right) dP(\omega')$$

$$= \int_{B \cap \{\tau \le s\}} \left(\int_A m_s(\omega'') \, dP^\tau(\omega'', \omega') \right) dP(\omega')$$

for $0 \le s \le t \le T$ and $A \in \mathscr{F}_s$, $B \in \mathscr{F}_\tau$, and use the fact that $\mathscr{F}_s$ is countably generated. Conclude that for $s \ge \tau(\omega')$, outside possibly of a set $E_{s,t}$ of P-measure zero,

$$E^{P^\tau(\cdot, \omega)}(m_t \mid \mathscr{F}_s) = m_s \quad P^\tau(\omega)\text{-a.s.}$$

Choose a dense countable subset D of $[0, T]$ and show that the family $\{\|m_t\|_{\mathbb{R}^n}, t \in D\}$ is uniformly integrable.

Now we will use the compact embedding argument, previously discussed in Sect. 3.8.

Theorem 4.2 *Let the coefficients A and B of (4.2) satisfy conditions (4.3), (4.4), (4.5), and (4.6). Consider the family of measures μ_*^n on $C([0, T], V^*)$ with support in $C([0, T], H)$, defined by*

$$\mu_*^n(Y) = \mu^n \left\{ x \in C([0, T], \mathbb{R}^n) : \sum_{i=1}^n x_i(t)\varphi_i \in Y \right\}, \quad Y \subset C([0, T], V^*),$$

where μ^n are the measures constructed in Lemma 4.2. Assume that the embedding $H \hookrightarrow V^*$ is compact. Then the family of measures $\{\mu_*^n\}_{n=1}^\infty$ is tight on $C([0, T], V^*)$.

Proof We will use Theorem 3.17. Denote by $B_{C([0,T],H)}(R) \subset C([0, T], H)$ the closed ball of radius R centered at the origin. By the definition of measures μ_*^n and Lemma 4.2, for any $\eta > 0$, we can choose $R > 0$ such that

$$\mu_*^n \left\{ \left(B_{C([0,T],H)}(R) \right)^c \right\} = \mu^n \left\{ x \in C([0, T], \mathbb{R}^n) : \sup_{0 \le t \le T} \left\| \sum_{i=1}^n x_i(t)\varphi_i \right\|_H > R \right\}$$

$$= \mu^n \left\{ x \in C([0, T], \mathbb{R}^n) : \sup_{0 \le t \le T} \|x(t)\|_{\mathbb{R}^n} > R \right\} < \eta.$$

Denote the closed ball of radius R centered at zero in H by $B_H(R)$. Then its closure in V^*, denoted by $\overline{B_H(R)}^{V^*}$, is a compact subset of V^*, and we have

$$\mu_*^n \circ x(t)^{-1}\left(\overline{B_H(R)}^{V^*}\right) \geq 1 - \eta, \quad 0 \leq t \leq T,$$

fulfilling the first condition for tightness in Theorem 3.17.

Again, using the compactness of $\overline{B_H(R)}^{V^*}$ in V^*, for any $\varepsilon > 0$, we can find an index $n_0 \geq 1$ such that

$$\left\|\sum_{j=n_0+1}^{\infty} x_j\varphi_j\right\|_{V^*} < \varepsilon/4 \quad \text{if } \|x\|_H \leq R. \tag{4.27}$$

Since the embedding $H \hookrightarrow V^*$ is continuous and linear, we have $\|x\|_{V^*} \leq C\|x\|_H$ for $x \in H$ and some constant C, independent of η and R, so that

$$\left\|\sum_{j=1}^{n_0}(x_j - y_j)\varphi_j\right\|_{V^*} \leq C\left\|\sum_{j=1}^{n_0}(x_j - y_j)\varphi_j\right\|_H.$$

Recall the modulus of continuity (3.106) and indicate the space, e.g., V^*, in the subscript in $w_{V^*}(x, \delta)$ if the V^* norm is to be used. Then, with $B_{C([0,T],\mathbb{R}^n)}(R)$ denoting the closed ball with radius R centered at the origin in $C([0, T], \mathbb{R}^n)$ and $n > n_0$,

$$\mu_*^n\left\{x \in C([0, T], V^*) : x \in B_{C([0,T],H)}(R), w_{V^*}(x, \delta) > \varepsilon\right\}$$

$$\leq \mu^n\left\{x \in B_{C([0,T],\mathbb{R}^n)}(R) : w_{V^*}\left(\sum_{j=1}^{n}(x(\cdot))_j\varphi_j, \delta\right) > \varepsilon\right\}$$

$$\leq \mu^n\left\{x \in B_{C([0,T],\mathbb{R}^n)}(R) : \sup_{\substack{0 \leq s,t \leq T \\ |s-t|<\delta}}\left\|\sum_{j=1}^{n_0}\left((x(t))_j - (x(s))_j\right)\varphi_j\right\|_{V^*}\right.$$

$$\left. + \sup_{\substack{0 \leq s,t \leq T \\ |s-t|<\delta}}\left\|\sum_{j=n_0+1}^{n}\left((x(t))_j - (x(s))_j\right)\varphi_j\right\|_{V^*} > \varepsilon\right\}$$

$$\leq \mu^n\left\{x \in B_{C([0,T],\mathbb{R}^n)}(R) : C\sup_{\substack{0 \leq s,t \leq T \\ |s-t|<\delta}}\left\|\sum_{j=1}^{n_0}\left((x(t))_j - (x(s))_j\right)\varphi_j\right\|_H + \varepsilon/4 > \varepsilon\right\}$$

$$= \mu^n\left\{x \in B_{C([0,T],\mathbb{R}^n)}(R) : \sup_{\substack{0 \leq s,t \leq T \\ |s-t|<\delta}}\left(\sum_{j=1}^{n_0}\left((x(t))_j - (x(s))_j\right)^2\right)^{1/2} > 3\varepsilon/(4C)\right\}.$$

For the stopping time $\tau_R = \inf\{0 \leq t \leq T : x(t) \notin B_{C([0,T],\mathbb{R}^n)}(R)\}$ or T, the $\mathbb{R}^n$-valued martingale

$$m_t^R(x) = x(t \wedge \tau_R) - \int_0^{t \wedge \tau_R} a^n(s, x(s))\, ds$$

has the quadratic variation process given by

$$\langle\!\langle m^R(x) \rangle\!\rangle_t = \int_0^{t \wedge \tau_R} b^n(s, x(s))(b^n(s, x(s))^T\, ds$$

with the function $\mathrm{tr}(b^n(s, x(s))(b^n(s, x(s))^T)$ bounded on bounded subsets of $\mathbb{R}^n$ uniformly relative to the variable s, due to condition (4.10). Hence, for $t \geq s$,

$$\langle m^R(x) \rangle_t - \langle m^R(x) \rangle_s = \mathrm{tr}(\langle\!\langle m^R(x) \rangle\!\rangle_t - \langle\!\langle m^R(x) \rangle\!\rangle_s) \leq \beta(R)(t - s)$$

with the constant $\beta(R)$ not depending on n. Now, by Lemma 4.3, we have

$$\mu^n\big(x \in C([0, T], \mathbb{R}^n) : w_{\mathbb{R}^n}(m^R(x), \delta) > \varepsilon/(2C)\big) < \eta \qquad (4.28)$$

for sufficiently small δ independent of n.

Let $n \geq n_0$ and $\sup_{0 \leq t \leq T} \|x(t)\|_{\mathbb{R}^n} \leq R$. Using (4.14), we have

$$\left(\sum_{j=1}^{n_0} ((a^n(t, x(t)))_j)^2 \right)^{1/2} \leq \theta_{n_0}(1 + R^2);$$

hence, for a sufficiently small constant δ, we can write

$$\left(\sum_{j=1}^{n_0} \left(\int_s^t (a^n(t, x(t)))_j \right)^2 \right)^{1/2} \leq \varepsilon/(4C) \quad \text{whenever } |t - s| < \delta.$$

Also, whenever $\sup_{0 \leq t \leq T} \|x(t)\|_{\mathbb{R}^n} \leq R$,

$$m_t^n = x(t) - \int_0^t a^n(s, x(s))\, ds = m_t^R.$$

We can continue our calculations as follows:

$$\mu^n\left\{ x \in B_{C([0,T],\mathbb{R}^n)}(R) : \sup_{\substack{0 \leq s, t \leq T \\ |s-t| < \delta}} \left(\sum_{j=1}^{n_0} ((x(t))_j - (x(s))_j)^2 \right)^{1/2} > 3\varepsilon/(4C) \right\}$$

$$\leq \mu^n\left\{ x \in B_{C([0,T],\mathbb{R}^n)}(R) : w_{\mathbb{R}^n}(m^n, \delta) \right.$$

$$\left. + w_{\mathbb{R}^{n_0}}\left(\int_0^\cdot a^n(s, x(s))\, ds, \delta \right) > 3\varepsilon/(4C) \right\}$$

$$\leq \mu^n\left\{x \in B_{C([0,T],\mathbb{R}^n)}(R): w_{\mathbb{R}^n}(m^n,\delta) > \varepsilon/(2C)\right\}$$

$$\leq \mu^n\left\{w_{\mathbb{R}^n}(m^R,\delta) > \varepsilon/(2C)\right\} \leq \eta.$$

Summarizing, for any $\varepsilon, \eta > 0$ and sufficiently small $\delta > 0$, there exists n_0 such that for $n > n_0$,

$$\mu_*^n\left\{x \in C([0,T],V^*): w_{V^*}(x,\delta) > \varepsilon\right\}$$

$$\leq \mu_*^n\left\{\left(B_{C([0,T],H)}(R)\right)^c\right\} + \mu_*^n\left\{x \in B_{C([0,T],H)}(R): w_{V^*}(x,\delta) > \varepsilon\right\}$$

$$\leq 2\eta,$$

concluding the proof. □

We will now summarize the desired properties of the measures μ^n and μ_*^n.

Corollary 4.1 *Let $X^n(t)$ be solutions to (4.17), μ^n be their laws in $C([0,T],\mathbb{R}^n)$, and μ_*^n be the measures induced in $C([0,T],V^*)$ as in Theorem 4.2. Then for some constant C independent of n,*

$$\int_{C([0,T],\mathbb{R}^n)} \sup_{0 \leq t \leq T} \|x(t)\|_{\mathbb{R}^n}^2 \ln\ln\left(3 + \|x(t)\|_{\mathbb{R}^n}^2\right) \mu^n(dx) < C, \tag{4.29}$$

implying the uniform integrability of $\|X^n\|_{\mathbb{R}^n}^2$. The $\|\cdot\|_H$ norm of $x(t)$ satisfies the following properties:

$$\int_{C([0,T],V^*)} \sup_{0 \leq t \leq T} \|x(t)\|_H^2 \ln\ln\left(3 + \|x(t)\|_H^2\right) \mu_*^n(dx)$$

$$= E\left(\sup_{0 \leq t \leq T} \|J_n X^n(t)\|_H^2 \ln\ln\left(3 + \|J_n X^n(t)\|_H^2\right)\right) < C. \tag{4.30}$$

For a cluster point μ_* of the tight sequence μ_*^n,

$$\int_{C([0,T],V^*)} \sup_{0 \leq t \leq T} \|x(t)\|_H^2 \mu_*(dx) < C. \tag{4.31}$$

There exists a constant C such that for any $R > 0$,

$$\mu_*\left\{x \in C([0,T],V^*): \sup_{0 \leq t \leq T} \|x(t)\|_H > R\right\} < C/R^2, \tag{4.32}$$

and also,

$$\mu_*\left\{x \in C([0,T],V^*): \sup_{0 \leq t \leq T} \|x(t)\|_H < \infty\right\} = 1. \tag{4.33}$$

Finally, the $\|\cdot\|_V$ norm of $x(t)$ satisfies

$$\int_{C([0,T],V^*)} \int_0^T \|x(t)\|_V^2 \, dt \, \mu_*^n(dx) < C \tag{4.34}$$

and

$$\int_{C([0,T],V^*)} \int_0^T \|x(t)\|_V^2 \, dt \, \mu_*(dx) < \infty. \tag{4.35}$$

Proof Property (4.29) is just (4.19) and inequality (4.30) is just a restatement of (4.29)

To prove (4.31), assume, using the Skorokhod theorem, that $J_n X_n \to X$ a.s. in $C([0, T], V^*)$. We introduce the function $\alpha_H : V^* \to \mathbb{R}$ by

$$\alpha_H(u) = \sup\{\langle v, u \rangle, \ v \in V, \ \|v\|_H \leq 1\}.$$

Clearly $\alpha_H(u) = \|u\|_H$ if $u \in H$, and it is a lower semicontinuous function as a supremum of continuous functions. For $u \in V^* \setminus H$, $\alpha_H(u) = +\infty$ (Exercise 4.6). Thus, we can extend the norm $\| \cdot \|_H$ to a lower semicontinuous function on V^*.

By the Fatou lemma and (4.29),

$$\int_{C([0,T],V^*)} \sup_{0 \leq t \leq T} \|x(t)\|_H^2 \, \mu_*(dx)$$

$$= E\left(\sup_{0 \leq t \leq T} \|X(t)\|_H^2 \right)$$

$$\leq E \liminf_{n \to \infty} \left(\sup_{0 \leq t \leq T} \|J_n X^n(t)\|_H^2 \right)$$

$$\leq \liminf_{n \to \infty} E\left(\sup_{0 \leq t \leq T} \|J_n X^n(t)\|_H^2 \right)$$

$$= \liminf_{n \to \infty} \int_{C([0,T],V^*)} \sup_{0 \leq t \leq T} \|x(t)\|_H^2 \, \mu_*^n(dx) < C.$$

Property (4.32) follows from the Markov inequality, and (4.33) is a consequence of (4.31). To prove (4.34), we apply the Itô formula and (4.11) to obtain that

$$E\|J_n X^n(t)\|_H^2 = E\|J_n \xi_0^n\|_H^2 + 2E \int_0^t \langle a^n(s, X^n(s)), X^n(s) \rangle_{\mathbb{R}^n} \, ds$$

$$+ E \int_0^t \text{tr}\left(b^n(s, X^n(s))(b^n(s, X^n))^T \right) ds$$

$$\leq E\|J_n \xi_0\|_H^2 + \lambda \int_0^t E\|J_n X^n(s)\|_H^2 \, ds$$

$$- \alpha \int_0^t E\|J_n X^n(s)\|_V^2 \, ds + \gamma.$$

Using the bound in (4.30), we conclude that

$$\sup_n \int_0^T E\|J_n X^n(t)\|_V^2 \, dt < \infty.$$

Finally, we can extend the norm $\| \cdot \|_V$ to a lower semicontinuous function on V^* by introducing the lower semicontinuous function

$$\alpha_V(u) = \sup\{\langle v, u \rangle, \ v \in V, \ \|v\|_V \le 1\},$$

since $\alpha_V(u) = \|u\|_V$ if $u \in V$ and, for $u \in V^* \setminus V$, $\alpha_V(u) = +\infty$. Now (4.35) follows by the Fatou lemma. $\qquad\square$

Exercise 4.6 Justify the statements about α_H made in the proof of Corollary 4.1.

As we stated in the introduction, in order to identify the solution, weak or strong, as a continuous H-valued process, we will need the following deep result, which is a stochastic extension of a lemma of Lions. This result is included in [42], Theorem I.3.1, and a detailed proof is given in [64], Theorem 4.2.5.

Theorem 4.3 *Let* $X(0) \in L^2(\Omega, \mathscr{F}_0, P, H)$, *and* $Y \in L^2([0, T] \times \Omega, V^*)$ *and* $Z \in L^2([0, T] \times \Omega, \mathscr{L}_2(K_Q, H))$ *be both progressively measurable. Define the continuous V^*-valued process*

$$X(t) = X(0) + \int_0^t Y(s)\, ds + \int_0^t Z(s)\, dW_s, \quad t \in [0, T].$$

If for its $dt \otimes P$-equivalence class $\hat{X}$, we have $\hat{X} \in L^2([0, T] \times \Omega, V)$, then X is an H-valued continuous $\mathscr{F}_t$-adapted process,

$$E\left(\sup_{t \in [0,T]} \|X(t)\|_H^2 \right) < \infty \tag{4.36}$$

and the following Itô formula holds for the square of its H-norm P-a.s.:

$$\|X(t)\|_H^2 = \|X(0)\|_H^2 + \int_0^t \left(2\langle \bar{X}(s), Y(s) \rangle + \|Z(s)\|_{\mathscr{L}_2(K_Q,H)}^2 \right) ds$$

$$+ 2 \int_0^t \langle X(s), Z(s)\, dW_s \rangle_H, \quad t \in [0, T] \tag{4.37}$$

for any V-valued progressively measurable version $\bar{X}$ of $\hat{X}$.

Remark 4.1 Note that the process

$$\bar{X}(t) = 1_{\{\alpha_V(\hat{X}(t)) < \infty\}} \hat{X}(t)$$

serves as a V-valued progressively $dt \otimes dP$-measurable version of $\hat{X}$.

We are now ready to formulate the existence theorem.

Theorem 4.4 *Let $V \hookrightarrow H \hookrightarrow V^*$ be a Gelfand triplet of real separable Hilbert spaces with compact inclusions. Let the coefficients A and B of (4.2) satisfy conditions (4.3), (4.4), (4.5), and (4.6). Let the initial condition ξ_0 be an H-valued random variable satisfying (4.7). Then (4.2) has a weak solution $X(t)$ in $C([0, T], H)$ such that*

$$E\left(\sup_{0 \le t \le T} \|X(t)\|_H^2\right) < \infty \tag{4.38}$$

and

$$E \int_0^T \|X(t)\|_V^2 \, dt < \infty. \tag{4.39}$$

Proof Let $X^n(t)$ be solutions to (4.17), μ^n be their laws in $C([0, T], \mathbb{R}^n)$, and μ_*^n be the measures induced in $C([0, T], V^*)$ as in Theorem 4.2, with a cluster point μ_*. We need to show that μ_* is the law of a weak solution to (4.2). Again, using the Skorokhod theorem, assume that $J_n X^n(t)$ and $X(t)$ are processes with laws μ_*^n and μ_*, respectively, with $J_n X^n \to X$ P-a.s. By (4.30) and (4.33)–(4.35), $J_n X^n$ and X are P-a.s. in $C([0, T], V^*) \cap L^\infty([0, T], H) \cap L^2([0, T], V)$. Denote by $\{\varphi_j\}_{j=1}^\infty \subset V$ a complete orthogonal system in V, which is an ONB in H. Note that such a system always exists, see Exercise 4.7. Then, the vectors $\psi_j = \varphi_j / \|\varphi_j\|_V$ form an ONB in V. For $x \in C([0, T], V^*) \cap L^\infty([0, T], H) \cap L^2([0, T], V)$, consider

$$M_t(x) = x(t) - x(0) - \int_0^t A(s, x(s)) \, ds.$$

Using (4.4) and (4.30), we have, for any $v \in V$ and some constant C,

$$\int \left(\langle v, A(s, x(s)) \rangle^2 \ln \ln (3 + \|x(s)\|_H^2) \right) \mu_*^n(dx)$$

$$\le \int \left(\|A(s, x(s))\|_{V^*}^2 \|v\|_V^2 \ln \ln (3 + \|x(s)\|_H^2) \right) \mu_*^n(dx)$$

$$\le \int \theta \left(1 + \|x(s)\|_H^2 \right) \ln \ln (3 + \|x(s)\|_H^2) \|v\|_V^2 \, \mu_*^n(dx)$$

$$\le C\theta \|v\|_V^2 \tag{4.40}$$

and, in a similar fashion, adding (4.31) to the argument,

$$\int \langle v, A(s, x(s)) \rangle^2 \mu_*(dx) \le C\theta \|v\|_V^2. \tag{4.41}$$

Properties (4.38) and (4.39) are just restatements of (4.31) and (4.35). By involving (4.41) we conclude that the continuous process $\langle v, M_t(\cdot) \rangle$ is μ_*-square integrable. We will now show that for any $v \in V$, $s \le t$, and any bounded function g_s on $C([0, T], V^*)$ which is measurable with respect to the cylindrical σ-field generated by the cylinders with bases over $[0, s]$,

$$\int \left(\langle v, M_t(x) - M_s(x) \rangle g_s(x) \right) \mu_*(dx) = 0, \tag{4.42}$$

i.e., that $\langle v, M_t(\cdot) \rangle \in \mathcal{M}_T^2(\mathbb{R})$ (continuous square-integrable real-valued martingales). First, assume that g_s is continuous and extend the result to the general case by the monotone class theorem (functional form).

Let for $v \in V$, $v^m = \sum_{j=1}^{m} \langle v, \psi_j \rangle v \psi_j$. Then, as $m \to \infty$,

$$\int \left| g_s(x) \langle v - v^m, M_t(x) \rangle \right| \mu_*(dx) \to 0$$

by uniform integrability, since $|g_s(x)| \|M_t(x)\|_{V^*} \|v - v^m\|_V \to 0$. Hence,

$$\int \left| g_s(x) \left(\langle v, M_t(x) \rangle - \langle v, M_s(x) \rangle \right) \right.$$
$$\left. - g_s(x) \left(\langle v^m, M_t(x) \rangle - \langle v^m, M_s(x) \rangle \right) \right| \mu_*(dx) \to 0. \tag{4.43}$$

By the choice of the vectors φ_j and ψ_j, we have, for $x^n(t) = (x_1(t), \ldots, x_n(t)) \in \mathbb{R}^n$,

$$\langle v^m, J_n x^n(t) \rangle = \left\langle \sum_{j=1}^{n} x_j(t) \varphi_j \right\rangle = \sum_{j=1}^{n \wedge m} x_j(t) \langle v, \varphi_j \rangle_H.$$

For $n \geq m$, the process

$$\langle v^m, M_t(J_n x^n(\cdot)) \rangle = \langle v^m, J_n x^n(t) \rangle_H - \langle v^m, x(0) \rangle_H - \int_0^t \langle v^m, A(s, J_n x^n(s)) \rangle ds$$

$$= \sum_{j=1}^{m} \langle v^m, \varphi_j \rangle_H \left\{ (x^n(t))_j - (x(0))_j - \int_0^t (a^n(s, x^n(s)))_j ds \right\}$$

is a martingale relative to the measure μ^n. Hence, the above and the uniform integrability of $\langle M_t(J_n X^n), v \rangle$ (that follows from (4.30) and (4.40)) imply that

$$\int \left(g_s(x) \langle v^m, M_t(x) - M_s(x) \rangle \right) \mu_*(dx)$$

$$= E\left(g_s(X) \langle v^m, M_t(X) - M_s(X) \rangle \right)$$

$$= \lim_{n \to \infty} E\left(g_s(X^n) \langle v^m, M_t(J_n X^n) - M_s(J_n X^n) \rangle \right)$$

$$= \lim_{n \to \infty} \int \left(g_s(J_n x^n) \langle v^m, M_t(J_n x^n) - M_s(J_n x^n) \rangle \right) \mu^n(dx^n) = 0.$$

The above conclusion, together with (4.43), ensures (4.42). Next, we find the increasing process for the martingale $\langle v, M_t(x) \rangle$. We begin with some estimates. For

$x, v \in V$, we have

$$\langle v, (B(s, x(s))Q^{1/2})(B(s, x(s))Q^{1/2})^* v \rangle \leq \|v\|_H^2 \|B(s, x)\|_{\mathcal{L}_2(K_Q, H)}^2.$$

Hence,

$$\int \langle v, (B(s, x(s))Q^{1/2})(B(s, x(s))Q^{1/2})^* v \rangle \mu_*^n(dx)$$

$$\leq \|v\|_H^2 \int \theta(1 + \|x\|_H^2) \mu_*^n(dx)$$

$$\leq \theta(1 + C)\|v\|_H^2 \tag{4.44}$$

by (4.30), and by (4.31)

$$\int \langle v, (B(s, x(s))Q^{1/2})(B(s, x(s))Q^{1/2})^* v \rangle \mu_*(dx) \leq \theta(1 + C)\|v\|_H^2.$$

As a consequence, we obtain that

$$\left| \int\int_s^t \Big(\langle v^m, (B(u, x(u))Q^{1/2})(B(u, x(u))Q^{1/2})^* v^m \rangle \right.$$

$$\left. - \langle v, (B(u, x(u))Q^{1/2})(B(u, x(u))Q^{1/2})^* v \rangle \Big) g_s(x) \, du \, \mu_*(dx) \right|$$

$$\leq 2 \left| \sup_x (g_s(x)) \right| T\theta(1 + C) \|v^m - v\|_H \|v\|_H < \varepsilon/2 \tag{4.45}$$

for m sufficiently large. Next, observe that

$$\int \Big(\langle v^m, M_t(x) \rangle^2 - \langle v, M_t(x) \rangle^2 \Big) g_s(x) \mu_*(dx)$$

$$\leq \left| \sup_x (g_s(x)) \right| \left(\int \langle v^m - v, M_t(x) \rangle^2 \mu_*(dx) \right)^{1/2} \left(\int \langle v^m + v, M_t(x) \rangle^2 \mu_*(dx) \right)^{1/2}$$

$$< \varepsilon/2, \tag{4.46}$$

since by (4.31) and (4.41) the integrals above are bounded by $D\|v^m - v\|_V^2$ and $D\|v^m + v\|_V^2$, respectively, for some constant D.

By the uniform integrability of $\langle v, M_t(J_n X^n) \rangle^2$ (ensured by (4.30) and (4.40)), we have

$$\int \Big(\langle v^m, M_t(x) \rangle^2 - \langle v^m, M_s(x) \rangle^2 \Big) g_s(x) \mu_*(dx)$$

$$= E\Big(\big(\langle v^m, M_t(X) \rangle^2 - \langle v^m, M_s(X) \rangle^2 \big) g_s(X) \Big)$$

$$= \lim_{n \to \infty} E\Big(\big(\langle v^m, M_t(J_n X^n) \rangle^2 - \langle v^m, M_s(J_n X^n) \rangle^2 \big) g_s(J_n X^n) \Big)$$

$$= \lim_{n \to \infty} E \left(\left\{ \sum_{j=1}^{m} \left(X^n(t) - \xi_0^n - \int_0^t a^n\left(u, X^n(u)\right) du \right)_j \langle v, \varphi_j \rangle_H \right\}^2 g_s\left(J_n X^n\right) \right)$$

$$- \lim_{n \to \infty} E \left(\left\{ \sum_{j=1}^{m} \left(X^n(s) - \xi_0^n - \int_0^s a^n\left(u, X^n(u)\right) du \right)_j \langle v, \varphi_j \rangle_H \right\}^2 g_s\left(J_n X^n\right) \right)$$

$$= \lim_{n \to \infty} E \left(\int_s^t \left(\sum_{j=1}^{m} \left(b^n\left(u, X^n(u)\right)\left(b^n\left(u, X^n(u)\right)\right)^T\right)_{jj} \langle v, \varphi_j \rangle_H^2 g_s\left(J_n X^n\right) \right) du \right)$$

$$= \lim_{n \to \infty} E \left(\int_s^t \left(\sum_{j=1}^{m} \sum_{k=1}^{n} \langle\left(B\left(u, J_n X^n(u)\right) Q^{1/2}\right)^* \varphi_j, f_k \rangle_K^2 \langle v, \varphi_j \rangle_H^2 g_s\left(J_n X^n\right) \right) du \right).$$

Here, we have used the fact that the martingale

$$X^n(t) - \xi_0^n - \int_0^t a^n\left(s, X^n(s)\right) ds = \int_0^t b^n\left(s, X^n(s)\right) dB_s^n$$

has an increasing process given by $\int_0^t \text{tr}(b(s, X^n(s))(b(s, X^n(s))^T) \, ds$.

By using the positive and negative parts of $g_s(x)$ separately, we can assume, without any loss of generality, that $g_s(x) \geq 0$ in the following argument. Consider the last expectation above. It is dominated by

$$E \left(\int_s^t \left(\sum_{j=1}^{m} \sum_{k=1}^{\infty} \langle\left(B\left(u, J_n X^n(u)\right) Q^{1/2}\right)^* \varphi_j, f_k \rangle_K^2 \langle v, \varphi_j \rangle_H^2 g_s\left(J_n X^n\right) \right) du \right)$$

$$= E \left(\int_s^t \left(\sum_{j=1}^{m} \|\left(B\left(u, J_n X^n(u)\right) Q^{1/2}\right)^* \varphi_j \|_K^2 \langle v, \varphi_j \rangle_H^2 g_s\left(J_n X^n\right) \right) du \right)$$

$$= E \left(\int_s^t \left(\sum_{j=1}^{m} \langle\left(B\left(u, J_n X^n(u)\right) Q^{1/2}\right)\left(B\left(u, J_n X^n(u)\right) Q^{1/2}\right)^* \varphi_j, \varphi_j \rangle_H \right. \right.$$

$$\left. \left. \langle v, \varphi_j \rangle_H^2 g_s\left(J_n X^n\right) \right) du \right)$$

$$= E \left(\int_s^t \langle\left(\left(B\left(u, J_n X^n(u)\right) Q^{1/2}\right)\left(B\left(u, J_n X^n(u)\right) Q^{1/2}\right)^* v^m, v^m \rangle_H g_s\left(J_n X^n\right)\right) du \right)$$

$$\to E \left(\int_s^t \langle\left(B\left(u, X(u)\right) Q^{1/2}\right)\left(B\left(u, X(u)\right) Q^{1/2}\right)^* v^m, v^m \rangle_H g_s(X)\right) du \right)$$

$$= \int \int_s^t \langle\left(B\left(u, x(u)\right) Q^{1/2}\right)\left(B\left(u, x(u)\right) Q^{1/2}\right)^* v^m, v^m \rangle_H g_s(x) \, du \, \mu_*(dx),$$

using the weak convergence and uniform integrability of the integrand ensured by (4.5) and (4.31). Hence,

$$\lim_{n\to\infty} E\left(\int_0^t \left(\sum_{j=1}^m \sum_{k=1}^n \langle (B(u, J_n X^n(u))Q^{1/2})^* \varphi_j, f_k \rangle_K^2 \langle v, \varphi_j \rangle_H^2 g_s(J_n X^n)\right) du\right)$$

$$\leq \iint_s^t \langle \langle (B(u, x(u))Q^{1/2})(B(u, x(u))Q^{1/2})^* v^m, v^m \rangle_H g_s(x)) du\, \mu_*(dx).$$

To show the opposite inequality, note that if $n \geq r$, then

$$\liminf_{n\to\infty} \sum_{j=1}^m \sum_{k=1}^n \langle (B(u, J_n X^n(u))Q^{1/2})^* \varphi_j, f_k \rangle_K^2 \langle v, \varphi_j \rangle_H^2 g_s(J_n X^n)$$

$$\geq \sum_{j=1}^m \sum_{k=1}^r \liminf_{n\to\infty} \langle (B(u, J_n X^n(u))Q^{1/2})^* \varphi_j, f_k \rangle_K^2 \langle v, \varphi_j \rangle_H^2 g_s(J_n X^n)$$

$$= \sum_{j=1}^m \sum_{k=1}^r \langle (B(u, J_n X_n(u))Q^{1/2})^* \varphi_j, f_k \rangle_K^2 \langle v, \varphi_j \rangle_H^2 g_s(X)$$

$$\to \langle (B(u, X(u))Q^{1/2})(B(u, X(u))Q^{1/2})^* v^m, v^m \rangle_H g_s(X),$$

and an application of the Fatou lemma gives the equality

$$\int \left((\langle v^m, M_t(x) \rangle^2 - \langle v^m, M_s(x) \rangle^2) g_s(x) \right) \mu_*(dx)$$

$$= \iint_s^t \langle \langle (B(u, x(u))Q^{1/2})(B(u, x(u))Q^{1/2})^* v^m, v^m \rangle_H g_s(x)) du\, \mu_*(dx).$$

$$(4.47)$$

Summarizing, calculations in (4.45), (4.46), and (4.47) prove that for $v \in V$, the process $\langle v, M_t(x) \rangle$ is a square-integrable continuous martingale with the increasing process given by

$$\int_0^t \langle (B(u, x(u))Q^{1/2})(B(u, x(u))Q^{1/2})^* v, v \rangle du.$$

Let $\{\psi_j^*\}_{j=1}^\infty$ be the dual orthonormal basis in V^* defined by the duality

$$\langle u, \psi_j^* \rangle_{V^*} = \langle \psi_j, u \rangle, \quad u \in V^*.$$

Since by (4.4)

$$\|M_t(x)\|_{V^*}^2 \leq C\left(1 + \sup_{0\leq t\leq T} \|x(t)\|_H^2\right),$$

the martingale $M_t(x) \in \mathcal{M}_T^2(V^*)$, i.e., it is a continuous μ_*-square-integrable V^*-valued martingale.

Denote $M_t^j(x) = \langle M_t(x), \psi_j^* \rangle_{V^*}$. Using (2.5) and the property of the dual basis, its increasing process is given by

$$\langle\!\langle M(x) \rangle\!\rangle_t(u), v\rangle_{V^*} = \sum_{j,k=1}^{\infty} \langle M^j(x), M^k(x) \rangle_t \langle \psi_j^*, u \rangle_{V^*} \langle \psi_k^*, v \rangle_{V^*}$$

$$= \sum_{j,k=1}^{\infty} \langle M^j(x), M^k(x) \rangle_t \langle \psi_j, u \rangle \langle \psi_k, v \rangle \quad u, v \in V^*.$$

Since

$$M_t^j(x) M_t^k(x) = \langle \psi_j, M_t(x) \rangle \langle \psi_k, M_t(x) \rangle,$$

we can write

$$\langle M^j(x), M^k(x) \rangle_t = \langle\!\langle \psi_j, M(x) \rangle, \langle \psi_k, M(x) \rangle \rangle\!\rangle_t$$

$$= \int_0^t \langle (B(s, x(s)) Q^{1/2})(B(s, x(s)) Q^{1/2})^* \psi_j, \psi_k \rangle ds.$$

Define, for any $0 \le t \le T$, a map $\Phi(s) : K \to V^*$ by

$$\Phi(s)(k) = \sum_{j=1}^{\infty} \langle (B(s, X(s)) Q^{1/2})^* \psi_j, f_m \rangle_K \psi_j^*, \quad k \in K.$$

Then

$$\Phi^*(s)(u) = \sum_{j=1}^{\infty} \langle \psi_j, u \rangle (B(s, X(s)) Q^{1/2})^* \psi_j, \quad u \in V^*,$$

and we have, for $u, v \in V^*$,

$$\int_0^t \langle \Phi(s) \Phi^*(s) u, v \rangle_{V^*} ds$$

$$= \int_0^t \sum_{j,k=1}^{\infty} \langle (B(s, X(s)) Q^{1/2})^* \psi_j, (B(s, X(s)) Q^{1/2})^* \psi_k \rangle_K \langle \psi_j, u \rangle \langle \psi_k, v \rangle ds$$

$$= \int_0^t \sum_{j,k=1}^{\infty} \langle (B(s, X(s)) Q^{1/2})(B(s, X(s)) Q^{1/2})^* \psi_j, \psi_k \rangle \langle \psi_j, u \rangle \langle \psi_k, v \rangle ds$$

$$= \langle\!\langle M(X) \rangle\!\rangle_t(u), v \rangle_{V^*},$$

giving that

$$\langle\!\langle M(X) \rangle\!\rangle_t = \int_0^t \Phi(s) \Phi^*(s) \, ds.$$

Note that $\Phi(s) \in \mathscr{L}_2(K, V^*)$, since

$$\sum_{m=1}^{\infty} \|\Phi(s) f_m\|_{V^*}^2 = \sum_{m,j=1}^{\infty} \langle \psi_j^*, B(s, X(s)) Q^{1/2} f_m \rangle_{V^*}^2$$

$$= \sum_{m,j=1}^{\infty} \langle \psi_j, B(s, X(s)) Q^{1/2} f_m \rangle^2$$

$$\leq \sum_{m,j=1}^{\infty} \langle \varphi_j, B(s, X(s)) Q^{1/2} f_m \rangle_H^2$$

$$= \sum_{m=1}^{\infty} \|B(s, X(s)) Q^{1/2} f_m\|_H^2$$

$$= \|B(s, X(s))\|_{\mathscr{L}_2(K_Q, H)}^2 < \infty,$$

where we have used the assumption on the duality on Gelfand triplet and the fact that $\psi_j = \varphi_j / \|\varphi_j\|_V$ with the denominator greater than or equal to one. Consequently, the growth condition (4.5), together with (4.38), implies that

$$E \int_0^T \|\Phi(t)\|_{\mathscr{L}_2(K, V^*)}^2.$$

Using the cylindrical version of the martingale representation theorem, Corollary 2.2, we can write

$$M_t(X) = \int_0^t \Phi(s) \, d\widetilde{W}_s.$$

Define

$$W_t = \sum_{m=1}^{\infty} \widetilde{W}_t \left(Q^{1/2} f_m \right) f_m.$$

By Exercise 2.4, W_t is a K-valued Q-Wiener process. Using Lemma 2.8, we calculate

$$X(t) - \xi_0 - \int_0^t A(s, X(s)) \, ds = M_t(X)$$

$$= \int_0^t \Phi(s) \, d\widetilde{W}_s$$

$$= \sum_{m=1}^{\infty} \int_0^t \left(\Phi(s) f_m \right) d\widetilde{W}_s(f_m)$$

$$= \sum_{m=1}^{\infty} \int_0^t \sum_{j=1}^{\infty} \langle \psi_j, B(s, X(s)) Q^{1/2} f_m \rangle \psi_j^* \, d\widetilde{W}_s(f_m)$$

$$= \sum_{m=1}^{\infty} \int_0^t \sum_{j=1}^{\infty} \langle \psi_j^*, B(s, X(s)) f_m \rangle_{V^*} \psi_j^* \, d\widetilde{W}_s(Q^{1/2} f_m)$$

$$= \int_0^t \sum_{m=1}^{\infty} B(s, X(s)) f_m \, d\widetilde{W}_s(Q^{1/2} f_m)$$

$$= \int_0^t B(s, X(s)) \, dW_s.$$

We are now in a position to apply Theorem 4.3 to $X(t)$, $Y(t) = A(t, X(t))$, and $Z(t) = B(t, X(t))$ to obtain that $X \in C([0, T], H)$, completing the proof. $\qquad\square$

Exercise 4.7 Show that under the assumption of compact embedding in the Gelfand triplet, there exists a vector system $\{\varphi_j\}_{j=1}^{\infty} \subset V$ which is a complete orthogonal system in V and an ONB in H.

Hint: show that the canonical isomorphism $I : V^* \to V$ takes a unit ball in V^* to a subset of the unit ball in V, which is relatively compact in H. For the eigenvectors h_n of I, we have

$$\langle h_n, h_m \rangle_H = \langle h_n, h_m \rangle = \langle h_n, I h_m \rangle_V = \lambda_m \langle h_n, h_m \rangle_V.$$

We now address the problem of the existence and uniqueness of a strong solution using a version of the Yamada and Watanabe result in infinite dimensions. Recall the notion of pathwise uniqueness.

Definition 4.2 If for any two H-valued weak solutions (X_1, W) and (X_2, W) of (4.2) defined on the same filtered probability space $(\Omega, \mathscr{F}, \{\mathscr{F}_t\}_{0 \leq t \leq T}, P)$ with the same Q-Wiener process W and such that $X_1(0) = X_2(0)$ P-a.s., we have that

$$P\big(X_1(t) = X_2(t), \ 0 \leq t \leq T\big) = 1,$$

then we say that (4.2) has the pathwise uniqueness property.

We introduce here the *weak monotonicity* condition

(WM) There exists $\theta \in \mathbb{R}$ such that for all $u, v \in V$, $t \in [0, T]$,

$$2\langle u - v, A(t, u) - A(t, v) \rangle + \| B(t, u) - B(t, v) \|_{\mathscr{L}_2(K_Q, H)}^2 \leq \theta \| u - v \|_H^2.$$
$$\tag{4.48}$$

The weak monotonicity is crucial in proving the uniqueness of weak and strong solutions. In addition, it allows one to construct strong solutions in the absence of the compact embedding $V \hookrightarrow H$.

Theorem 4.5 *Let the conditions of Theorem 4.4 hold and assume the weak monotonicity condition (4.48). Then the solution to (4.2) is pathwise unique.*

Proof Let X_1, X_2 be two weak solutions as in Definition 4.2, $Y(t) = X_1(t) - X_2(t)$, and denote a V-valued progressively measurable version of the latter by $\bar{Y}$. Applying the Itô formula and the monotonicity condition (4.48) yields

$$
\begin{aligned}
e^{-\theta t} \|Y(t)\|_H^2 = & -\theta \int_0^t e^{-\theta s} \|Y(s)\|_H^2 \, ds \\
& + \int_0^t e^{-\theta s} \left(2 \langle \bar{Y}(s), A(s, X_1(s)) - A(s, X_2(s)) \rangle \right. \\
& \left. + \|B(s, X_1(s)) - B(s, X_2(s))\|_{\mathcal{L}_2(K_Q, H)}^2 \right) ds \\
& + 2 \int_0^t e^{-\theta s} \langle Y_s, (B(s, X_1(s)) - B(s, X_2(s))) \, dW_s \rangle_H \\
\leq & \; M_t,
\end{aligned}
$$

where M_t is a real-valued continuous local martingale represented by the stochastic integral above. The inequality above also shows that $M_t \geq 0$. Hence, by the Doob maximal inequality, $M_t = 0$. $\qquad\square$

As a consequence of an infinite-dimensional version of the result of Yamada and Watanabe [67], we have the following corollary.

Corollary 4.2 *Under the conditions of Theorem 4.5, (4.2) has a unique strong solution.*

4.3 Strong Variational Solutions

We will now study the existence and uniqueness problem for strong solutions. A monotonicity condition will be imposed on the coefficients of the SDE (4.2), and the compactness of embeddings $V \hookrightarrow H \hookrightarrow V^*$ will be dropped. We emphasize that the monotonicity condition will allow us to construct approximate strong solutions using projections of a single Q-Wiener process, as opposed to constructing finite-dimensional weak solutions in possibly different probability spaces. In the presence of monotonicity, we can weaken other assumptions on the coefficients of the variational SDE. A reader interested in exploring this topic in more depth is referred to a detailed presentation in [64], where the authors reduce the conditions even slightly further.

We assume that in the Gelfand triplet $V \hookrightarrow H \hookrightarrow V^*$, V is a real separable Hilbert space.

The joint continuity assumption is replaced by the spatial continuity:

(SC) For any $t \in [0, T]$, the mappings

$$V \ni v \to A(t, v) \in V^* \quad \text{and}$$
$$V \ni v \to \left(B(t, v)Q^{1/2}\right)\left(B(t, v)Q^{1/2}\right)^* \in \mathscr{L}_1(H) \tag{4.49}$$

are continuous.

We now assume that the coefficient A satisfies the following growth condition:

(G-A')

$$\left\|A(t, v)\right\|_{V^*}^2 \le \theta\left(1 + \|v\|_V^2\right), \quad v \in V. \tag{4.50}$$

The coercivity condition (4.6) remains in force, and, in addition, we assume the weak monotonicity condition (4.48).

Exercise 4.8 The coefficient B satisfies the following growth condition:

$$\left\|B(t, v)\right\|_{\mathscr{L}_2(K_Q, H)}^2 \le \lambda\|v\|_H^2 + \theta'\left(1 + \|v\|_V^2\right), \quad v \in V, \ \lambda \in \mathbb{R}, \theta' \ge 0.$$

We will rely on the following finite-dimensional result for an SDE (4.15) with the initial condition ξ_0. Its more refined version is stated as Theorem 3.1.1 in [64].

Theorem 4.6 *Assume that $a : [0, \infty] \times \mathbb{R}^n \to \mathbb{R}^n$, $b : [0, \infty] \times \mathbb{R}^n \to \mathbb{R}^n \otimes \mathbb{R}^n$, and $a(t, x), b(t, x)$ are continuous in $x \in \mathbb{R}^n$ for each fixed value of $t \ge 0$. Let*

$$\int_0^T \sup_{\|x\|_{\mathbb{R}^n} \le R} \left(\|a(t, x)\|_{\mathbb{R}^n} + \|b(t, x)\|^2\right) dt < \infty,$$

where $\|b(t, x)\|^2 = \text{tr}(b(t, x)b^T(t, x))$. Assume that for all $t \ge 0$ and $R > 0$, on the set $\{\|x\|_{\mathbb{R}^n} \le R, \|y\|_{\mathbb{R}^n} \le R\}$, we have

$$2\langle x, a(t, x)\rangle_{\mathbb{R}^n} + \|b(t, x)\|^2 \le \theta\left(1 + \|x\|_{\mathbb{R}^n}^2\right)$$

and

$$2\langle x - y, a(t, x) - a(t, y)\rangle_{\mathbb{R}^n} + \|b(t, x) - b(t, y)\|^2 \le \theta\|x - y\|_{\mathbb{R}^n}^2.$$

Let $E\|\xi_0\|_{\mathbb{R}^n} < \infty$.

Then there exists a unique strong solution $X(t)$ to (4.15) such that for some constant C,

$$E\left(\|X(t)\|_{\mathbb{R}^n}^2\right) \le C\left(1 + E\|\xi_0\|_{\mathbb{R}^n}^2\right).$$

Here is the variational existence and uniqueness theorem for strong solutions.

Theorem 4.7 *Let $V \hookrightarrow H \hookrightarrow V^*$ be a Gelfand triplet of real separable Hilbert spaces, and let the coefficients A and B of (4.2) satisfy conditions (4.49), (4.50), (4.6), and (4.48). Let the initial condition ξ_0 be an H-valued $\mathscr{F}_0$-measurable random variable satisfying $E \|\xi_0\|_H^2 < c_0$ for some constant c_0. Then (4.2) has a unique strong solution $X(t)$ in $C([0, T], H)$ such that*

$$E\left(\sup_{0 \le t \le T} \|X(t)\|_H^2 \right) < \infty \tag{4.51}$$

and

$$E \int_0^T \|X(t)\|_V^2 \, dt < \infty. \tag{4.52}$$

Proof Let $\{\varphi_i\}_{i=1}^\infty$ be an orthonormal basis in H obtained by the Gramm–Schmidt orthonormalization process from a dense linearly independent subset of V. Define $P_n : V^* \to H_n \subset V$ by

$$P_n u = \sum_{i=1}^n \langle \varphi_i, u \rangle \varphi_i, \quad u \in V^*.$$

By the assumption of the Gelfand triplet, P_n is the orthogonal projection of H onto H_n. Also, let

$$W_t^n = \sum_{i=1}^n \lambda_i^{1/2} w_i(t) f_i$$

as in (2.3). Consider the following SDE on H_n:

$$dX^n(t) = P_n A\big(t, X^n(t)\big) \, dt + P_n B\big(t, X^n(t)\big) \, dW_t^n \tag{4.53}$$

with the initial condition $X^n(0) = P_n \xi_0$ and identify it with the SDE (4.15) in $\mathbb{R}^n$. It is a simple exercise to show that the conditions of Theorem 4.6 hold (Exercise 4.9); hence, we have a unique strong finite-dimensional solution $X^n(t) \in H_n$. We will now show its boundedness in the proper L^2 spaces. Identifying $X^n(t)$ with an $\mathbb{R}^n$-valued process and applying the finite-dimensional Itô formula yield

$$\|X^n(t)\|_H^2 = \|X^n(0)\|_H^2 + \int_0^t \Big(2\langle X^n(s), A\big(s, X^n(s)\big) \rangle$$

$$+ \big\| P_n B\big(s, X^n(s)\big) \tilde{P}_n \big\|_{\mathscr{L}_2(K_Q, H)}^2 \Big) \, ds + M^n(t),$$

where

$$M^n(t) = \int_0^t 2\langle X^n(s), P_n B\big(s, X^n(s)\big) \, dW_s^n \rangle_H$$

is a local martingale, and $\tilde{P}_n$ denotes the orthogonal projection on span$\{f_1, \ldots, f_k\}$ $\subset K$.

Let σ_l, $l = 1, 2, \ldots$, be stopping times localizing M^n, and $\eta_l = \inf\{t : \|X^n(t)\|_H \geq l\}$. Then $\tau_l = \sigma_l \wedge \eta_l$ localizes M^n, $\tau_l \to \infty$, and we can apply expectations as follows:

$$E\left(\left\|X^n(t \wedge \tau_l)\right\|_H^2\right) = E\left(\left\|X^n(0)\right\|_H^2\right) + \int_0^t E\left(1_{[0,\tau_l]}(s)2\langle X^n(s), P_n A\left(s, X^n(s)\right)\rangle\right.$$

$$\left. + \left\|P_n B\left(s, X^n(s)\right)\tilde{P}_n\right\|_{\mathscr{L}_2(K_Q, H)}^2\right) ds.$$

Integration by parts and coercivity (4.6) yield

$$E\left(e^{-\lambda t}\left\|X^n(t \wedge \tau_l)\right\|_H^2\right) - E\left(\left\|X^n(0)\right\|_H^2\right)$$

$$= \int_0^t E\left(-\lambda e^{-\lambda s}\left\|X^n(s \wedge \tau_l)\right\|_H^2\right) ds$$

$$+ \int_0^t e^{-\lambda s} E\left(1_{[0,\tau_l]}(s)2\langle X^n(s), P_n A\left(s, X^n(s)\right)\rangle\right.$$

$$\left. + \left\|P_n B\left(s, X^n(s)\right)\tilde{P}_n\right\|_{\mathscr{L}_2(K_Q, H)}^2\right) ds$$

$$\leq \int_0^t E\left(-\lambda e^{-\lambda s}\left\|X^n(s \wedge \tau_l)\right\|_H^2\right) ds$$

$$+ \int_0^t E\left(\lambda e^{-\lambda s}\left\|X^n(s)\right\|_H^2\right) ds$$

$$- \int_0^t E\left(1_{[0,\tau_l]}(s)\alpha e^{-\lambda s}\left\|X^n(s)\right\|_V^2\right) ds + \gamma T. \tag{4.54}$$

Rearranging and applying the Fatou lemma as $l \to \infty$, we obtain

$$E\left(e^{-\lambda t}\left\|X^n(t)\right\|_H^2\right) - E\left(\left\|X^n(0)\right\|_H^2\right) + \int_0^t E\left(\alpha e^{-\lambda s}\left\|X^n(s)\right\|_V^2\right) ds \leq C.$$

Hence, X^n is bounded in $L^2([0, T] \times \Omega, V)$ and in $L^2([0, T] \times \Omega, H)$. Using (4.6), we also have the boundedness of $A(\cdot, X^n(\cdot))$ in $L^2([0, T] \times \Omega, V^*) = (L^2([0, T] \times \Omega, V))^*$. In addition, by Exercise 4.8, $P_n B(\cdot, X^n(\cdot))$ is bounded in $L^2([0, T] \times \Omega, \mathscr{L}_2(K_Q, H))$. Therefore, by using the Alaoglu theorem and passing to a subsequence if necessary, we can assume that there exist X, Y, and Z such that

$$X^n \to X \text{ weakly in } L^2\left([0, T] \times \Omega, V\right), \text{ and weakly in } L^2\left([0, T] \times \Omega, H\right),$$

$$P_n A\left(\cdot, X^n(\cdot)\right) \to Y \text{ weakly in } L^2\left([0, T] \times \Omega, V^*\right), \tag{4.55}$$

$$P_n B\left(\cdot, X^n(\cdot)\right) \to Z \text{ weakly in } L^2\left([0, T] \times \Omega, \mathscr{L}_2(K_Q, H)\right).$$

Also $P_n B(\cdot, X^n(\cdot))\tilde{P}_n \to Z$ weakly in $L^2([0, T] \times \Omega, \mathscr{L}_2(K_Q, H))$ and

$$\int_0^t P_n B\left(s, X^n(s)\right)\tilde{P}_n \, dW_s = \int_0^t P_n B\left(s, X^n(s)\right) dW_s^n,$$

we can claim that

$$\int_0^t P_n B\big(s, X^n(s)\big) dW_s^n \rightarrow \int_0^t Z(s) dW_s$$

weakly in $L^2([0, T] \times \Omega, H)$, the reason being that the stochastic integral is a continuous transformation from $\Lambda_2(K_Q, H)$ to $L^2([0, T] \times \Omega, H)$, and so it is also continuous with respect to weak topologies in those spaces (see Exercise 4.10).

For any $v \in \bigcup_{n \geq 1} H_n$ and $g \in L^2([0, T] \times \Omega, \mathbb{R})$, using the assumption on the duality, we obtain

$$E\left\{\int_0^T \langle g(t)v, X(t)\rangle dt\right\}$$

$$= \lim_{n \to \infty} E\left(\int_0^T \langle g(t)v, X^n(t)\rangle dt\right)$$

$$= \lim_{n \to \infty} E\left\{\int_0^T \left(\langle g(t)v, X^n(0)\rangle + \int_0^t \langle g(t)v, P_n A\big(s, X^n(s)\big)\rangle ds\right.\right.$$

$$\left.\left. + \left\langle \int_0^t P_n B\big(s, X^n(s)\big) dW_s^n, g(t)v\right\rangle_H\right)\right\} dt$$

$$= \lim_{n \to \infty} E\left\{\langle v, X^n(0)\rangle_H \int_0^T g(t) dt + \int_0^T \left\langle \int_s^T g(t)v\, dt, P_n A\big(s, X^n(s)\big)\right\rangle ds\right.$$

$$\left. + \int_0^T \left\langle \int_0^t P_n B\big(s, X^n(s)\big) dW_s^n, g(t)v\right\rangle_H\right\} dt$$

$$= E\left\{\int_0^T \left\langle g(t)v, X(0) + \int_0^t Y(s) ds + \int_0^t Z(s) dW_s\right\rangle dt\right\}.$$

Therefore,

$$X(t) = X(0) + \int_0^t Y(s) ds + \int_0^t Z(s) dW_s, \quad dt \otimes dP\text{-a.e.},$$

and applying Theorem 4.3, we conclude that X is a continuous H-valued process with

$$E\left(\sup_{0 \leq t \leq T} \|X(t)\|_H^2\right) < \infty.$$

We now verify that $Y(t) = A(t, X(t))$ and $Z(t) = B(t, X(t))$, $dt \otimes dP$-a.e. For a nonnegative function $\psi \in L^\infty([0, T], \mathbb{R})$, we have

$$E \int_0^T \langle \psi(t) X(t), X^n(t)\rangle_H dt$$

$$\leq E \int_0^T \left(\sqrt{\psi(t)} \|X(t)\|_H\right)\left(\sqrt{\psi(t)} \|X^n(t)\|_H\right) dt$$

$$\leq \left(E \int_0^T \psi(t) \|X(t)\|_H^2 \, dt \right)^{1/2} \left(E \int_0^T \psi(t) \|X^n(t)\|_H^2 \, dt \right)^{1/2}.$$

Hence, by the weak convergence of X^n to X,

$$E \int_0^T \psi(t) \|X(t)\|_H^2 \, dt = \lim_{n\to\infty} E \int_0^T \langle \psi(t) X(t), X^n(t) \rangle_H \, dt$$

$$\leq \left(E \int_0^T \psi(t) \|X(t)\|_H^2 \, dt \right)^{1/2} \liminf_{n\to\infty} \left(E \int_0^T \psi(t) \|X^n(t)\|_H^2 \, dt \right)^{1/2} < \infty,$$

giving

$$E \int_0^T \psi(t) \|X(t)\|_H^2 \, dt \leq \liminf_{n\to\infty} E \int_0^T \psi(t) \|X^n(t)\|_H^2 \, dt. \tag{4.56}$$

Let $\phi \in L^2([0,T] \times \Omega, V)$. Revisiting the calculations in (4.54), with the constant c in the weak monotonicity condition (4.48), replacing λ, and taking the limit as $l \to \infty$ yield

$$E\left(e^{-ct} \|X^n(t)\|_H^2 \right) - E\left(\|X^n(0)\|_H^2 \right)$$

$$= \int_0^t e^{-cs} E\left(-c \|X^n(s)\|_H^2 + \|P_n B(s, X^n(s)) \tilde{P}_n\|_{\mathscr{L}_2(K_Q, H)}^2 \right) ds$$

$$+ \int_0^t e^{-cs} E\left(2\langle X^n(s), P_n A(s, X^n(s)) \rangle \right) ds$$

$$\leq \int_0^t e^{-cs} E\left(-c \|X^n(s)\|_H^2 + \|B(s, X^n(s))\|_{\mathscr{L}_2(K_Q, H)}^2 \right) ds$$

$$+ \int_0^t e^{-cs} E\left(2\langle X^n(s), P_n A(s, X^n(s)) \rangle \right) ds$$

$$= \int_0^t e^{-cs} E\left(-c \|X^n(s) - \phi(s)\|_H^2 + \|B(s, X^n(s)) - B(s, \phi(s))\|_{\mathscr{L}_2(K_Q, H)}^2 \right.$$

$$+ 2\langle X^n(s) - \phi(s), P_n A(s, X^n(s)) - A(s, \phi(s)) \rangle \right) ds$$

$$+ \int_0^t e^{-cs} E\left(c \|\phi(s)\|_H^2 - 2c\langle X^n(s), \phi(s) \rangle_H \right.$$

$$+ 2\langle B(s, X^n(s)), B(s, \phi(s)) \rangle_{\mathscr{L}_2(K_Q, H)} - \|B(s, \phi(s))\|_{\mathscr{L}_2(K_Q, H)}^2$$

$$+ 2\langle X^n(s), A(s, \phi(s)) \rangle + 2\langle \phi(s), P_n A(s, X^n(s)) - A(s, \phi(s)) \rangle \right) ds.$$

Since by (4.48) the first of the last two integrals is negative, by letting $n \to \infty$, using the weak convergence (4.55) in L^2, and applying (4.56), we conclude that for any function ψ as above,

$$\int_0^T \left(\psi(t) E \left(e^{-ct} \|X(t)\|_H^2 - \|X(0)\|_H^2 \right) \right) dt$$

$$\leq \int_0^T \left(\psi(t) \int_0^t \left(e^{-cs} E \left(c \|\phi(s)\|_H^2 - 2c\langle X(s), \phi(s)\rangle_H \right.\right.\right.$$

$$+ 2\langle Z(s), B(s, \phi(s))\rangle_{\mathscr{L}_2(K_Q, H)} - \|B(s, \phi(s))\|_{\mathscr{L}_2(K_Q, H)}^2$$

$$\left.\left.\left. + 2\langle \bar{X}(s), A(s, \phi(s))\rangle + 2\langle \phi(s), Y(s) - A(s, \phi(s))\rangle \right) \right) ds \right) dt. \quad (4.57)$$

Recall the Itô formula (4.37). With stopping times τ_l localizing the local martingale represented by the stochastic integral, we have

$$E\left(\|X(t \wedge \tau_l)\|_H^2 \right) - E\left(\|X(0)\|_H^2 \right)$$

$$= \int_0^t E\left(1_{[0, \tau_l]}(s)\left(2\langle \bar{X}(s), Y(s)\rangle + \|Z(s)\|_{\mathscr{L}_2(K_Q, H)}^2 \right) \right) ds.$$

Since, by (4.36) and by the square integrability of Y and Z, we can pass to the limit using the Lebesgue DCT, the above equality yields

$$E\left(\|X(t)\|_H^2 \right) - E\left(\|X(0)\|_H^2 \right)$$

$$= \int_0^t E\left(2\langle \bar{X}(s), Y(s)\rangle + \|Z(s)\|_{\mathscr{L}_2(K_Q, H)}^2 \right) ds. \quad (4.58)$$

Applying integration by parts, we get

$$E\left(e^{-ct} \|X(t)\|_H^2 \right) - E\left(\|X(0)\|_H^2 \right)$$

$$= \int_0^t e^{-cs} E\left(2\langle \bar{X}(s), Y(s)\rangle + \|Z(s)\|_{\mathscr{L}_2(K_Q, H)}^2 - c e^{-cs} \|X(s)\|_H^2 \right) ds. \quad (4.59)$$

We now substitute the expression for the left-hand side of (4.59) into the left-hand side of (4.57) and arrive at

$$\int_0^T E\left(\psi(t) \int_0^t e^{-cs} \left(2\langle \bar{X}(s) - \phi(s), Y(s) - A(s, \phi(s))\rangle \right.\right. \quad (4.60)$$

$$\left.\left. + \|B(s, \phi(s)) - Z(s)\|_{\mathscr{L}_2(K_Q, H)}^2 - c \|X(s) - \phi(s)\|_H^2 \right) ds \right) dt \leq 0. \quad (4.61)$$

Substituting $\phi = \bar{X}$ gives that $Z = B(\cdot, \bar{X})$. Now let $\phi = \bar{X} - \varepsilon \tilde{\phi} v$ with $\varepsilon > 0$, $\tilde{\phi} \in L^\infty([0, T] \times \Omega, \mathbb{R})$, and $v \in V$. Let us divide (4.60) by ε and pass to the limit as

$\varepsilon \to 0$ using the Lebesgue DCT. Utilizing (4.49) and (4.48), we obtain that

$$\int_0^T E\left(\psi(t) \int_0^t e^{-cs} \tilde{\phi}(s)\langle v, Y(s) - A(s, \bar{X}(s))\rangle ds\right) dt \leq 0.$$

This proves that $Y = A(\cdot, \bar{X})$ due to the choice of ψ and $\tilde{\phi}$.

The argument used in the proof of Theorem 4.5 can now be applied to show the uniqueness of the solution. □

Exercise 4.9 Show that the coefficients of (4.53) identified with the coefficients of (4.15) satisfy the conditions of Theorem 4.6.

Exercise 4.10 Show that if $T : X \to Y$ is a continuous linear operator between Banach spaces X and Y, then T is also continuous with respect to weak topologies on X and Y.

Exercise 4.11 Let X and Y be two solutions to (4.2). Using (4.58), show that under the conditions of Theorem 4.5 or Theorem 4.7,

$$E\|X(t) - Y(t)\|_H^2 \leq e^{ct} E\|X(0) - Y(0)\|_H^2, \quad 0 \leq t \leq T.$$

Note that this implies the uniqueness of the solution, providing an alternative argument to the one used in text.

4.4 Markov and Strong Markov Properties

Similarly as in Sect. 3.4, we now consider strong solutions to (4.2) on the interval $[s, T]$. The process $\bar{W}_t = W_{t+s} - W_s$ is a Q-Wiener process with respect to $\bar{\mathscr{F}}_t = \mathscr{F}_{t+s}$, $t \geq 0$, and its increments on $[0, T - s]$ are identical with the increments of W_t on $[s, T]$. Consider (4.2) with $\bar{W}_t$ replacing W_t and $\bar{\mathscr{F}}_0 = \mathscr{F}_s$ replacing $\mathscr{F}_0$. By Theorem 4.7, there exists a unique strong solution $X(t)$ of (4.2), so that for any $0 \leq s \leq T$ and an $\mathscr{F}_s$-measurable random variable ξ, there exists a unique process $X(\cdot, s, \xi)$ such that

$$X(t, s, \xi) = \xi + \int_s^t A\big(r, X(r, s, \xi)\big) dr + \int_s^t B\big(r, X(r, s, \xi)\big) dW_r. \tag{4.62}$$

As before, for a real bounded measurable function φ on H and $x \in H$,

$$(P_{s,t}\varphi)(x) = E\big(\varphi\big(X(t, s; x)\big)\big), \tag{4.63}$$

and this definition can be extended to functions φ such that $\varphi(X(t, s; x)) \in L^1(\Omega, \mathbb{R})$ for arbitrary $s \leq t$. As usual, for a random variable η,

$$(P_{s,t}\varphi)(\eta) = E\big(\varphi\big(X(t, s; x)\big)\big)\big|_{x=\eta}.$$

The Markov property (3.52) (and consequently (3.57)) of the solution now follows almost word by word by the arguments used in the proof of Theorem 3.6 and by Exercise 4.11. A proof given in [64] employs similar ideas.

Theorem 4.8 *The unique strong solution to* (4.2) *obtained in Theorem 4.7 is a Markov process.*

Remark 4.2 In the case where the coefficients A and B are independent of t and with $x \in H$,

$$X(t + s, t; x) = x + \int_t^{t+s} A\big(X(u, t; x)\big) du + \int_t^{t+s} B\big(X(u, t, x)\big) dW_u$$

$$= x + \int_0^s A\big(X(t + u, t; x)\big) du + \int_0^s B\big(X(t + u, t, x)\big) d\bar{W}_u,$$

where $\bar{W}_u = W_{t+u} - W_t$. Repeating the arguments in Sect. 3.4, we argue that

$$\{X(t + s, t; x), s \geq 0\} \stackrel{d}{=} \{X(s, 0; x), s \geq 0\},$$

i.e., the solution is a homogeneous Markov process with

$$P_{s,t}(\varphi) = P_{0,t-s}(\varphi), \quad 0 \leq s \leq t,$$

for all bounded measurable functions φ on H.

As before, we denote

$$P_t = P_{0,t}.$$

Due to the continuity of the solution with respect to the initial condition, P_t is a Feller semigroup, and $X(t)$ is a Feller process.

In Sect. 7.5 we will need the strong Markov property for strong variational solutions. We prove this in the next theorem. Consider the following variational SDE:

$$dX(t) = A\big(X(t)\big) dt + B\big(X(t)\big) dW_t \tag{4.64}$$

with the coefficients

$$A : V \to V^* \quad \text{and} \quad B : V \to \mathscr{L}_2(K_Q, H)$$

and an H-valued $\mathscr{F}_0$-measurable initial condition $\xi \in L^2(\Omega, H)$.

Definition 4.3 Let τ be a stopping time with respect to a filtration $\{\mathscr{F}_t\}_{t \geq 0}$ (an $\mathscr{F}_t$-stopping time for short). We define

$$\mathscr{F}_\tau = \sigma\{A \in \mathscr{F} : A \cap \{\tau \leq t\} \in \mathscr{F}_t, \, t \geq 0\}.$$

Exercise 4.12 Show that $\mathscr{F}_\tau^W = \sigma\{W_{s \wedge \tau}, \ s \geq 0\}$ and that $\mathscr{F}_\tau^X = \sigma\{X_{s \wedge \tau}, \ s \geq 0\}$ for a strong solution to (4.2).

Exercise 4.13 Let $X(t)$ be a progressively measurable process with respect to filtration $\{\mathscr{F}_t\}_{t \geq 0}$, and τ be an $\mathscr{F}_t$-stopping time. Show that $X_\tau 1_{\{\tau < \infty\}}$ is $\mathscr{F}_\tau$-measurable.

Definition 4.4 A solution $X(t)$ of (4.64) in $C([0, T], H)$ is called a *strong Markov process* if it satisfies the following *strong Markov property*:

$$E\big(\varphi\big(X(\tau + s; \xi)\big)\big|\mathscr{F}_\tau^{W, \xi}\big) = (P_s\varphi)\big(X(\tau; \xi)\big) \quad P\text{-a.s. on } \{\tau < \infty\}, \qquad (4.65)$$

for any real-valued function φ such that $\varphi(X(t; \xi)) \in L^1(\Omega, \mathbb{R})$ and an $\mathscr{F}_t^{W, \xi}$-stopping time τ.

Theorem 4.9 *Under the assumptions of Theorem 4.7, the unique strong solution $X(t)$ of (4.64) in $C([0, T], H)$ is a strong Markov process.*

Proof By the monotone class theorem (functional form) we only need to show that for any bounded continuous function $\varphi : H \to \mathbb{R}$ and $A \in \mathscr{F}_\tau^{W, \xi}$,

$$E\big(\varphi\big(X(\tau + s; \xi)\big)1_{A \cap \{\tau < \infty\}}\big) = E\big((P_s\varphi)\big(X(\tau; \xi)\big)1_{A \cap \{\tau < \infty\}}\big). \qquad (4.66)$$

If τ takes finitely many values, then $A \in \mathscr{F}_{\max\{\tau(\omega)\}}$, and (4.66) is a consequence of Theorem 4.8.

Let τ_n be a sequence of $\mathscr{F}_t^{W, \xi}$-stopping times, each taking finitely many values, and $\tau_n \downarrow \tau$ on $\tau < \infty$ (see Exercise 4.14). Since $\tau_n \geq \tau$, we have $\mathscr{F}_{\tau_n}^{W, \xi} \supset \mathscr{F}_\tau^{W, \xi}$ and $A \subset \mathscr{F}_{\tau_n}^{W, \xi}$ for all n. Consequently, (4.66) holds for τ_n,

$$E\big(\varphi\big(X(\tau_n + s; \xi)\big)1_{A \cap \{\tau < \infty\}}\big) = E\big((P_s\varphi)\big(X(\tau_n; \xi)\big)1_{A \cap \{\tau < \infty\}}\big).$$

By the continuity of φ and $X(t)$,

$$E\big(\varphi\big(X(\tau_n + s; \xi)\big)1_{A \cap \{\tau < \infty\}}\big) \to E\big(\varphi\big(X(\tau + s; \xi)\big)1_{A \cap \{\tau < \infty\}}\big).$$

By the Feller property of the semigroup P_s,

$$E\big((P_s\varphi)\big(X(\tau_n; \xi)\big)1_{A \cap \{\tau < \infty\}}\big) \to E\big((P_s\varphi)\big(X(\tau; \xi)\big)1_{A \cap \{\tau < \infty\}}\big).$$

This completes the proof. $\qquad \qquad \square$

Exercise 4.14 For an $\mathscr{F}_t$-stopping time τ, construct a sequence of $\mathscr{F}_t$-stopping times such that $\tau_n \downarrow \tau$ on $\{\tau < \infty\}$.

Corollary 4.3 *Under the assumptions of Theorem 4.9, the unique strong solution $X(t)$ of (4.64) in $C([0, T], H)$ has the following strong Markov property:*

$$E\big(\varphi\big(X(\tau + s; \xi)\big)\big|\mathscr{F}_\tau^X\big) = (P_s\varphi)\big(X(\tau; \xi)\big) \quad P\text{-a.s. on } \{\tau < \infty\} \qquad (4.67)$$

for any real-valued function φ such that $\varphi(X(t;\xi)) \in L^1(\Omega, \mathbb{R})$ with $\mathscr{F}_s^X = \sigma\{X(r;\xi), r \leq s\}$ and an $\mathscr{F}_t^X$-stopping time τ.

Exercise 4.15 Show that if $X(t) = X(t;\xi)$ is as a solution to (4.64) as in Theorem 4.9, then the strong Markov property (4.67) implies that

$$E\big(\varphi\big(X(\tau+s)\big)\big|\mathscr{F}_\tau^X\big) = E\big(\varphi\big(X(\tau+s)\big)\big|X_\tau\big)$$

for any a real-valued function φ such that $\varphi(X(t;\xi)) \in L^1(\Omega, \mathbb{R})$ and any $\mathscr{F}_t^X$-stopping time τ.

We now refer the reader to Sect. 4.1 in [64], where several examples and further references are provided.

Chapter 5
Stochastic Differential Equations with Discontinuous Drift

5.1 Introduction

In this chapter, we consider genuine infinite-dimensional stochastic differential equations not connected to SPDEs.

This problem has been discussed in Albeverio's work on solutions to infinite-dimensional stochastic differential equations with values in $C([0, T], \mathbb{R}^{\mathbb{Z}^d})$, which was motivated by applications to quantum lattice models in statistical mechanics [2, 3]. Leha and Ritter [47] have studied the existence problem in $C([0, T], H^w)$, where H is a real separable Hilbert space endowed with its weak topology and applied their results to modeling unbounded spin systems. Our purpose here is to extend the results obtained in [2, 3], and [47] using techniques developed in Chaps. 3 and 4.

We begin with the problem of the existence of weak solutions for SDEs with discontinuous drift in a Hilbert space H discussed in [47]. The discontinuity is modeled by a countable family of real-valued functions. The solution has finite-dimensional Galerkin approximation and is realized in $C([0, T], H^w)$. The result in [47] is generalized, and we also show that, under the assumptions in [47], both the Galerkin approximation and the infinite-dimensional approximation of [47] produce solutions with identical laws.

Next, we study the solutions in $C([0, T], \mathbb{R}^{\mathbb{Z}^d})$ using ideas in [3] and the technique of compact embedding in [22].

5.2 Unbounded Spin Systems, Solutions in $C([0, T], H^w)$

In Chap. 3, it was shown that with the usual continuity and growth assumptions on the coefficients of a stochastic differential equation in an infinite-dimensional real separable Hilbert space H, the solution can be obtained in a larger Hilbert space H_{-1} such that the embedding $H \hookrightarrow H_{-1}$ is compact. The space H_{-1} was arbitrary, however needed due to a (deterministic) example of Godunov [26]. This forced us to extend continuously the coefficients of the SDE to H_{-1}.

L. Gawarecki, V. Mandrekar, *Stochastic Differential Equations in Infinite Dimensions*,   185
Probability and Its Applications,
DOI 10.1007/978-3-642-16194-0_5, © Springer-Verlag Berlin Heidelberg 2011

We exploit this idea to obtain solutions in $C([0, T], H^w)$ (H^w denotes H with its weak topology) using the technique of compact embedding in Chap. 3. As a consequence of this approach, we are able to give an extension of an interesting result on equations with discontinuous drift due to Leha and Ritter [47], by eliminating local Lipschitz assumptions. When the drift is Lipschitz continuous, we relate the solution constructed here to that obtained by Leha and Ritter. The advantage of the construction presented here is that the weak solution has finite-dimensional Galerkin approximation as opposed to the infinite-dimensional approximation given in [47].

Let us now consider H as a space isomorphic to

$$l_2 = \left\{ x \in \mathbb{R}^\infty : \sum_{i=1}^\infty (x^i)^2 < \infty \right\}$$

with the canonical isomorphism

$$h \rightarrow (x^1, x^2, \ldots) = (\langle h, e_1 \rangle_H, \langle h, e_2 \rangle_H, \ldots),$$

where $\{e_k\}_{k=1}^\infty$ is an ONB in H. Then the natural choice of the larger space is $(\mathbb{R}^\infty, \rho_{\mathbb{R}^\infty})$ with its metric defined for coordinate-wise convergence

$$\rho_{\mathbb{R}^\infty}(x, y) = \sum_{k=1}^\infty \frac{1}{2^k} \frac{|x^k - y^k|}{1 + |x^k - y^k|},$$

as the embedding $J : l_2 \hookrightarrow \mathbb{R}^\infty$ is continuous and compact, see Exercise 5.1.

Exercise 5.1 Prove that the embedding $J : l_2 \hookrightarrow \mathbb{R}^\infty$ is continuous and compact. Show that for some constant C and $x, y \in l_2$,

$$\rho_{\mathbb{R}^\infty}(Jx, Jy) \leq C\|x - y\|_H. \tag{5.1}$$

Let $(\Omega, \mathcal{F}, \{\mathcal{F}_t\}_{t \leq T}, P)$ be a filtered probability space, and W_t be an H-valued Wiener process with covariance Q, a nonnegative trace-class operator on H, and with eigenvalues $\lambda_i > 0$ and the associated eigenvectors e_i, $i = 1, 2, \ldots$. Denote by H^τ the space H endowed with the topology induced by $\mathbb{R}^\infty$ (under the identification with l_2).

Theorem 5.1 *Let* $F : H \rightarrow H$, *and assume that* $F^k(x) := \langle F(x), e_k \rangle_H : H^\tau \rightarrow \mathbb{R}$ *and* $q^k : \mathbb{R} \rightarrow \mathbb{R}$, $k = 1, 2, \ldots$, *are continuous. Assume that* $\gamma_k > 0$ *are constants with* $\sum_{k=1}^\infty \gamma_k < \infty$. *Let the following conditions hold:*

(GF)

$$\|F(x)\|_H^2 \leq \ell(1 + \|x\|_H^2), \quad x \in H; \tag{5.2}$$

(Gq1)

$$uq^k(u) \leq \ell(\gamma_k + u^2), \quad u \in \mathbb{R}; \tag{5.3}$$

(Gq2) Let $\bar{q}_n(u^1, \ldots, u^n) = (q^1(u^1), \ldots, q^n(u^n)) \in \mathbb{R}^n$. There exists a positive integer m, independent on n, such that

$$\left\| \bar{q}_n(u_n) \right\|_{\mathbb{R}^n}^2 \leq C\left(1 + \|u_n\|_{\mathbb{R}^n}^{2m}\right). \tag{5.4}$$

Then, there exists a weak solution X in $C([0, T], H^w)$ to the equation

$$X(t) = x + \int_0^t \left(F\big(X(s)\big) + q\big(X(s)\big)\right) ds + W_t$$

in the following sense. There exists an H-valued Q-Wiener process W_t and a process $X(\cdot) \in C([0, T], H^w)$ such that for every $k = 1, 2, \ldots,$

$$X^k(t) = x^k + \int_0^t \left(F^k\big(X(s)\big) + q^k\big(X^k(s)\big)\right) ds + W_t^k. \tag{5.5}$$

Here, $y^k = \langle y, e_k \rangle_H$ is the kth coordinate of $y \in H$.

Proof Consider the following sequence of equations:

$$X_n(t) = P_n x + \int_0^t \left(P_n F\big(P_n X_n(s)\big) + q_n\big(P_n X_n(s)\big)\right) ds + W_t^n. \tag{5.6}$$

Here, P_n is the projection of H onto $\text{span}\{e_1, \ldots, e_n\}$, $q_n : P_n H \to P_n H$, $q_n(y) = \sum_{k=1}^n q^k(y^k) e_k$, $y \in P_n H$, and W_t^n is an H-valued Wiener process with covariance $Q_n = P_n Q P_n$.

We can consider (5.6) in $\mathbb{R}^n$ by identifying $P_n H$ with $\mathbb{R}^n$ and treating W_t^n as an $\mathbb{R}^n$-valued Wiener process. Denote

$$G_n(x) = \left(F^1(x_n) + q^1(x^1), \ldots, F^n(x_n) + q^n(x^n)\right) \in \mathbb{R}^n, \quad x \in \mathbb{R}^n, \quad x_n = \sum_{k=1}^n x^k e_k.$$

Note that conditions (5.2) and (5.3) imply that

$$\left\langle x, G_n(x) \right\rangle_{\mathbb{R}^n} \leq C\left(1 + \|x\|_{\mathbb{R}^n}^2\right), \tag{5.7}$$

so that, by Theorem 4.1, there exists a weak solution $\xi_n(\cdot) \in C([0, T], \mathbb{R}^n)$.

We now establish estimates for the moments of $\xi_n(t)$. Denote

$$\tau_R = \inf_{0 \leq t \leq T} \left\{ \left\| \xi_n(t) \right\|_{\mathbb{R}^n} > R \right\}$$

or $\tau_R = T$ if the infimum is taken over an empty set. Using the Itô formula for the function $\|x\|_{\mathbb{R}^n}^2$ on $\mathbb{R}^n$, we get

$$\left\| \xi_n(t \wedge \tau_R) \right\|_{\mathbb{R}^n}^2 = \|P_n x\|_H^2 + 2 \int_0^{t \wedge \tau_R} \left\langle \xi_n(s), G_n\big(\xi_n(s)\big) \right\rangle_{\mathbb{R}^n} ds$$

$$+ (t \wedge \tau_R)\,\mathrm{tr}(Q_n) + 2 \int_0^{t \wedge \tau_R} \langle \xi_n(s), dW_s^n \rangle_{\mathbb{R}^n}.$$

For the stochastic integral term, we use that $2a \leq 1 + a^2$; hence, by the Doob inequality, Theorem 2.2, and (5.7), we have

$$E \sup_{0 \leq s \leq t} \| \xi_n(s \wedge \tau_R) \|_{\mathbb{R}^n}^2 \leq C_1 + C_2 \int_0^t \| \xi_n(s \wedge \tau_R) \|_{\mathbb{R}^n}^2 \, ds.$$

Now the Gronwall lemma implies that

$$E \sup_{0 \leq t \leq T} \| \xi_n(t \wedge \tau_R) \|_{\mathbb{R}^n}^2 < C_3.$$

Taking $R \to \infty$ and using the monotone convergence theorem imply

$$\sup_n E \sup_{0 \leq t \leq T} \| \xi_n(t) \|_{\mathbb{R}^n}^2 < \infty. \tag{5.8}$$

Let l be a positive integer. Using the Itô formula for the function $\|x\|_{\mathbb{R}^n}^{2l}$ on $\mathbb{R}^n$, we get

$$\| \xi_n(t \wedge \tau_R) \|_{\mathbb{R}^n}^{2l} = \| P_n x \|_H^{2l} + 2l \int_0^{t \wedge \tau_R} \| \xi_n(s) \|_{\mathbb{R}^n}^{2(l-1)} \langle \xi_n(s), G_n(\xi_n(s)) \rangle_{\mathbb{R}^n} \, ds$$

$$+ 2l(l-1) \int_0^{t \wedge \tau_R} \| \xi_n(s) \|_{\mathbb{R}^n}^{2(l-2)} \| Q_n^{1/2} \xi_n(s) \|_{\mathbb{R}^n}^2 \, ds$$

$$+ l \int_0^{t \wedge \tau_R} \| \xi_n(s) \|_{\mathbb{R}^n}^{2(l-1)} \mathrm{tr}(Q_n) \, ds$$

$$+ 2l \int_0^{t \wedge \tau_R} \| \xi_n(s) \|_{\mathbb{R}^n}^{2(l-1)} \langle \xi_n(s), dW_s^n \rangle_{\mathbb{R}^n}.$$

Taking the expectation to both sides and using (5.7) yield

$$E \| \xi_n(t \wedge \tau_R) \|_{\mathbb{R}^n}^{2l} \leq C_1 + C_2 \int_0^t E \| \xi_n(t \wedge \tau_R) \|_{\mathbb{R}^n}^{2(l-1)} \, ds$$

$$+ C_3 \int_0^t E \| \xi_n(t \wedge \tau_R) \|_{\mathbb{R}^n}^{2l} \, ds$$

$$\leq (C_1 + C_2 T) + (C_2 + C_3) \int_0^t E \| \xi_n(t \wedge \tau_R) \|_{\mathbb{R}^n}^{2l} \, ds,$$

where we have used the fact that $a^{2(l-1)} \leq 1 + a^{2l}$. By Gronwall's lemma, for some constant C,

$$E \| \xi_n(t \wedge \tau_R) \|_{\mathbb{R}^n}^{2l} \leq C,$$

which, as $R \to \infty$, leads to

$$E \sup_{0 \le t \le T} \|\xi_n(t)\|_{\mathbb{R}^n}^{2l} \le C. \tag{5.9}$$

Using (5.8) and (5.9) and essentially repeating the argument in Lemma 3.10, we now obtain an estimate for the fourth moment of the increment of the process $\xi_n(t \wedge \tau_R)$. Applying the Itô formula for the function $\|x\|_{\mathbb{R}^n}^4$ on $\mathbb{R}^n$, we get

$$\|\xi_n(t + h) - \xi_n(t)\|_{\mathbb{R}^n}^4$$
$$= 4 \int_t^{t+h} \|\xi_n(u) - \xi_n(t)\|_{\mathbb{R}^n}^2 \langle \xi_n(u) - \xi_n(t), G_n(\xi_n(u)) \rangle_{\mathbb{R}^n} du$$
$$+ 2 \int_t^{t+h} \|\xi_n(u) - \xi_n(t)\|_{\mathbb{R}^n}^2 \operatorname{tr}(Q_n) du$$
$$+ 2 \int_t^{t+h} \langle Q_n(\xi_n(u) - \xi_n(t)), \xi_n(u) - \xi_n(t) \rangle_{\mathbb{R}^n} du$$
$$+ 4 \int_t^{t+h} \|\xi_n(u) - \xi_n(t)\|_{\mathbb{R}^n}^2 \langle \xi_n(u) - \xi_n(t), dW_u^n \rangle_{\mathbb{R}^n}.$$

Taking the expectation of both sides and using assumptions (5.2) and (5.4), which imply the polynomial growth of G_n, we calculate

$$E \|\xi_n(t + h) - \xi_n(t)\|_{\mathbb{R}^n}^4$$
$$\le C_1 E \int_t^{t+h} \|\xi_n(u) - \xi_n(t)\|_{\mathbb{R}^n}^3 (1 + \|\xi_n(u)\|) du$$
$$+ C_2 E \int_t^{t+h} \|\xi_n(u) - \xi_n(t)\|_{\mathbb{R}^n}^2 du$$
$$\le C_3 \left(\left[E \int_t^{t+h} \|\xi_n(u) - \xi_n(t)\|_{\mathbb{R}^n}^4 du \right]^{3/4} h^{1/4} \right.$$
$$\left. + \left[E \int_0^{t+h} \|\xi_n(u) - \xi_n(t)\|_{\mathbb{R}^n}^4 du \right]^{1/2} h^{1/2} \right) \le Ch$$

for a suitable constant C. Substituting repeatedly, starting with $C(u - t)$ for $E \|\xi_n(u) - \xi_n(t)\|_{\mathbb{R}^n}^4$, leads to the following estimate for the fourth moment of the increment of ξ_n:

$$E \|\xi_n(t + h) - \xi_n(t)\|_{\mathbb{R}^n}^4 \le Ch^2$$

for some constant C independent of n.

By (5.1), the compactness of the embedding J and Lemma 3.14, the measures $\mu_n = P \circ X_n^{-1}$ are tight on $C([0, T], \mathbb{R}^\infty)$. Let μ be a limit point of the sequence

μ_n. Since the function $C([0, T], \mathbb{R}^\infty) \ni x \mapsto \sup_{0 \leq t \leq T} \sum_{k=1}^\infty (x^k(t))^2 \in \mathbb{R}$ is lower semicontinuous, we conclude that

$$E_\mu \left(\sup_{0 \leq t \leq T} \sum_{k=1}^\infty (x^k(t))^2 \right) \leq \liminf_n E_{\mu_n} \left(\sup_{0 \leq t \leq T} \sum_{k=1}^\infty (x^k(t))^2 \right)$$

$$= \liminf_n E_P \left(\sup_{0 \leq t \leq T} \| X_n(t) \|_H^2 \right) < \infty.$$

Here, $X_n(t) = \sum_{k=1}^n \xi_n^k(t) e_k$.

Using the Skorokhod theorem, by changing the underlying probability space, we can assume that

$$X_n \to X, \quad P\text{-a.s. in } C([0, T], \mathbb{R}^\infty)$$

(not in $C([0, T], H^w)$, which is not metrizable) with $X_n(t)$, $X(t) \in H$, $t \in [0, T]$, P-a.s., since

$$P \left(\sup_{0 \leq t \leq T} \sum_{k=1}^\infty (X^k(t))^2 < \infty \right) = 1.$$

The process $X(t)$ is in H; hence the measure μ is concentrated on $C([0, T], H^w)$, as the topologies on H^τ and on H^w coincide on norm-bounded sets of H, and a.e. path is such a set.

Now, the random variables $(X_n^k(t))^2$ and $\int_0^t [F_n^k(X_n(s)) + q^k(X_n^k(s))]^2 \, ds$ are P-uniformly integrable in view of (5.2), (5.4), and (5.9). In addition, for $n \geq k$, $F_n^k(X_n(s)) = F^k(X_n(s))$, and hence,

$$E_P \int_0^t |F_n^k(X_n(s)) - F^k(X(s))|^2 \, ds \to 0$$

due to the continuity of F^k on H^τ. Consequently, the sequence of Brownian motions

$$Y_n^k(t) = X_n^k(t) - x^k - \int_0^t [F_n^k(X_n(s)) + q^k(X_n^k(s))] \, ds$$

converges in $L^2(\Omega, P)$ to the Brownian motion

$$Y^k(t) = X^k(t) - x^k - \int_0^t [F^k(X(s)) + q^k(X^k(s))] \, ds.$$

Define an H-valued Brownian motion

$$W_t = \sum_{k=1}^\infty Y^k(t) e_k,$$

then $X(t)$, W_t satisfy (5.5). $\qquad\square$

Let us relate these results to the work of Leha and Ritter [47]. We begin with the general uniqueness and existence theorem in [46].

Theorem 5.2 *Let H be real separable Hilbert space, and W_t be a Q-Wiener process. Assume that $A : H \to \mathcal{L}(H)$ and $B \in \mathcal{L}(H)$ satisfy the following growth and local Lipschitz conditions. There exist constants C and $C_n, n = 1, 2, \ldots$, such that*

(1) $\operatorname{tr}(B(x)QB^*(x)) \le C(1 + \|x\|_H^2), x \in H$;
(2) $\langle x, A(x)\rangle_H \le C(1 + \|x\|_H^2), x \in H$;
(3) $\|A(x) - A(y)\|_{\mathcal{L}(H)} + \|B(x) - B(y)\|_H \le C_n \|x - y\|_H$ *for* $\|x\|_H \le n$ *and* $\|y\|_H \le n$.

Then there exists a unique strong solution to the equation

$$X(t) = x + \int_0^t A\big(X(s)\big)\,ds + \int_0^t B\big(X(s)\big)\,dW_s, \quad t > 0.$$

Exercise 5.2 Prove Theorem 5.2.
   Hint: assume global Lipschitz condition first to obtain the usual growth conditions on A and B and produce solution $X(t)$ as in Theorem 3.3. Define

$$A_n(x) = \begin{cases} A(x), & \|x\|_H \le n, \\ A(nx/\|x\|_H), & \|x\|_H > n, \end{cases}$$

and define B_n in a similar way. Show that A_n and B_n are globally Lipschitz and show that for the corresponding solutions $X_n(t)$, there exists a process $X(t)$ such that $X(t \wedge \tau_n) = X_n(t \wedge \tau_n)$ with τ_n denoting the first exit time of X_n from the ball of radius n centered at the origin. Finally, show that $P(\tau_n < t) \to 0$ by using the estimate for the second moment of X_n obtained from the application of the Itô formula to the function $\|x\|_H^2$.

Following [47], for a finite subset V of the set of positive integers, define q^V : $H \to H$ by

$$\big(q^V(y), e_k\big)_H = \begin{cases} q^k(y^k), & k \in V, \\ 0, & k \notin V. \end{cases}$$

From Theorem 5.2 we know that for any fixed V, under conditions (5.2), (5.3) and the local Lipschitz condition on the coefficients F and q^V, there exists a unique strong solution to the equation

$$\xi^V(t) = x + \int_0^t \big(F\big(\xi^V(s)\big) + q^V\big(\xi^V(s)\big)\big)\,ds + W_t$$

with the solution $\xi^V(\cdot) \in C([0, T], H)$.
   Leha and Ritter proved further in Theorem 2.4 in [47] that, under conditions (5.2), (5.3) and the local Lipschitz condition (3) of Theorem 5.2 on F and

q^V (the Lipschitz constant possibly dependent on V), there exists a weak solution ξ in $C([0, T], H^w)$ constructed as a weak limit of ξ^{V_n} satisfying, for each k,

$$\xi^k(t) = x^k + \int_0^t \left(F^k(\xi(s)) + q^k(\xi^k(s)) \right) ds + W^k(t).$$

We now show that, under the global Lipschitz condition independent on V, the laws of the above solution $\xi(t)$ and the solution $X(t)$ obtained in Theorem 5.1 coincide.

Theorem 5.3 *Assume that the functions $F + q^V : H \to H$ satisfy global Lipschitz conditions for every V, a finite subset of positive integers, with the Lipschitz constants independent on V. Assume that conditions (5.2) and (5.3) hold. Then the weak solution ξ of Theorem 2.4 in [47] and the weak solution X constructed in Theorem 5.1, using the compact embedding argument, have the same law.*

Proof We first note that by Theorem 5.2 under the Lipschitz condition, both approximating sequences ξ_n^V of Theorem 2.4 in [47] and X_n of Theorem 5.1 can be constructed as strong solutions on the same probability space. Let $F_n = P_n \circ F \circ P_n$. Then

$$X_n(t) = x_n + \int_0^t \left(F_n(X_n(s)) + q_n(X_n(s)) \right) ds + W_t^n,$$

$$\xi^{V_n}(t) = x + \int_0^t \left(F(\xi^{V_n}(s)) + q^{V_n}(\xi^{V_n}(s)) \right) ds + W_t.$$

The laws $\mathscr{L}(\xi^V)$ are tight (see the proof of Theorem 2.4 in [47]). Therefore, for a sequence $V_n = \{1, 2, \ldots, n\}$, there is a subsequence V_{n_k} such that $\mathscr{L}(\xi^{V_{n_k}}) \to \mathscr{L}(\xi)$. Therefore, for simplicity, we assume as in Theorem 2.4 in [47] that $V_n = \{1, \ldots, n\}$ and that $\xi^{V_n} \to \xi$ weakly. Denote $Y_n(t) = P_n \xi^{V_n}(t)$. Then

$$Y_n(t) = x_n + \int_0^t \left(P_n F(\xi^{V_n}(s)) + q_n(Y_n(s)) \right) ds + W^n(t).$$

We obtain

$$E \left\| X_n(t) - Y_n(t) \right\|_H^2 \leq CE \int_0^t \left(\left\| P_n F(X_n(s)) - P_n F(\xi^{V_n}(s)) \right\|_H^2 \right.$$

$$+ \left. \left\| q_n(X_n(s)) - q_n(Y_n(s)) \right\|_H^2 \right) ds$$

$$\leq C_1 E \int_0^t \left(\left\| X_n(s) - \xi^{V_n}(s) \right\|_H^2 + \left\| X_n(s) - Y_n(s) \right\|_H^2 \right) ds$$

$$\leq C_1 E \int_0^t \left(\left\| Y_n(s) - \xi^{V_n}(s) \right\|_H^2 + 2 \left\| X_n(s) - Y_n(s) \right\|_H^2 \right) ds,$$

where we have used the Lipschitz condition.

Note that (5.9) also holds for ξ^{V_n} (the same proof works, with H replacing $\mathbb{R}^n$). Hence, by changing the underlying probability space to ensure the a.s. convergence by means of Skorokhod's theorem, we have that

$$E \int_0^t \left\| Y_n(s) - \xi^{V_n}(s) \right\|_H^2 ds \leq 3 \left(E \int_0^t \left\| Y_n(s) - P_n \xi(s) \right\|_H^2 ds \right.$$

$$\left. + E \int_0^t \left\| P_n \xi(s) - \xi(s) \right\|_H^2 ds + E \int_0^t \left\| \xi(s) - \xi^{V_n}(s) \right\|_H^2 ds \right) \to 0$$

as $n \to \infty$, due to the uniform integrability, implying that with some $\varepsilon_n \to 0$,

$$E \left\| X_n(t) - Y_n(t) \right\|_H^2 \leq 2C_1 \varepsilon_n + C_1 \int_0^t E \left\| X_n(s) - Y_n(s) \right\|_H^2 ds.$$

Using Gronwall's lemma, we obtain that

$$E \left\| X_n(t) - \xi^{V_n}(t) \right\|_H^2 \leq \varepsilon_n e^{C_1 t}$$

with the expression on the right-hand side converging to zero as $n \to \infty$.

We conclude that $X_n(t) - \xi^{V_n}(t) \to 0$ in $L^2(\Omega)$. Thus, $\xi^{V_n} \to \xi$ implies that $X_n(t) \to \xi(t)$ weakly. Therefore, $\text{Law}(X) = \text{Law}(\xi)$ on $C([0, T], \mathbb{R}^\infty)$, and consequently the laws coincide on $C([0, T], H^w)$. □

Example 5.1 (Unbounded Spin Systems) In statistical mechanics an unbounded spin system can be described by a family of interaction potentials and on-site energy functions. For a family of subsets $V \subset \mathbb{Z}_+$ with $|V| < \infty$, consider interaction potentials $\varphi_V : \mathbb{R}^V \to \mathbb{R}$. A typical example is pair potentials $\varphi_{\{k,l\}} : \mathbb{R}^2 \to \mathbb{R}$,

$$\varphi_{\{k,l\}}(u, v) = -J_{k,l} u v, \quad k \neq l.$$

On-site energy is modeled by functions $\varphi_k : \mathbb{R} \to \mathbb{R}$.

The energy function H defined on the space of configurations with finite support in $\mathbb{R}^\infty$ into $\mathbb{R}^\infty$ is now defined at every site "k" as

$$H^k(x) = \phi_k(x^k) - \sum_{l \neq k} J_{k,l} x^k x^l.$$

The components of the drift coefficient of an SDE takes the following form:

$$F^k(x) = \frac{\partial}{\partial x^k} \sum_{l \neq k} J_{k,l} x^k x^l = \sum_{l \neq k} J_{k,l} x^l,$$

$$q^k(u) = -\phi_k'(u),$$

so that

$$F^k(x) + q^k(x) = -\frac{\partial}{\partial x_k} H^k(x).$$

The growth condition (5.2) is satisfied if, for example,

$$J_{k,l} = \begin{cases} 1 & \text{if } |k - l| = 1, \\ 0 & \text{otherwise,} \end{cases}$$

that is, there is only the closest neighbor interaction.

Conditions (5.3) and (5.4) hold if, for example,

$$\varphi_k(u) = P(u) = a_n u^{2n} + a_{n-1} u^{2(n-1)} + \cdots + a_1 u^2 + a_0 \qquad (5.10)$$

with $a_n > 0$.

The results show that there exists a weak solution only under the assumption that $q^k(u) : \mathbb{R} \to \mathbb{R}$ satisfy growth conditions and are continuous functions. It should be noted that even if $J_{k,l} = 0$ for all k, l, the growth condition (5.3) is necessary for the existence of a solution without explosion, and continuity is needed in the proof of the Peano theorem.

In Euclidian quantum field theory continuous spin models serve as lattice approximations (see [60] for details) with $\mathbb{R}^\infty$ replaced by $\mathbb{R}^{\mathbb{Z}^d}$. In [60],

$$J_{k,l} = \begin{cases} 1 & \text{if } \sum_{j=1}^d |k_j - l_j| = 1, \\ 0 & \text{otherwise,} \end{cases}$$

$$\varphi_k(u) = (d + m^2/2) u^2 + P(u),$$

where $k, j \in \mathbb{Z}^d$.

We will study such models in the next section.

5.3 Locally Interacting Particle Systems, Solutions in $C([0, T], \mathbb{R}^{\mathbb{Z}^d})$

We use recent ideas from Albeverio et al. [3] to study the dynamics of an infinite particle system corresponding to a Gibbs measure on the lattice. Our technique is to study weak solutions of an infinite system of SDEs using the work in [22, 36], and the methods in Chap. 4 related to the case of a SDE in the dual to a nuclear space. This allows us to extend the existence result in [3] by removing the dissipativity condition.

The work [3] provides results for the existence and uniqueness of solutions to a system of SDEs describing a lattice spin-system model with spins taking values in "loop spaces." The space of configurations $\Omega_\beta = C(S_\beta)^{\mathbb{Z}^d}$, where S_β is a circle with circumference $\beta > 0$.

We consider a lattice system of locally interacting diffusions, which is a special case of the system studied in [3], when the continuous parameter in S_β is absent. The resulting infinite-dimensional process is of extensive interest (see [3], Sect. 2, for references).

The fundamental problem is to study the dynamics corresponding to a given Gibbs measure on $\mathbb{R}^{\mathbb{Z}^d}$. We consider here a lattice system of locally interacting diffusions and study weak solutions in $C([0, T], \mathbb{R}^{\mathbb{Z}^d})$. The space $\mathbb{Z}^d$ is equipped with the Euclidian norm and $\mathbb{R}^{\mathbb{Z}^d}$ has the metric for coordinate-wise convergence, similar as in Sect. 5.2.

We begin with the system of SDEs describing the lattice system of interest,

$$X^k(t) = \int_0^t \left(F^k(X^k(s)) + q^k(X^k(s)) \right) ds + W_t^k. \tag{5.11}$$

Here, for $k \in \mathbb{Z}^d$ and $x \in \mathbb{R}^{\mathbb{Z}^d}$, x^k denotes the kth coordinate of x. The coefficient $F : \mathbb{R}^{\mathbb{Z}^d} \to \mathbb{R}^{\mathbb{Z}^d}$ is defined by

$$F(x) = \left\{ F^k(x) \right\}_{k \in \mathbb{Z}^d} = \left\{ -\frac{1}{2} \sum_{j \in B_{\mathbb{Z}^d}(k,\rho)} a(k - j) x^j \right\}_{k \in \mathbb{Z}^d} \tag{5.12}$$

with $B_{\mathbb{Z}^d}(k, \rho)$ denoting a sphere of radius ρ centered in $k \in \mathbb{Z}^d$.

Let $l_2(\mathbb{Z}^d)$ denote the Hilbert space of square-integrable sequences indexed by elements of $\mathbb{Z}^d$. The "dynamical matrix" $A = (a_{k,j})_{k,j \in \mathbb{Z}^d} \in \mathscr{L}(l_2(\mathbb{Z}^d))$ is lattice translation invariant, and the interactions are local, i.e., $a_{k,j} = a(k - j)$, and $a_{k,j} = 0$ for $|k - j|_{\mathbb{Z}^d} > \rho$.

The family of drifts $q(x) = \{q_k(x^k)\}_{k \in \mathbb{Z}^d} : \mathbb{R}^{\mathbb{Z}^d} \to \mathbb{R}^{\mathbb{Z}^d}$ is in general a singular mapping on the scales of the Hilbert space $l_2(\mathbb{Z}^d)$. The functions $q_k : \mathbb{R} \to \mathbb{R}$ are the derivatives of potentials $V_k(u)$. In [3], $V_k(u) = \lambda P(u)$ with $P(u)$ as in (5.10), which is the case in an important class of the so-called $P(\varphi)$ models.

We note that

$$\left| F^k(x) \right| \leq \frac{1}{2} \|A\| \left(\sum_{j \in B_{\mathbb{Z}^d}(k,\rho)} (x^j)^2 \right)^{1/2}, \tag{5.13}$$

where

$$\|A\| = \left(\sum_{j \in B_{\mathbb{Z}^d}(k,\rho)} a^2(k - j) \right)^{1/2}.$$

The assumptions we impose on functions $q_k : \mathbb{R} \to \mathbb{R}$ are the same as in Sect. 5.2. Denote

$$l_2^n = l_2^n(\mathbb{Z}^d) = \left\{ x \in \mathbb{R}^{\mathbb{Z}^d} : \sum_{k \in \mathbb{Z}^d} (1 + |k|_{\mathbb{Z}^d})^{2n} (x^k)^2 < \infty \right\},$$

$$\Phi = \bigcap_{n=1}^{\infty} l_2^n \hookrightarrow \cdots \hookrightarrow l_2^1 \hookrightarrow l_2 = l_2^0 \hookrightarrow l_2^{-1} \hookrightarrow \cdots \hookrightarrow \bigcup_{n=1}^{\infty} l_2^{-n} =: \Phi'.$$

The embeddings between the Hilbert spaces $l_2^n \hookrightarrow l_2^m$ with $m + \frac{d}{2} < n$ are compact (in fact, Hilbert–Schmidt) operators. The space Φ endowed with the projective limit topology is a nuclear space of fast decreasing sequences, and the space Φ', endowed with the inductive limit topology, is dual to Φ.

Let Q be a continuous quadratic form on Φ and denote its extension (which always exists) to a nuclear form on some l_2^{-m}, $m > 0$, by the same symbol.

Theorem 5.4 *Let $F : \mathbb{R}^{\mathbb{Z}^d} \to \mathbb{R}^{\mathbb{Z}^d}$ be as in (5.12), defined by a lattice translation-invariant "dynamical matrix" A, and the drifts $q^k : \mathbb{R} \to \mathbb{R}$, $k \in \mathbb{Z}^d$, be continuous and satisfy conditions (5.3) and (5.4). Then there exists a weak solution $X(\cdot) \in C([0, T], l_2^{-p})$, for some $p > 0$, to the equation*

$$X(t) = x + \int_0^t \left(F\big(X(s)\big) + q\big(X(s)\big) \right) ds + W_t, \quad x \in l_2, \tag{5.14}$$

in the following sense. There exist an l_2^{-p}-valued Q-Wiener process W_t and a process $X(\cdot) \in C([0, T], l_2^{-p})$ such that for every $k \in \mathbb{Z}^d$,

$$X^k(t) = x^k + \int_0^t \left(F^k\big(X(s)\big) + q^k\big(X^k(s)\big) \right) ds + W_t^k. \tag{5.15}$$

Proof Let us show that for $m > 0$, $F : l_2^{-m} \to l_2^{-m}$ is Lipschitz continuous. If $x, y \in l_2^{-m}$, $m > 0$, then

$$\left\| F(x) - F(y) \right\|_{l_2^{-m}}^2 = \sum_{k \in \mathbb{Z}^d} \left(1 + |k|_{\mathbb{Z}^d} \right)^{-2m} \left(F^k(x) - F^k(y) \right)^2$$

$$\leq \sum_{k \in \mathbb{Z}^d} \left(1 + |k|_{\mathbb{Z}^d} \right)^{-2m} \left(\frac{1}{4} \|A\|^2 \sum_{j \in B_{\mathbb{Z}^d}(k, \rho)} \left(x^j - y^j \right)^2 \right)$$

$$\leq C_1 \| x - y \|_{l_2^{-m}}^2.$$

Note that F and q^k satisfy all conditions of Theorem 5.1. Following its proof, we first construct solutions ξ_n of (5.6). Let $|B_n|$ denote the cardinality of the ball $B_{\mathbb{Z}^d}(0, n)$ of radius n centered at 0 in $\mathbb{Z}^d$. Observe that

$$\xi_n \in C\big([0, T], \mathbb{R}^{|B_n|}\big).$$

Next, we obtain approximations $X_n(t) = \sum_{j \in B_{\mathbb{Z}^d}(0,n)} \xi_n^j h_j$, where we denote by $\{h_k\}_{k \in \mathbb{Z}^d}$ the canonical basis in l_2. We have

$$X_n \in C\big([0, T], l_2\big).$$

Since for $m > d/2$, the embedding $l_2 \hookrightarrow l_2^{-m}$ is compact and

$$\| \cdot \|_{l_2^{-m}} \leq C \| \cdot \|_{l_2},$$

Lemma 3.14 guarantees that the measures $\mu_n = \mathcal{L}(X_n)$ are tight on $C([0, T], l_2^{-m})$.

Let μ be a limit point of the sequence μ_n. Using the Skorokhod theorem, we can assume that

$$X_n \to X, \quad P\text{-a.s. in } C([0, T], l_2^{-m}).$$

The random variables $(X_n^k(t))^2$ and $\int_0^t [F_n^k(X_n(s)) + q^k(X_n^k(s))]^2 \, ds$ are P-uniformly integrable. By the same arguments as in the proof of Theorem 5.1, we conclude that the sequence of Brownian motions

$$Y_n^k(t) = X_n^k(t) - x^k - \int_0^t \left[F_n^k(X_n(s)) + q^k(X_n^k(s)) \right] ds$$

converges in $L^2(\Omega, P)$ to the Brownian motion

$$Y^k(t) = X^k(t) - x^k - \int_0^t \left[F^k(X(s)) + q^k(X^k(s)) \right] ds,$$

and we define the l_2^{-m}-valued Brownian motion

$$W_t = \sum_{k \in \mathbb{Z}^d} Y^k(t) h_k^{-m},$$

where $\{h_k^{-m}\}_{k \in \mathbb{Z}^d}$ is the basis in l_2^{-m} obtained by applying the Gramm–Schmidt orthonormalization of the vectors h_k. Then $X(t)$, W_t satisfy (5.15). $\qquad\square$

When the assumptions on the drifts q^k are more restrictive, (5.14) can be considered in a Hilbert space. We will require that the functions $q^k(u)$ have linear growth.

Theorem 5.5 *Let $F : \mathbb{R}^{\mathbb{Z}^d} \to \mathbb{R}^{\mathbb{Z}^d}$ be as in (5.12), defined by a lattice translation-invariant "dynamical matrix" A, and the drifts $q^k : \mathbb{R} \to \mathbb{R}$, $k \in \mathbb{Z}^d$, be continuous and satisfy the linear growth condition*

$$\left| q^k(u) \right| \leq C(1 + |u|), \quad u \in \mathbb{R}.$$

Assume that $B : \Phi' \to \Phi'$, $B(x) \in \mathscr{L}(l_2^{-m})$ if $x \in l_2^{-m}$ for $m > 0$, and it is continuous in x. In addition,

$$\operatorname{tr}\left(B(x) Q (B(x))^* \right) \leq \theta \left(1 + \|x\|_{l_2^{-m}}^2 \right).$$

Then there exists a weak solution $X(\cdot) \in C([0, T], l_2^{-p})$, for some $p > 0$, to the equation

$$X(t) = x + \int_0^t \left(F(X(s)) + q(X(s)) \right) ds + \int_0^t B(X(s)) \, dW_s, \quad x \in l_2, \quad (5.16)$$

where W_t is an l_2^{-p}-valued Q-Wiener process.

Proof If $x \in l_2^{-m}$, $m > 0$, then

$$
\begin{aligned}
\left\| F(x) + q(x) \right\|_{l_2^{-m}}^2 &= \sum_{k \in \mathbb{Z}^d} \left(1 + |k|_{\mathbb{Z}^d}\right)^{-2m} \left(F^k(x) + q_k(x^k)\right)^2 \\
&\leq 2 \sum_{k \in \mathbb{Z}^d} \left(1 + |k|_{\mathbb{Z}^d}\right)^{-2m} \left((F^k(x))^2 + (q_k(x^k))^2\right) \\
&\leq 2 \sum_{k \in \mathbb{Z}^d} \left(1 + |k|_{\mathbb{Z}^d}\right)^{-2m} \left(\frac{1}{4}\|A\|^2 \sum_{j \in B_{\mathbb{Z}^d}(k,\rho)} (x^j)^2 + C^2(1 + |x^k|)^2\right) \\
&\leq C_1 \left(1 + \|x\|_{l_2^{-m}}^2\right),
\end{aligned}
$$

showing that $G(x) = F(x) + q(x) : l_2^{-m} \to l_2^{-m}$, $m > d/2$ (otherwise, we face a divergent series, see Exercise 5.3), and providing the estimate

$$
\left\| G(x) \right\|_{l_2^{-m}}^2 \leq C_1 \left(1 + \|x\|_{l_2^{-m}}^2\right).
$$

Moreover, we know from the proof of Theorem 5.4 that $F : l_2^{-m} \to l_2^{-m}$, $m > 0$, is Lipschitz continuous.

We will now use the approach presented in Sect. 3.8. With $\{e_k^{-r}\}_{k \in \mathbb{Z}^d}$, $r > 0$, denoting the ONB in l_2^{-r} consisting of the eigenvectors of Q, let $P_n : l_2^{-r} \to l_2^{-r}$ be defined by

$$
P_n x = \sum_{k \in B_{\mathbb{Z}^d}(0,n)} \langle x, e_k^{-r} \rangle_{l_2^{-r}} e_k^{-r}.
$$

Note that for $0 < m < r$ and $P_n : l_2^{-m} \to l_2^{-m}$,

$$
P_n x = \sum_{k \in B_{\mathbb{Z}^d}(0,n)} \langle x, e_k^{-m} \rangle_{l_2^{-m}} e_k^{-m}, \quad x \in l_2^{-m}.
$$

Let us consider a weak solution of the equation

$$
dX_n(t) = P_n G\left(P_n X_n(t)\right) dt + d P_n B\left(P_n X_n(t)\right) W_t^n \tag{5.17}
$$

with a $P_n Q P_n$-Wiener process W_t in l_2^{-m}.

Equation (5.17) can be considered in an $|B_n|$-dimensional subspace of l_2^{-m} ($|B_n|$ denotes the cardinality of $B_{\mathbb{Z}^d}(0, n)$). In addition, we have the following estimates:

$$
\left\| P_n G(P_n x) \right\|_{l_2^{-m}}^2 \leq C_1 \left(1 + \|P_n x\|_{l_2^{-m}}^2\right),
$$

$$
\mathrm{tr}\left(P_n B(P_n x)(P_n Q P_n)\left(P_n B(P_n x)\right)^*\right) \leq \theta\left(1 + \|P_n x\|_{l_2^{-m}}^2\right).
$$

Since $P_n H_{l_2^{-m}}$ is finite-dimensional, the embedding $J : P_n l_2^{-m} \hookrightarrow P_n l_2^{-m}$ is compact, and we conclude by Theorem 3.12 (or by Theorem 4.1) that (5.17) has a weak solution $X_n(t)$ in the space $C([0, T], P_n l_2^{-m})$ satisfying the inequalities

$$
E \sup_{0 \le t \le T} \|X_n(t)\|_{l_2^{-m}}^{2j} < C', \quad j \ge 1,
$$

$$
E \|X_n(t + h) - X_n(t)\|_{l_2^{-m}}^4 \le C' h^2,
$$

(5.18)

with the constant C' independent of n. Consider now the sequence of measures on $C([0, T], l_2^{-m})$ induced by $\{X_n(t)\}_{t\in[0,T]}$ and denote them by μ_n. Again, using the fact that for $p > m + \frac{d}{2}$, the embedding $l_2^{-m} \hookrightarrow l_2^{-p}$ is Hilbert–Schmidt, we obtain the relative weak compactness of the sequence μ_n on $C([0, T], l_2^{-p})$ for some $p > 0$.

Because $X_n(t) \in C([0, T], l_2^{-p})$, there exists $X(t) \in C([0, T], l_2^{-p})$ such that $X_n \Rightarrow X$ in $C([0, T], l_2^{-p})$. By Skorokhod's theorem, we can assume that $X_n \to X$ a.s. For $x \in l_2^{-p}$, using the continuity of F on l_2^{-p} and uniform integrability, we obtain

$$
\langle X_n(t), x \rangle_{l_2^{-p}} - \int_0^t \langle F(X_n(s)), x \rangle_{l_2^{-p}} ds \to \langle X(t), x \rangle_{l_2^{-p}} - \int_0^t \langle F(X(s)), x \rangle_{l_2^{-p}} ds
$$

a.s. and in $L^1(\Omega, P)$. Next, let $\{x_l\}_{l=1}^\infty$ be a sequence converging to x in l_2^{-p}, $x_l^j = 0$, $j \notin B_{\mathbb{Z}^d}(0, l)$. We consider

$$
E \big| \langle q(X_n(t)), x \rangle_{l_2^{-p}} - \langle q(X(t)), x \rangle_{l_2^{-p}} \big| \le E \big\{ \big| \langle q(X_n(t)), x \rangle_{l_2^{-p}} - \langle q(X_n(t)), x_l \rangle_{l_2^{-p}} \big|
$$

$$
+ \big| \langle q(X_n(t)), x_l \rangle_{l_2^{-p}} - \langle q(X(t)), x_l \rangle_{l_2^{-p}} \big| + \big| \langle q(X(t)), x_l \rangle_{l_2^{-p}} - \langle q(X(t)), x \rangle_{l_2^{-p}} \big| \big\}
$$

$$
\le E \| q(X_n(t)) \|_{l_2^{-p}} \|x - x_l\|_{l_2^{-p}} + E \sum_{j \in B_{\mathbb{Z}^d}(0,l)} |q^j(X_n^j(t)) - q^j(X^j(t))| |x_l^j|
$$

$$
+ E \| q(X(t)) \|_{l_2^{-p}} \|x_l - x\|_{l_2^{-p}}
$$

$$
\le C E \sup_n \big(1 + \|X_n(t)\|_{l_2^{-p}} + \|X(t)\|_{l_2^{-p}}\big) \|x - x_l\|_{l_2^{-p}}
$$

$$
+ E \sum_{j \in B_{\mathbb{Z}^d}(0,l)} |q^j(X_n^j(t)) - q^j(X^j(t))| |x_l^j|.
$$

Using the estimate in (5.18), we can choose l, independent of n, such that the first summand is arbitrarily small. By choosing n large enough and using the continuity of q_k on $\mathbf{R}$, we can make the second summand arbitrarily small. Using the uniform integrability for the term involving q, we conclude that

$$
M_n(t) = \langle X_n(t), x \rangle_{l_2^{-p}} - \int_0^t \langle G(X_n(s)), x \rangle_{l_2^{-p}} ds
$$

$$
\to \langle X(t), x \rangle_{l_2^{-p}} - \int_0^t \langle G(X(s)), x \rangle_{l_2^{-p}} ds = M(t)
$$

(5.19)

in $L^1(\Omega)$ and a.s. for some subsequence.

Since

$$\mathrm{tr}\left(P_n B\left(X_n(s)\right) P_n Q P_n B^*\left(X_n(s)\right) P_n\right) \leq \theta\left(1 + \left\|X_n(s)\right\|_{l_2^{-p}}^2\right),$$

the LHS is uniformly integrable with respect to the measure $dP \times ds$, and it converges $dP \times ds$-a.e. to $\mathrm{tr}(B(X(s))QB^*(X(s)))$ (see Exercise 5.4), implying that

$$E \int_0^t \left(P_n Q P_n\left(B^*\left(X_n(s)\right)P_n x, B^*\left(X_n(s)\right)P_n x\right)\right) ds$$

$$\rightarrow E \int_0^t \left(Q\left(B^*(X(s))x, B^*(X(s))x\right)\right) ds,$$

i.e., $\langle M_n \rangle_t \rightarrow \langle M \rangle_t$ in $L^1(\Omega)$. In conclusion, the process on the right-hand side of (5.19) is a real continuous square-integrable martingale whose increasing process is given by

$$\int_0^t Q\left(B^*(X(s))x, B^*(X(s))x\right) ds,$$

An appeal to Lemma 2.1 and to the martingale representation theorem (Theorem 2.7) completes the proof. □

Exercise 5.3 Show that for $p > d/2$, $\sum_{k \in \mathbb{Z}^d}(1 + |k|)^{2p} < \infty$, and the series diverges otherwise.

Exercise 5.4 Show the convergence

$$\mathrm{tr}\left(P_n B\left(X_n(s)\right) P_n Q P_n B^*\left(X_n(s)\right) P_n\right) \rightarrow \mathrm{tr}\left(B\left(X(s)\right)QB^*\left(X(s)\right)\right)$$

claimed in the proof of Theorem 5.5.

Part II
Stability, Boundedness, and Invariant Measures

Chapter 6
Stability Theory for Strong and Mild Solutions

6.1 Introduction

Let $(\mathfrak{X}, \|\cdot\|_{\mathfrak{X}})$ be a Banach space, and let us consider the Cauchy problem

$$\begin{cases} \dfrac{du(t)}{dt} = Au(t), & 0 < t < T, \\ u(0) = x \in \mathfrak{X}. \end{cases} \tag{6.1}$$

We know that if A generates a C_0-semigroup $\{S(t), t \geq 0\}$, then the mild solution of the Cauchy problem (6.1) is given by

$$u^x(t) = S(t)x.$$

If $\mathfrak{X}$ is finite-dimensional, with a scalar product $\langle \cdot, \cdot \rangle_{\mathfrak{X}}$, Lyapunov proved the equivalence of the following three conditions:

(1) $\|u^x(t)\|_{\mathfrak{X}} \leq c_0 \|x\|_{\mathfrak{X}} e^{-rt}$, $r, c_0 > 0$.
(2) $\max\{\mathrm{Re}(\lambda) : \det(\lambda I - A) = 0\} < 0$.
(3) There exists a positive definite matrix R satisfying
    (i) $c_1 \|x\|_{\mathfrak{X}}^2 \leq \langle Rx, x \rangle_{\mathfrak{X}} \leq c_2 \|x\|_{\mathfrak{X}}^2$, $x \in \mathfrak{X}$, $c_1, c_2 > 0$,
    (ii) $A^*R + RA = -I$.

If condition (1) is satisfied, then the mild solution $\{u^x(t), t \geq 0\}$ of the Cauchy problem (6.1) is said to be *exponentially stable*.

To prove that (1) implies (3), the matrix R is constructed using the equation

$$\langle Rx, x \rangle_{\mathfrak{X}} = \int_0^{\infty} \|u^x(t)\|_{\mathfrak{X}}^2 \, dt.$$

When $\mathfrak{X}$ is infinite-dimensional, then the interesting examples of PDEs result in an unbounded operator A. In this case, if we replace condition (2) by

(2') $\max\{\mathrm{Re}(\lambda) : \lambda \in \sigma(A)\} < 0$,

with $\sigma(A)$ denoting the spectrum of A, the equivalence of (1) and (2') fails
$((2') \nRightarrow (1))$ due to the failure of the spectral mapping theorem (refer to [63], p. 117),

L. Gawarecki, V. Mandrekar, *Stochastic Differential Equations in Infinite Dimensions*,
Probability and Its Applications,
DOI 10.1007/978-3-642-16194-0_6, © Springer-Verlag Berlin Heidelberg 2011

unless we make more restrictive assumptions on A (e.g., A is analytic). A sufficient condition for exponential stability is given in [63], p. 116:

$$\int_0^\infty \|S(t)x\|_{\mathfrak{X}}^p \, dt < \infty, \quad \text{for } p > 1.$$

In our PDE examples, we need $p = 2$ and $\mathfrak{X} = H$, a real separable Hilbert space. In this case, condition (1) alone implies that R, given by

$$\langle Rx, y \rangle_H = \int_0^\infty \langle u^x(t), u^y(t) \rangle_H \, dt,$$

exists as a bilinear form, and in fact, the equivalence of conditions (1) and (3) above can be proved (see [13]). We now consider the Cauchy problem in a real Hilbert space,

$$\begin{cases} \dfrac{du(t)}{dt} = Au(t), & 0 < t < T, \\ u(0) = x \in H. \end{cases} \tag{6.2}$$

Theorem 6.1 *Let $(H, \langle \cdot, \cdot \rangle_H)$ be a real Hilbert space. The following conditions are equivalent:*

(1) *The solution of the Cauchy problem (6.2) $\{u^x(t), t \geq 0\}$ is exponentially stable.*
(2) *There exists a nonnegative symmetric operator R such that for $x \in \mathscr{D}(A)$,*

$$A^* Rx + RAx = -x.$$

Proof Define $\langle Rx, y \rangle_H$ as above. Using condition (1), we have

$$\langle Rx, x \rangle_H = \int_0^\infty \|S(t)x\|_H^2 \, dt < \infty. \tag{6.3}$$

Clearly, R is nonnegative definite and symmetric. Now, for $x, y \in H$,

$$\frac{d}{dt} \langle RS(t)x, S(t)y \rangle = \langle RAS(t)x, S(t)y \rangle_H + \langle RS(t)x, AS(t)y \rangle_H.$$

But

$$\langle RS(t)x, S(t)y \rangle_H = \int_t^\infty \langle S(u)x, S(u)y \rangle_H \, du$$

by the semigroup property. Hence, we obtain

$$\langle RAS(t)x, S(t)y \rangle_H + \langle RS(t)x, AS(t)y \rangle_H$$
$$= \frac{d}{dt} \int_t^\infty \langle S(u)x, S(u)y \rangle_H \, du$$
$$= -\langle S(t)x, S(t)y \rangle_H,$$

since $S(t)$ is strongly continuous. Thus, if $x \in \mathscr{D}(A)$, then

$$\langle (RA + A^*R)x, y \rangle_H = -\langle x, y \rangle_H,$$

giving (2).

From the above calculations condition (2) implies that

$$\frac{d}{dt} \langle RS(t)x, S(t)x \rangle_H = -\|S(t)x\|_H^2.$$

Hence,

$$\int_0^t \|S(u)x\|_H^2 \, du = \langle Rx, x \rangle_H - \langle RS(t)x, S(t)x \rangle_H$$

$$\leq \langle Rx, x \rangle_H.$$

Thus, $\int_0^\infty \|S(t)x\|_H^2 \, dt < \infty$.

We know that $S(t)x \to 0$ as $t \to \infty$ for each x (see Exercise 6.1). Hence, by the uniform boundedness principle, for some constant M, we have $\|S(t)\|_{\mathscr{L}(H)} \leq M$ for all $t \geq 0$.

Consider the map $T : H \to L^2(\mathbb{R}_+, H)$, $Tx = S(t)x$. Then T is a closed linear operator on H. Using the closed graph theorem, we have

$$\int_0^\infty \|S(t)x\|_H^2 \, dt \leq c^2 \|x\|_H^2.$$

Let $0 < \rho < M^{-1}$ and define

$$t_x(\rho) = \sup\{t : \|S(s)x\|_H > \rho\|x\|_H, \text{ for all } 0 \leq s \leq t\}.$$

Since $\|S(t)x\|_H \to 0$ as $t \to \infty$, we have that $t_x(\rho) < \infty$ for each $x \in H$, $t_x(\rho)$ is clearly positive, and

$$t_x(\rho)\rho^2\|x\|_H^2 \leq \int_0^{t_x(\rho)} \|S(t)x\|_H^2 \, dt \leq c^2\|x\|_H^2,$$

giving $t_x(\rho) \leq (c/\rho)^2 = t_0$.

For $t > t_0$, using the definition of $t_x(\rho)$, we have

$$\|S(t)x\|_H \leq \|S(t - t_x(\rho))\|_{\mathscr{L}(H)}\|S(t_x(\rho))x\|_H$$

$$\leq M\rho\|x\|_H.$$

Let $\beta = M\rho < 1$ and $t_1 > t_0$ be fixed. For $0 < s < t_1$, let $t = nt_1 + s$. Then

$$\|S(t)\|_{\mathscr{L}(H)} \leq \|S(nt_1)\|_{\mathscr{L}(H)}\|S(s)\|_{\mathscr{L}(H)}$$

$$\leq M\|S(t_1)\|_{\mathscr{L}(H)}^n \leq M\beta^n \leq M'e^{-\mu t},$$

where $M' = M/\beta$ and $\mu = -(1/t_1)\log\beta > 0$. $\qquad\qquad\square$

In particular, we have proved the following corollary.

Corollary 6.1 *If $S(\cdot)x \in L^2(\mathbb{R}_+, H)$ for all x in a real separable Hilbert space H, then*

$$\|S(t)\|_{\mathscr{L}(H)} \le c\,e^{-rt}, \quad \text{for some } r > 0.$$

Exercise 6.1 (a) Find a continuous function $f(t)$ such that $\int_0^\infty (f(t))^2\,dt < \infty$ but $\lim_{t\to\infty} f(t) \ne 0$.

(b) Show that if $\int_0^\infty \|S(t)x\|_H^2\,dt < \infty$ for every $x \in H$, then $\lim_{t\to\infty}\|S(t)x\|_H = 0$ for every $x \in H$.

Hint: recall that $\|S(t)\|_{\mathscr{L}(H)} \le Me^{\alpha t}$. Assume that $\|S(t_j)x\|_H > \delta$ for some sequence $t_j \to \infty$. Then, $\|S(t)x\|_H \ge \delta(Me)^{-1}$ on $[t_j - \alpha^{-1}, t_j]$.

We note that $\langle Rx, x\rangle_H$ does not play the role of the Lyapunov function, since in the infinite-dimensional case, $\langle Rx, x\rangle_H \ge c_1\|x\|^2$ with $c_1 > 0$ does not hold (see Example 6.1). We shall show that if A generates a pseudo-contraction semigroup, then we can produce a Lyapunov function related to R. The function Λ in Theorems 6.2 and 6.3 is called the *Lyapunov function*. Let us recall that $\{S(t),\ t \ge 0\}$ is a pseudo-contraction semigroup if there exists $\omega \in \mathbb{R}$ such that

$$\|S(t)\|_{\mathscr{L}(H)} \le e^{\omega t}.$$

Theorem 6.2 (a) *Let $\{u^*(t)\, t \ge 0\}$ be a mild solution to the Cauchy problem (6.2). Suppose that there exists a real-valued function Λ on H satisfying the following conditions:*

(1) $c_1\|x\|_H^2 \le \Lambda(x) \le c_2\|x\|_H^2$ *for $x \in H$,*
(2) $\langle \Lambda'(x), Ax\rangle_H \le -c_3\Lambda(x)$ *for $x \in \mathscr{D}(A)$,*

where c_1, c_2, c_3 are positive constants. Then the solution $u^*(t)$ is exponentially stable.

(b) *If the solution $\{u^*(t)\, t \ge 0\}$ to the Cauchy problem (6.2) is exponentially stable and A generates a pseudo-contraction semigroup, then there exists a real-valued function Λ on H satisfying conditions* (1) *and* (2) *in part* (a).

Proof (a) Consider $e^{c_3 t}\Lambda(u^*(t))$. We have

$$\frac{d}{dt}\left(e^{c_3 t}\Lambda\left(u^*(t)\right)\right) = c_3 e^{c_3 t}\Lambda\left(u^*(t)\right) + e^{c_3 t}\left\langle \Lambda'\left(u^*(t)\right), Au^*(t)\right\rangle_H.$$

Hence,

$$e^{c_3 t}\Lambda\left(u^*(t)\right) - \Lambda(x) = \int_0^t e^{c_3 s}\left\{c_3\Lambda\left(u^*(s)\right) + \left\langle \Lambda'\left(u^*(s)\right), Au^*(s)\right\rangle_H\right\}ds.$$

It follows, by condition (2), that

$$e^{c_3 t}\Lambda\left(u^*(t)\right) \le \Lambda(x).$$

Using (1), we have

$$c_1 \left\| u^x(t) \right\|_H^2 \le \mathrm{e}^{-c_3 t} \Lambda(x) \le c_2 \mathrm{e}^{-c_3 t} \|x\|_H^2,$$

proving (a).

(b) Conversely, we first observe that for $\Psi(x) = \langle Rx, x \rangle_H$ with R defined in (6.3), we have $\Psi'(x) = 2Rx$ by the symmetry of R. Since $R = R^*$, we can write

$$\left\langle \Psi'(x), Ax \right\rangle_H = \langle Rx, Ax \rangle_H + \langle Rx, Ax \rangle_H = \left\langle A^* Rx, x \right\rangle_H + \langle x, RAx \rangle_H$$

$$= \left\langle A^* Rx + RAx, x \right\rangle_H = -\|x\|_H^2.$$

Consider now, for some $\alpha > 0$ (to be determined later),

$$\Lambda(x) = \langle Rx, x \rangle_H + \alpha \|x\|_H^2.$$

Clearly $\Lambda(x)$ satisfies condition (1) in (a). Since $S(t)$ is a pseudo-contraction semigroup, there exists a constant λ (assumed positive WLOG) such that (see Exercise 3.5)

$$\langle x, Ax \rangle_H \le \lambda \|x\|_H^2, \quad x \in \mathscr{D}(A). \tag{6.4}$$

We calculate

$$\left\langle \Lambda'(x), Ax \right\rangle_H = \left\langle \Psi'(x), Ax \right\rangle_H + 2\alpha \langle x, Ax \rangle_H = \|x\|_H^2 (2\alpha\lambda - 1).$$

Choosing α small enough, so that $2\alpha\lambda < 1$, and using condition (1), we obtain (2) in (a).   □

Let us now consider the case of a coercive operator A (see condition (6.5)), with a view towards applications to PDEs. For this, we recall some concepts from Part I.

We have a Gelfand triplet of real separable Hilbert spaces

$$V \hookrightarrow H \hookrightarrow V^*,$$

where the embeddings are continuous. The space V^* is the continuous dual of V, with the duality on $V \times V^*$ denoted by $\langle \cdot, \cdot \rangle$ and satisfying

$$\langle v, h \rangle = \langle v, h \rangle_H$$

if $h \in H$.

Assume that V is dense in H. We shall now construct a Lyapunov function for determining the exponential stability of the solution of the Cauchy problem (6.2), where $A : V \to V^*$ is a linear bounded operator satisfying the coercivity condition

$$2\langle v, Av \rangle \le \lambda \|v\|_H^2 - \alpha \|v\|_V^2, \quad v \in V, \lambda \in \mathbb{R}, \alpha > 0. \tag{6.5}$$

We note that the following *energy equality* in [72] holds for solutions $u^x(t) \in L^2([0, T], V) \cap C([0, T], H)$:

$$\left\| u^x(t) \right\|_H^2 - \|x\|_H^2 = 2 \int_0^t \langle u^x(s), Au^x(s) \rangle ds. \tag{6.6}$$

We now state our theorem.

Theorem 6.3 (a) *The solution of the Cauchy problem (6.2) with a coercive coefficient A is exponentially stable if there exists a real-valued function Λ that is Fréchet differentiable on H, with Λ and Λ' continuous, locally bounded on H, and satisfying the following conditions:*

(1) $c_1 \|x\|_H^2 \le \Lambda(x) \le c_2 \|x\|_H^2$.
(2) *For $x \in V$, $\Lambda'(x) \in V$, and the function*

$$V \ni x \to \langle \Lambda'(x), v^* \rangle \in \mathbb{R}$$

  is continuous for any $v^* \in V^*$.
(3) *For $x \in V$, $\langle \Lambda'(x), Ax \rangle \le -c_3 \Lambda(x)$, where c_1, c_2, c_3 are positive constants.*

  In particular, if

$$2\langle \Lambda'(x), Ax \rangle_H = -\|x\|_V^2,$$

then condition (3) is satisfied.

  (b) *Conversely, if the solution to the Cauchy problem (6.2) is exponentially stable, then the real-valued function*

$$\Lambda(x) = \int_0^\infty \left\| u^x(t) \right\|_V^2 dt \tag{6.7}$$

satisfies conditions (1)–(3) in part (a).

Proof Note that for $t, t' \ge 0$,

$$\Lambda\big(u^x(t)\big) - \Lambda\big(u^x(t')\big) = \int_{t'}^t \frac{d}{ds} \Lambda\big(u^x(s)\big) ds.$$

But, using (2) and (3), we have

$$\frac{d}{ds} \Lambda\big(u^x(s)\big) = \langle \Lambda'\big(u^x(s)\big), Au^x(s) \rangle \le -c_3 \Lambda\big(u^x(s)\big).$$

Denoting $\Phi(t) = \Lambda(u^x(t))$, we can then write

$$\Phi'(t) \le -c_3 \Phi(t)$$

or, equivalently, $d(\Phi(t)e^{c_3 t})/dt \le 0$, giving $\Phi(t)e^{c_3 t} \le \Phi(0)$.

Using condition (1), we have

$$c_1 \left\| u^x(t) \right\|_H^2 \leq \Lambda\left(u^x(t)\right) \leq \Lambda(x) e^{-c_3 t} \leq c_2 \|x\|_H^2 e^{-c_3 t}.$$

To prove (b), we observe that, by the energy equality,

$$\left\| u^x(t) \right\|_H^2 = \|x\|_H^2 + 2 \int_0^t \left\langle A u^x(s), u^x(s) \right\rangle ds$$

$$\leq \|x\|_H^2 + |\lambda| \int_0^t \left\| u^x(s) \right\|_H^2 ds - \alpha \int_0^t \left\| u^x(s) \right\|_V^2 ds.$$

Hence,

$$\left\| u^x(t) \right\|_H^2 + \alpha \int_0^t \left\| u^x(s) \right\|_V^2 ds \leq \|x\|_H^2 + |\lambda| \int_0^t \left\| u^x(s) \right\|_H^2 ds.$$

Letting $t \to \infty$ and using the fact that

$$\left\| u^x(t) \right\|_H^2 \leq c \|x\|_H^2 e^{-\gamma t} \quad (\gamma > 0),$$

we obtain

$$\int_0^\infty \left\| u^x(s) \right\|_V^2 ds \leq \frac{1}{\alpha} \left(1 + \frac{|\lambda| c}{2\gamma} \right) \|x\|_H^2.$$

Define

$$\Lambda(x) = \int_0^\infty \left\| u^x(s) \right\|_V^2 ds,$$

then $\Lambda(x) \leq c_2 \|x\|_H^2$. Let x, $y \in H$ and consider

$$T(x, y) = \int_0^\infty \left(u^x(t), u^y(t) \right)_V dt.$$

Using the fact that $u^x(s) \in L^2([0, \infty), V)$ and the Schwarz inequality, we can see that $T(x, y)$ is a continuous bilinear form on V, which is continuous on H. Hence, $T(x, y) = \langle \tilde{C}x, y \rangle_H$. Since $\Lambda'(x) = 2\tilde{C}x$ (by identifying H with H^*), we can see that Λ and Λ' are locally bounded and continuous on H. By the continuity of the embedding $V \hookrightarrow H$, we have that for v, $v' \in V$, $T(v, v') = \langle Cv, v' \rangle_V$ for some bounded linear operator C on V, and property (2) in (a) follows. Now,

$$\left\| u^x(t) \right\|_H^2 - \|x\|_H^2 = 2 \int_0^t \left\langle A u^x(s), u^x(s) \right\rangle ds.$$

But $|\langle u^x(s), A u^x(s) \rangle| \leq c_2' \|u^x(s)\|_V^2$, giving

$$\left\| u^x(t) \right\|_H^2 - \|x\|_H^2 \geq -2c_2' \int_0^t \left\| u^x(s) \right\|_V^2 ds.$$

Let $t \to \infty$; then $\|u^x(t)\|_H^2 \to 0$, so that

$$-\|x\|_H^2 \geq -2c_2' \Lambda(x),$$

implying $\Lambda(x) \geq c_1 \|x\|_H^2$ for $c_1 = 1/(2c_2')$.

It remains to prove that $\Lambda(x)$ satisfies condition (3) in (a).

Note that

$$\Lambda\left(u^x(t)\right) = \int_0^\infty \left\|u^{u^x(t)}(s)\right\|_V^2 ds.$$

By the uniqueness of solution,

$$u^{u^x(t)}(s) = u^x(t+s).$$

Hence,

$$\Lambda\left(u^x(t)\right) = \int_0^\infty \left\|u^x(t+s)\right\|_V^2 ds = \int_t^\infty \left\|u^x(s)\right\|_V^2 ds.$$

Observe that

$$\frac{d}{ds} \Lambda\left(u^x(s)\right) = \left\langle \Lambda' u^x(s), A u^x(s) \right\rangle.$$

Since the map $\Lambda' : V \to H$ is continuous, we can write

$$\Lambda\left(u^x(t)\right) - \Lambda(x) = \int_0^t \left\langle \Lambda'\left(u^x(s)\right), A u^x(s) \right\rangle ds = -\int_0^t \left\|u^x(s)\right\|_V^2 ds.$$

By the continuity of the embedding $V \hookrightarrow H$, we have $\|x\|_H \leq c_0 \|x\|_V$, $x \in V$, $c_0 > 0$, and hence,

$$\int_0^t \left\langle \Lambda'\left(u^x(s)\right), A u^x(s) \right\rangle ds \leq -\frac{1}{c_0^2} \int_0^t \left\|u^x(s)\right\|_H^2 ds.$$

Now divide both sides by t and let $t \to 0$. Since Λ' is continuous and $u^x(\cdot) \in C([0,T], H)$, we get

$$\left\langle \Lambda'(x), Ax \right\rangle \leq -\frac{1}{c_0^2} \|x\|_H^2. \qquad \square$$

The following example shows that in the infinite-dimensional case, if we define

$$\Lambda(x) = \int_0^\infty \left\|u^x(t)\right\|_H^2 dt,$$

then $\Lambda(x)$ does not satisfy the lower bound in condition (2) of (a) of Theorem 6.3.

Example 6.1 Consider a solution of the following equation

$$\begin{cases} d_t u(t,x) = a^2 \dfrac{\partial^2 u}{\partial x^2} dt + \left(b \dfrac{\partial u}{\partial t} + cu\right) dt, \\ u(0,x) = \varphi(x) \in L^2(\mathbb{R}) \cap L^1(\mathbb{R}). \end{cases} \qquad (6.8)$$

Here, $H = L^2(\mathbb{R})$, and V is the Sobolev space $W^{1,2}(\mathbb{R})$. We denote by $\hat{\varphi}(\lambda)$ the Fourier transform of $\varphi(x)$ and use the similar notation $\hat{u}(t, \lambda)$ for the Fourier transform of $u(t, x)$. Then (6.8) can be written as follows:

$$\begin{cases} \dfrac{d\hat{u}(t, \lambda)}{dt} = -a^2\lambda^2\hat{u}(t, \lambda) + (ib\lambda + c)\hat{u}(t, \lambda), \\ \hat{u}(0, \lambda) = \hat{\varphi}(\lambda). \end{cases} \tag{6.9}$$

The solution is

$$\hat{u}^\varphi(t, \lambda) = \hat{\varphi}(\lambda)\exp\{(-a^2\lambda^2 + ib\lambda + c)t\}.$$

By Plancherel's theorem, $\|u^\varphi(t, \cdot)\|_H = \|\hat{u}^\varphi(t, \cdot)\|_H$, so that

$$\begin{aligned} \|u^\varphi(t, \cdot)\|_H^2 &= \int_{-\infty}^\infty |\hat{\varphi}(\lambda)|^2 \exp\{(-2a^2\lambda^2 + 2c)t\}\, d\lambda \\ &\leq \|\varphi\|_H^2 \exp\{\gamma t\} \quad (\gamma = 2c). \end{aligned}$$

For $c < 0$, we obtain an exponentially stable solution.

Take $A = -2a^2$, $B = 2c$. Then

$$\Lambda(\varphi) = \int_0^\infty \int_{-\infty}^\infty |\hat{\varphi}(\lambda)|^2 \exp\{-(A\lambda^2 + B)t\}\, d\lambda\, dt = \int_{-\infty}^\infty \frac{|\hat{\varphi}(\lambda)|^2}{A\lambda^2 + B}\, d\lambda$$

does not satisfy $\Lambda(\varphi) \geq c_1\|\varphi\|_H^2$ (see condition (1) in part (a) of Theorem 6.3).

In the next section, we consider the stability problem for infinite-dimensional stochastic differential equations using the Lyapunov function approach. We shall show that the fact that a Lyapunov function for the linear case is bounded below can be used to study the stability for nonlinear stochastic PDEs.

6.2 Exponential Stability for Stochastic Differential Equations

We recall some facts from Part I. Consider the following stochastic differential equation in H:

$$\begin{cases} dX(t) = (AX(t) + F(X(t)))\, dt + B(X(t))\, dW_t, \\ X(0) = x \in H, \end{cases} \tag{6.10}$$

where

(1) A is the generator of a C_0-semigroup $\{S(t), t \geq 0\}$ on H.
(2) W_t is a K-valued $\mathscr{F}_t$-Wiener process with covariance Q.
(3) $F : H \to H$ and $B : H \to \mathscr{L}(K, H)$ are Bochner-measurable functions satisfying

$$\|F(x)\|_H^2 + \mathrm{tr}(B(x)QB^*(x)) \leq \ell(1 + \|x\|_H^2),$$

$$\|F(x) - F(y)\|_H^2 + \mathrm{tr}((B(x) - B(y))Q(B(x) - B(y))^*) \leq \mathscr{K}\|x - y\|_H^2.$$

Then (6.10) has a unique $\mathscr{F}_t$-adapted mild solution (Chap. 3, Theorem 3.5), which is a Markov process (Chap. 3, Theorem 3.6) and depends continuously on the initial condition (Chap. 3, Theorem 3.7). That is, the integral equation

$$X(t) = S(t)x + \int_0^t S(t-s)F\big(X(s)\big)\,ds + \int_0^t S(t-s)B\big(X(s)\big)\,dW_s \quad (6.11)$$

has a solution in $C([0,T], L^{2p}((\Omega, \mathscr{F}, P), H))$, $p \geq 1$. Here $\mathscr{F} = \sigma(\bigcup_{t \geq 0} \mathscr{F}_t)$.

In addition, the solution of (6.11) can be approximated by solutions X_n obtained by using Yosida approximations of the operator A in the following manner.

Recall from (1.22), Chap. 1, that for $n \in \rho(A)$, the resolvent set of A, $R(n,A)$ denotes the resolvent of A at n, and if $R_n = nR(n, A)$, then $A_n = AR_n$ are the Yosida approximations of A. The approximating semigroup is $S_n(t) = e^{tA_n}$. Consider the strong solution X_n^x of

$$\begin{cases} dX(t) = (A_n X(t) + F(X(t)))\,dt + B(X(t))\,dW_t, \\ X(0) = x \in H. \end{cases} \quad (6.12)$$

Then $X_n^x \in C([0,T], L^{2p}((\Omega, \mathscr{F}, P), H))$, $p \geq 1$, by Theorem 3.5 in Chap. 3. By Proposition 3.2 in Chap. 3, for $p \geq 1$,

$$\lim_{n \to \infty} \sup_{0 \leq t \leq T} E\big(\|X_n^x(t) - X^x(t)\|_H^{2p}\big) = 0, \quad (6.13)$$

where $X^x(t)$ is the solution of (6.11).

We also recall the Itô formula, Theorem 2.9 in Chap. 2, for strong solutions of (6.10). Let $C_{b,\text{loc}}^2([0,T] \times H)$ denote the space of twice differentiable functions $\Psi : [0,T] \times H \to \mathbb{R}$ with locally bounded and continuous partial derivatives Ψ_t, Ψ_x, and Ψ_{xx}. Let $X^x(t)$ be a strong solution of (6.10), and $\Psi \in C_{b,\text{loc}}^2([0,T] \times H)$. Then, with $x \in \mathscr{D}(A)$,

$$\Psi\big(t, X^x(t)\big) - \Psi(0, x) = \int_0^t \big(\Psi_t\big(s, X^x(s)\big) + \mathscr{L}\Psi\big(s, X^x(s)\big)\big)\,ds$$

$$+ \int_0^t \big\langle \Psi_x\big(s, X^x(s)\big), B\big(X^x(s)\big)\,dW_s\big\rangle_H, \quad (6.14)$$

where

$$\mathscr{L}\Psi(t, x) = \big\langle \Psi_x(t, x), Ax + F(x)\big\rangle_H + \frac{1}{2}\,\text{tr}\big(\Psi_{xx}(t, x)B(x)QB^*(x)\big). \quad (6.15)$$

Clearly (6.14) is valid for strong solutions of (6.12), with $x \in H$ and A replaced by A_n in (6.15).

We are ready to discuss the stability of mild solutions of (6.10).

Definition 6.1 Let $\{X^x(t), t \geq 0\}$ be a mild solution of (6.10). We say that $X^x(t)$ is *exponentially stable in the mean square sense (m.s.s.)* if for all $t \geq 0$ and $x \in H$,

$$E\|X^x(t)\|_H^2 \leq ce^{-\beta t}\|x\|_H^2, \quad c, \beta > 0. \tag{6.16}$$

It is convenient to denote by $C_{2p}^2(H)$, with $p \geq 1$, the subspace of $C^2(H)$ consisting of functions $f : H \to \mathbb{R}$ whose first two derivatives satisfy the following growth condition:

$$\|f'(x)\|_H \leq C\|x\|_H^{2p} \quad \text{and} \quad \|f''(x)\|_{\mathscr{L}(H)} \leq C\|x\|_H^{2p}$$

for some constant $C \geq 0$.

Theorem 6.4 *The mild solution of (6.10) is exponentially stable in the m.s.s. if there exists a function* $\Lambda : H \to \mathbb{R}$ *satisfying the following conditions:*

(1) $\Lambda \in C_{2p}^2(H)$.
(2) *There exist constants* $c_1, c_2 > 0$ *such that*

$$c_1\|x\|_H^2 \leq \Lambda(x) \leq c_2\|x\|_H^2 \quad \text{for all } x \in H.$$

(3) *There exists a constant* $c_3 > 0$ *such that*

$$\mathscr{L}\Lambda(x) \leq -c_3\Lambda(x) \quad \text{for all } x \in \mathscr{D}(A)$$

with $\mathscr{L}\Lambda(x)$ *defined in (6.15).*

Proof Assume first that the initial condition $x \in \mathscr{D}(A)$. Let $X_n^x(t)$ be the mild solution of Theorem 3.5 in Chap. 3 to the approximating equation

$$\begin{cases} dX(t) = AX(t) + R_n F(X(t))\,dt + R_n B(X(t))\,dW_t, \\ X(0) = x \in \mathscr{D}(A), \end{cases} \tag{6.17}$$

that is,

$$X_n^x(t) = S(t)x + \int_0^t S(t-s)R_n F\big(X_n^x(s)\big)\,ds + \int_0^t S(t-s)R_n B\big(X_n^x(s)\big)\,dW_s$$

with R_n defined in (1.21). We note that (6.17) is an alternative to (6.12) in approximating the mild solution of (6.10) with strong solutions. This technique preserves the operator A and we have used it in the proof of Theorem 3.11.

Since $x \in \mathscr{D}(A)$ and $R_n : H \to \mathscr{D}(A)$, the solution $X_n^x(t) \in \mathscr{D}(A)$. Moreover, since the initial condition is deterministic, Theorem 3.5 in Chap. 3 guarantees that $X_n^x \in \tilde{\mathscr{H}}_2$. Then, the linear growth of B and the boundedness of R_n and $S(t)$ implies that the conditions of Theorem 3.2 are met, so that $X_n^x(t)$ is a strong solution of (6.17). We apply Itô's formula (6.14) to $e^{c_3 t}\Lambda(X_n^x(t))$ and take the expectations, to obtain

$$e^{c_3 t}E\Lambda\big(X_n^x(t)\big) - \Lambda\big(X_n^x(0)\big) = E\int_0^t e^{c_3 s}\big(c_3\Lambda\big(X_n^x(s)\big) + \mathscr{L}_n\Lambda\big(X_n^x(s)\big)\big)\,ds,$$

where

$$\mathscr{L}_n \Lambda(x) = \langle \Lambda'(x), Ax + R_n F(x) \rangle_H + \frac{1}{2} \operatorname{tr}(\Lambda''(x)(R_n B(x)) Q(R_n B(x))^*). \quad (6.18)$$

By condition (3),

$$c_3 \Lambda(x) + \mathscr{L}_n \Lambda(x) \le -\mathscr{L} \Lambda(x) + \mathscr{L}_n \Lambda(x).$$

The RHS of the above equals to

$$\langle \Lambda'(x), (R_n - I) F(x) \rangle_H$$
$$+ \frac{1}{2} \operatorname{tr}\{\Lambda''(x)[(R_n B(x)) Q(R_n B(x))^* - B(x) Q(B(x))^*]\}.$$

Hence,

$$e^{c_3 t} E \Lambda(X_n^x(t)) - \Lambda(x)$$

$$\le E \int_0^t e^{c_3 s} \Big\{ \langle \Lambda'(X_n^x(s)), (R_n - I) F(X_n^x(s)) \rangle_H$$

$$+ \frac{1}{2} \operatorname{tr}\{\Lambda''(X_n^x(s))[(R_n B(X_n^x(s))) Q(R_n B(X_n^x(s)))^*$$

$$- B(X_n^x(s)) Q(B(X_n^x(s)))^*]\} \Big\} ds. \quad (6.19)$$

In order to pass to the limit, we need to show that

$$\sup_{0 \le t \le T} E \|X_n^x(t) - X^x(t)\|_H^2 \to 0. \quad (6.20)$$

Consider

$$E \|X_n^x(t) - X^x(t)\|_H^2$$

$$\le E \Big\| \int_0^t S(t - s)(R_n F(X_n^x(s)) - F(X^x(s))) ds$$

$$+ \int_0^t S(t - s)(R_n B(X_n^x(s)) - B(X^x(s))) dW_s \Big\|_H^2$$

$$\le C \Big\{ E \Big\| \int_0^t S(t - s) R_n (F(X_n^x(s)) - F(X^x(s))) ds \Big\|_H^2$$

$$+ E \int_0^t \|S(t - s) R_n (B(X_n^x(s)) - B(X^x(s)))\|_{\mathscr{L}_2(K_Q, H)}^2 ds$$

$$+ E \Big\| \int_0^t S(t - s)(R_n - I) F(X^x(s)) ds \Big\|_H^2$$

$$+ E \int_0^t \|S(t - s)(R_n - I) B(X^x(s))\|_{\mathscr{L}_2(K_Q, H)}^2 ds \Big\}.$$

The first two summands are bounded by $C \mathcal{K} E \int_0^t \|X_n^x(s) - X^x(s)\|_H^2$ for $n > n_0$ (n_0 sufficiently large), where C depends on $\sup_{0 \le t \le T} \|S(t)\|_{\mathscr{L}(H)}$ and $\sup_{n > n_0} \|R_n\|_{\mathscr{L}(H)}$, and $\mathcal{K}$ is the Lipschitz constant.

By the properties of R_n, the integrand in the third summand converges to zero, and, by (2.17) in Lemma 2.2, Chap. 2, the integrand in the fourth summand converges to zero. Both integrands are bounded by $C\ell \|X^x(s)\|_H^2$ for some constant C depending on the norms of $S(t)$ and R_n, similar as above, and the constant ℓ in the linear growth condition. By the Lebesgue DCT, the third and fourth summands can be bounded uniformly in t by $\varepsilon_n(T) \to 0$.

An appeal to Gronwall's lemma completes the argument.

The convergence in (6.20) allows us to choose a subsequence $X_{n_k}^x$ such that

$$X_{n_k}^x(t) \to X^x(t), \quad 0 \le t \le T, \quad P\text{-a.s.}$$

We will denote such a subsequence again by X_n^x.

Now we use assumption (1), the continuity and local boundedness of Λ', the continuity of F, the uniform boundedness of $\|R_n\|_{\mathscr{L}(H)}$, and the convergence $(R_n - I)x \to 0$ to conclude that

$$E \int_0^t e^{c_3 s} \langle \Lambda'(X_n^x(s)), (R_n - I)F(X_n^x(s)) \rangle_H \, ds \to 0$$

by the Lebesgue DCT. Now, using Exercise 2.19, we have

$$\mathrm{tr}\{\Lambda''(X_n^x(s))(R_n B(X_n^x(s)))Q(R_n B(X_n^x(s)))^*\}$$
$$= \mathrm{tr}\{(R_n B(X_n^x(s)))^* \Lambda''(X_n^x(s))(R_n B(X_n^x(s)))Q\}$$
$$= \sum_{j=1}^\infty \lambda_j \langle \Lambda''(X_n^x(s))(R_n B(X_n^x(s)))f_j, (R_n B(X_n^x(s)))f_j \rangle_H$$

with

$$\langle \Lambda''(X_n^x(s))(R_n B(X_n^x(s)))f_j, (R_n B(X_n^x(s)))f_j \rangle_H$$
$$\to \langle \Lambda''(X^x(s))B(X^x(s))f_j, B(X^x(s))f_j \rangle_H.$$

Hence,

$$\mathrm{tr}\{\Lambda''(X_n^x(s))(R_n B(X_n^x(s)))Q(R_n B(X_n^x(s)))^*\}$$
$$\to \mathrm{tr}\{\Lambda''(X^x(s))B(X^x(s))Q(B(X^x(s)))^*\}.$$

Obviously,

$$\mathrm{tr}\{\Lambda''(X_n^x(s))B(X_n^x(s))Q(B(X_n^x(s)))^*\}$$
$$\to \mathrm{tr}\{\Lambda''(X^x(s))B(X^x(s))Q(B(X^x(s)))^*\}.$$

Now we use assumption (1), the continuity and local boundedness of Λ' and Λ'', the growth condition on F and B, and the fact that

$$\sup_{0 \leq t \leq T} E \| X_n^x(s) \|_H^2 < \infty,$$

and apply Lebesgue's DCT to conclude that the right-hand side in (6.19) converges to zero.

By the continuity of Λ and (6.20), we obtain

$$e^{c_3 t} E \Lambda \big(X^x(t) \big) \leq \Lambda(x),$$

and finally, by condition (2),

$$E \| X^x(t) \|_H^2 \leq \frac{c_2}{c_1} e^{-c_3 t} \| x \|_H^2, \quad x \in \mathscr{D}(A). \tag{6.21}$$

We recall that the mild solution $X^x(t)$ depends continuously on the initial condition $x \in H$ in the following way (Lemma 3.7):

$$\sup_{t \leq T} E \| X^x(t) - X^y(t) \|_H^2 \leq c_T \| x - y \|_H^2, \quad T > 0.$$

Then for $t \leq T$,

$$E \| X^x(t) \|_H^2 \leq E \| X^y(t) \|_H^2 + E \| X^x(t) - X^y(t) \|_H^2$$

$$\leq \frac{c_2}{c_1} e^{-c_3 t} \| y \|_H^2 + c_T \| x - y \|_H^2$$

$$\leq \frac{c_2}{c_1} e^{-c_3 t} 2 \| x - y \|_H^2 + \frac{c_2}{c_1} e^{-c_3 t} 2 \| x \|_H^2 + c_T \| x - y \|_H^2$$

for all $y \in \mathscr{D}(A)$, forcing inequality (6.21) to hold for all $x \in H$, since $\mathscr{D}(A)$ is dense in H. □

The function Λ defined in Theorem 6.4, satisfying conditions (1)–(3), is called a *Lyapunov function*.

We now consider the linear case of (6.10) with $F \equiv 0$ and $B(x) = B_0 x$, where $B_0 \in \mathscr{L}(H, \mathscr{L}(K, H))$, and $\| B_0 x \| \leq d \| x \|_H$,

$$\begin{cases} dX(t) = AX(t)\, dt + B_0 X(t)\, dW_t, \\ X(0) = x \in H. \end{cases} \tag{6.22}$$

Mild solutions are solutions of the corresponding integral equation

$$X(t) = S(t) x + \int_0^t S(t - s) B_0 X(s)\, dW_s. \tag{6.23}$$

The concept of exponential stability in the m.s.s. for mild solutions of (6.22) obviously transfers to this case. We show that the existence of a Lyapunov function is a necessary condition for stability of mild solutions of (6.22). The following notation will be used:

$$\mathscr{L}_0 \Psi(x) = \langle \Psi'(x), Ax \rangle_H + \frac{1}{2} \mathrm{tr}\big(\Psi''(x)(B_0 x) Q (B_0 x)^*\big). \tag{6.24}$$

Theorem 6.5 *Assume that A generates a pseudo-contraction semigroup of operators $\{S(t), t \geq 0\}$ on H and that the mild solution of (6.22) is exponentially stable in the m.s.s. Then there exists a function $\Lambda_0(x)$ satisfying conditions (1) and (2) of Theorem 6.4 and the condition*

(3') $\mathscr{L}_0 \Lambda_0(x) \leq -c_3 \Lambda_0(x), x \in \mathscr{D}(A),$ *for some $c_3 > 0$.*

Proof Let

$$\Lambda_0(x) = \int_0^\infty E \|X^x(t)\|_H^2 \, dt + \alpha \|x\|_H^2,$$

where the value of the constant $\alpha > 0$ will be determined later. Note that $X^x(t)$ depends on x linearly. The exponential stability in the m.s.s. implies that

$$\int_0^\infty E \|X^x(t)\|_H^2 \, dt < \infty.$$

Hence, by the Schwarz inequality,

$$T(x, y) = \int_0^\infty E \langle X^x(t), X^y(t) \rangle_H \, dt$$

defines a continuous bilinear form on $H \times H$, and there exists a symmetric bounded linear operator $\tilde{T} : H \to H$ such that

$$\langle \tilde{T} x, x \rangle_H = \int_0^\infty E \|X^x(t)\|_H^2 \, dt.$$

Let

$$\Psi(x) = \langle \tilde{T} x, x \rangle_H.$$

Using the same arguments, we define bounded linear operators on H by

$$\langle \tilde{T}(t) x, x \rangle_H = \int_0^t E \|X^x(s)\|_H^2 \, ds.$$

Consider solutions $\{X_n^x(t), t \geq 0\}$ to the following equation:

$$\begin{cases} dX(t) = A_n X(t) \, dt + B_0 X(t) \, dW_t, \\ X(0) = x \in H, \end{cases}$$

obtained using the Yosida approximations of A. Just as above, we have continuous bilinear forms T_n, symmetric linear operators $\tilde{T}_n(t)$, and real-valued continuous functions $\Psi_n(t)$, defined for X_n,

$$T_n(t)(x, y) = \int_0^t E\langle X_n^x(u), X_n^y(u)\rangle_H \, du,$$

$$\langle \tilde{T}_n(t)x, x\rangle_H = \int_0^t E\|X_n^x(u)\|_H^2 \, du,$$

$$\Psi_n(t)(x) = \langle \tilde{T}_n(t)x, x\rangle_H.$$

We have

$$\Psi_n(t)\big(X_n^x(s)\big) = \left(\int_0^t E\|X_n^y(u)\|_H^2 \, du \right)\bigg|_{y=X_n^x(s)}.$$

Let $\varphi : H \to \mathbb{R}$, $\varphi(h) = \|h\|_H^2$, and

$$\big(\tilde{P}_t\varphi\big)(x) = E\varphi\big(X^x(t)\big)$$

be the transition semigroup. Using the uniqueness of the solution, the Markov property (3.59) yields

$$E\Psi_n(t)\big(X_n^x(s)\big) = E\int_0^t \big(\tilde{P}_u\varphi\big)\big(X_n^x(s)\big)\, du$$

$$= E\int_0^t E\big(\varphi\big(X_n^x(u+s)\big)\big|\mathscr{F}_s^{X_n^x}\big)\, du$$

$$= \int_0^t E\|X_n^x(u+s)\|_H^2 \, du$$

$$= \Psi_n(t+s)(x) - \Psi_n(s)(x). \tag{6.25}$$

With t and n fixed, we use the Itô formula for the function $\Psi_n(t)(x)$, then take the expectation of both sides to arrive at

$$E\big(\Psi_n(t)\big(X_n^x(s)\big)\big) = \Psi_n(t)(x) + \int_0^s E\big(\mathscr{L}_n\Psi_n(t)\big(X_n^x(u)\big)\big)\, du, \tag{6.26}$$

where

$$\mathscr{L}_n\Psi_n(t)(x) = 2\langle \tilde{T}_n(t)x, A_nx\rangle_H + \mathrm{tr}\big(\tilde{T}_n(t)(B_0x)Q(B_0x)^*\big).$$

Putting (6.25) and (6.26) together, we have

$$\Psi_n(t+s)(x) - \Psi_n(s)(x) = \int_0^s E\big(\mathscr{L}_n\Psi_n(t)\big(X_n^x(u)\big)\big)\, du + \Psi_n(t)(x).$$

Rearranging the above, dividing by s, and taking the limit as $s \to 0$ give

$$\frac{\Psi_n(t+s)(x) - \Psi_n(t)(x)}{s} = \frac{1}{s}\int_0^s E\big(\mathscr{L}_n\Psi_n(t)\big(X_n^x(u)\big)\big)\, du + \frac{\Psi_n(s)(x)}{s}. \tag{6.27}$$

We fix n and t, and intend to take the limit in (6.27) as $s \to 0$.

The processes $X_n^x(u)$ are continuous in the mean-square, since

$$E \| X_n^x(u) - X_n^x(v) \|_H^2$$

$$\leq 2 \big(\| A_n \|_{\mathscr{L}(H)}^2 + \| B_0 \|_{\mathscr{L}(H,\mathscr{L}(K,H))}^2 \big) \operatorname{tr}(Q) \int_u^v E \| X_n^x(r) \|_H^2 \, dr.$$

Hence,

$$\lim_{s \to 0} \frac{\Psi_n(s)(x)}{s} = \lim_{s \to 0} \frac{1}{s} \int_0^s E \| X_n^x(u) \|_H^2 \, du = \| x \|_H^2. \tag{6.28}$$

Now consider

$$E \mathscr{L}_n \Psi_n(t) \big(X_n^x(u) \big)$$

$$= E \big(2 \langle \tilde{T}_n(t) X_n^x(u), A_n X_n^x(u) \rangle_H \big) + E \big(\operatorname{tr} \big(\tilde{T}_n(t) \big(B_0 X_n^x(u) \big) Q \big(B_0 X_n^x(u) \big)^* \big) \big).$$

Since

$$\lim_{u \to 0} A_n X_n^x(u) = A_n x, \qquad \lim_{u \to 0} \tilde{T}_n(t) X_n^x(u) = \tilde{T}_n(t) x,$$

and

$$\big| \langle \tilde{T}_n(t) X_n^x(u), A_n X_n^x(u) \rangle \big| \leq \| \tilde{T}_n(t) \|_{\mathscr{L}(H)} \| A_n \|_{\mathscr{L}(H)} \| X_n^x(u) \|_H^2 \in L^1(\Omega),$$

the Lebesgue DCT gives

$$\lim_{u \to 0} E \big(2 \langle \tilde{T}_n(t) X_n^x(u), A_n X_n^x(u) \rangle_H \big) = 2 \langle \tilde{T}_n(t) x, A_n x \rangle_H.$$

For the term involving the *trace*, we simplify the notation and denote

$$\Phi_n(u) = B_0 X_n^x(u) \quad \text{and} \quad x_n^j(u) = \Phi_n(u) f_j,$$

where $\{ f_j \}_{j=1}^\infty$ is an ONB in K that diagonalizes the covariance operator Q. Using Exercise 2.19, we have

$$\operatorname{tr} \big(\tilde{T}_n(t) \Phi_n(u) Q \big(\Phi_n(u) \big)^* \big) = \operatorname{tr} \big(\big(\Phi_n(u) \big)^* \tilde{T}_n(t) \Phi_n(u) Q \big)$$

$$= \sum_{j=1}^\infty \lambda_j \langle \tilde{T}_n(t) \Phi_n(u) f_j, \Phi_n(u) f_j \rangle_H$$

$$= \sum_{j=1}^\infty \lambda_j \langle \tilde{T}_n(t) x_n^j(u), x_n^j(u) \rangle_H$$

$$= \sum_{j=1}^\infty \lambda_j \int_0^t E \| X_n^{x_n^j(u)}(s) \|_H^2 \, ds. \tag{6.29}$$

Denote $x^j = (B_0 x) f_j$. Since B_0 is continuous, as $u \to 0$, $B_0 X_n^x(u) \to B_0 x$ in $\mathscr{L}(K, H)$, so that $x_n^j(u) \to x^j$ in H. By the continuity of the solution X_n with respect to the initial condition (Chap. 3, Lemma 3.7),

$$\sup_{0 \le s \le T} E \| X_n^{x_n^j(u)}(s) - X_n^{x^j}(s) \|_H^2 \to 0 \quad \text{as } u \to 0,$$

so that, by Lebesgue's DCT and by reversing the calculations in (6.29),

$$\sum_{j=1}^{\infty} \lambda_j \int_0^t E \| X_n^{x_n^j(u)}(s) \|_H^2 \, ds$$

$$\to \sum_{j=1}^{\infty} \lambda_j \int_0^t E \| X_n^{x^j}(s) \|_H^2 \, ds = \text{tr}\big(\tilde{T}_n(t)(B_0 x) Q (B_0 x)^*\big).$$

Summarizing, we proved that

$$\frac{d\Psi_n(t)(x)}{dt} = \mathscr{L}_n \Psi_n(t)(x) + \|x\|_H^2.$$

In the next step, we fix t and allow $n \to \infty$. By the mean-square continuity of $X_n^x(t)$ and the definition of $\langle \tilde{T}_n(t)x, x \rangle_H$ and $\langle \tilde{T}(t)x, x \rangle_H$, we can calculate the derivatives below, and the convergence follows from condition (6.13):

$$\frac{d\Psi_n(t)(x)}{dt} = E \| X_n^x(t) \|_H^2 \to E \| X^x(t) \|_H^2 = \frac{d\Psi(t)(x)}{dt}.$$

Now, we need to show that as $n \to \infty$, for $x \in \mathscr{D}(A)$,

$$\mathscr{L}_n \langle \tilde{T}_n(t)x, x \rangle_H \to \mathscr{L}_0 \langle \tilde{T}(t)x, x \rangle_H. \tag{6.30}$$

Consider

$$\left| \langle \tilde{T}_n(t)x, A_n x \rangle_H - \langle \tilde{T}(t)x, Ax \rangle_H \right|$$

$$\le \| \tilde{T}_n(t)x \|_H \| (A_n - A)x \|_H + \left| \langle (\tilde{T}_n(t) - \tilde{T}(t))x, Ax \rangle_H \right| \to 0.$$

Since (6.13) implies that

$$\lim_{n \to \infty} E \int_0^T \| X_n^x(u) - X^x(u) \|_H^2 \, du = 0, \tag{6.31}$$

we thus have the weak convergence of $T_n(t)x$ to $T(t)x$, and, further, by the Banach–Steinhaus theorem, we deduce that $\sup_n \|T_n(t)\|_{\mathscr{L}(H)} < \infty$. Using calculations similar as in (6.29), we have

$$\mathrm{tr}\big(\tilde{T}_n(t)(B_0 x)Q(B_0 x)^*\big) = \sum_{j=1}^{\infty} \lambda_j \big\langle \tilde{T}_n(t)x^j, x^j \big\rangle_H$$

$$\to \sum_{j=1}^{\infty} \lambda_j \big\langle \tilde{T}(t)x^j, x^j \big\rangle_H$$

$$= \mathrm{tr}\big(\tilde{T}(t)B_0 x Q(B_0 x)^*\big),$$

by Lebesgue's DCT, proving the convergence in (6.30). Summarizing, we have

$$\frac{d\langle \tilde{T}(t)x, x \rangle_H}{dt} = \mathscr{L}_0 \langle \tilde{T}(t)x, x \rangle_H + \|x\|_H^2.$$

We will now let $t \to \infty$. Then, by the exponential stability condition,

$$\frac{d\langle \tilde{T}(t)x, x \rangle_H}{dt} = E\big\|X^x(t)\big\|_H^2 \to 0.$$

Since $\langle \tilde{T}(t)x, x \rangle_H \to \langle \tilde{T}x, x \rangle_H$, using the weak convergence of $\tilde{T}(t)x$ to $\tilde{T}x$ and the Lebesgue DCT, exactly as above, we obtain that

$$\mathscr{L}_0 \langle \tilde{T}(t)x, x \rangle_H = 2\langle \tilde{T}(t)x, Ax \rangle_H + \mathrm{tr}\big(\tilde{T}(t)B_0 x Q(B_0 x)^*\big)$$

$$\to 2\langle \tilde{T}x, Ax \rangle_H + \mathrm{tr}\big(\tilde{T}B_0 x Q(B_0 x)^*\big) = \mathscr{L}_0 \langle \tilde{T}x, x \rangle_H.$$

In conclusion,

$$\mathscr{L}_0 \Psi(x) = -\|x\|_H^2, \quad x \in \mathscr{D}(A).$$

Now, Λ_0 satisfies conditions (1) and (2). To prove condition (3'), let us note that, as in Sect. 6.1, since $\|S(t)\| \le e^{\omega t}$, inequality (6.4) is valid for some constant $\lambda > 0$. Hence,

$$\mathscr{L}_0 \|x\|_H^2 = 2\langle x, Ax \rangle_H + \mathrm{tr}\big((B_0 x)Q(B_0 x)^*\big) \le \big(2\lambda + d^2 \mathrm{tr}\, Q\big)\|x\|_H^2 \qquad (6.32)$$

gives

$$\mathscr{L}_0 \Lambda_0(x) \le -\|x\|_H^2 + \alpha\big(2\lambda + d^2 \mathrm{tr}(Q)\big)\|x\|_H^2 \le -c_3 \Lambda_0(x),$$

$c_3 > 0$, by choosing α small enough. □

Remark 6.1 For the nonlinear equation (6.10), we need to assume $F(0) = 0$ and $B(0) = 0$ to assure that zero is a solution. In this case, if the solution $\{X^x(t), t \ge 0\}$ is exponentially stable in the m.s.s., we can still construct

$$\Lambda(x) = \int_0^{\infty} E\big\|X^x(t)\big\|_H^2 \, dt + \alpha \|x\|_H^2.$$

We however do not know if it satisfies condition (1) of Theorem 6.4. If we assume that it does, then one can show, as in Theorem 6.5, that it satisfies condition (2). Then we can prove that $\Lambda(x)$ also satisfies condition (3).

First, observe that for $\Psi(x) = \langle Rx, x \rangle_H$, as before,

$$\mathscr{L}\Psi(x) = -\|x\|_H^2$$

and

$$\mathscr{L}\Lambda(x) = \mathscr{L}\Psi(x) + \alpha\mathscr{L}\|x\|_H^2$$
$$= -\|x\|_H^2 + \alpha\big(2\langle x, Ax + F(x)\rangle_H + \mathrm{tr}\big(B(x)QB^*(x)\big)\big),$$

noting the form of the infinitesimal generator $\mathscr{L}$ of the Markov process $X^x(t)$. We obtain

$$\mathscr{L}\Lambda(x) \le -\|x\|_H^2 + 2\alpha\lambda\|x\|_H^2 + \alpha\big(2\langle x, F(x)\rangle_H + \mathrm{tr}\big(B(x)QB^*(x)\big)\big).$$

Now using the fact that $F(0) = 0$, $B(0) = 0$, and the Lipschitz property of F and B, we obtain

$$\mathscr{L}\Lambda(x) \le -\|x\|_H^2 + \alpha\big(2\lambda + 2\mathscr{K} + \mathscr{K}^2\,\mathrm{tr}(Q)\big)\|x\|_H^2.$$

Hence, for α small enough, condition (3) follows from condition (2).

As shown in Part I, the differentiability with respect to the initial value requires stringent assumptions on the coefficients F and B. In order to make the result more applicable, we provide another technique that uses first-order approximation. We use trace norm of a difference of nonnegative definite operators in the approximation condition. Recall that for any trace-class operator T, we defined the trace norm in (2.1) by

$$\tau(T) = \mathrm{tr}\big((TT^*)^{1/2}\big).$$

Note (see [68]) that for a trace-class operator T and a bounded operator S,

(a) $|\mathrm{tr}(T)| \le \tau(T)$,
(b) $\tau(ST) \le \|S\|\tau(T)$ and $\tau(TS) \le \|S\|\tau(T)$.

Theorem 6.6 *Assume that A generates a pseudo-contraction semigroup of operators $\{S(t), t \ge 0\}$ on H. Suppose that the solution $\{X_0^x(t), t \ge 0\}$ of the linear equation (6.22) is exponentially stable in the m.s.s. Then the solution $\{X^x(t), t \ge 0\}$ of (6.10) is exponentially stable in the m.s.s. if*

$$2\|x\|_H\|F(x)\|_H + \tau\big(B(x)QB^*(x) - B_0x Q(B_0x)^*\big) \le \frac{\beta}{2c}\|x\|_H^2. \tag{6.33}$$

Proof Let $\Lambda_0(x) = \langle \tilde{T}x, x \rangle_H + \alpha \|x\|_H^2$, as in the proof of Theorem 6.5. Note that $\langle \tilde{T}x, x \rangle_H = E \int_0^\infty \|X_0^x(t)\|_H^2 \, dt$, so that

$$\langle \tilde{T}x, x \rangle_H \leq \int_0^\infty ce^{-\beta t} \|x\|_H^2 \, dt = \frac{c}{\beta} \|x\|_H^2.$$

Hence, $\|\tilde{T}\|_{\mathscr{L}(H)} \leq c/\beta$. Clearly Λ_0 satisfies conditions (1) and (2) of Theorem 6.4. It remains to prove that

$$\mathscr{L}\Lambda_0(x) \leq -c_3 \Lambda_0(x).$$

Consider

$$\mathscr{L}\Lambda_0(x) - \mathscr{L}_0\Lambda_0(x)$$

$$= \langle \Lambda_0'(x), F(x) \rangle_H + \frac{1}{2} \operatorname{tr}\left(\Lambda_0''(x)\left(B(x)QB^*(x) - (B_0x)Q(B_0x)^*\right)\right)$$

$$\leq 2\langle (\tilde{T}+\alpha)x, F(x) \rangle_H + \tau\left((\tilde{T}+\alpha)\left(B(x)QB^*(x) - (B_0x)Q(B_0x)^*\right)\right)$$

$$\leq \left(\|\tilde{T}\|_{\mathscr{L}(H)} + \alpha\right)\left(2\|x\|_H \|F(x)\|_H + \tau\left(B(x)QB^*(x) - (B_0x)Q(B_0x)^*\right)\right)$$

$$\leq \left(\frac{1}{2} + \alpha\frac{\beta}{2c}\right)\|x\|_H^2.$$

It follows that

$$\mathscr{L}\Lambda_0(x) \leq \mathscr{L}_0\Lambda_0(x) + \left(\frac{1}{2} + \alpha\frac{\beta}{2c}\right)\|x\|_H^2$$

$$\leq -\|x\|_H^2 + \alpha\left(2\lambda + d^2\operatorname{tr}(Q)\right)\|x\|_H^2 + \left(\frac{1}{2} + \frac{\alpha\beta}{2c}\right)\|x\|_H^2.$$

For α small enough, we obtain condition (3) in Theorem 6.4 using condition (2). $\square$

We now consider stability in probability of the zero solution of (6.10).

Definition 6.2 Let $\{X^x(t)\}_{t \geq 0}$ be the mild solution of (6.10) with $F(0) = 0$ and $B(0) = 0$ (assuring that zero is a solution). The *zero solution* of (6.10) is called *stable in probability* if for any $\varepsilon > 0$,

$$\lim_{\|x\|_H \to 0} P\left(\sup_{t \geq 0} \|X^x(t)\|_H > \varepsilon \right) = 0. \tag{6.34}$$

Once a Lyapunov function satisfying conditions (1) and (2) of Theorem 6.4 is constructed, the following theorem provides a technique for proving condition (6.34).

Theorem 6.7 *Let $X^x(t)$ be the solution of (6.10). Assume that there exists a function $\Psi \in C_{2p}^2(H)$ having the following properties:*

(1) $\Psi(x) \to 0$ as $\|x\|_H \to 0$.
(2) $\inf_{\|x\|_H > \varepsilon} \Psi(x) = \lambda_\varepsilon > 0$.
(3) $\mathscr{L}\Psi(x) \le 0$, when $x \in \mathscr{D}(A)$ and $\|x\|_H < \delta$ for some $\delta > 0$.

Then, $\{X^x(t), \, t \ge 0\}$ satisfies condition (6.34).

Proof The proof is similar to the proof of Theorem 6.4. We assume first that the initial condition $x \in \mathscr{D}(A)$ and consider strong solutions $X_n^x(t)$ of the approximating equations (6.17), $n = 1, 2, \dots$,

$$X_n^x(t) = S(t)x + \int_0^t S(t-s) R_n F\left(X_n^x(s)\right) ds + \int_0^t S(t-s) R_n B\left(X_n^x(s)\right) dW_s.$$

Denote $B_\varepsilon = \{x \in H : \|x\| < \varepsilon\}$ and let

$$\tau_\varepsilon = \inf\{t : \left\|X^x(t)\right\|_H > \varepsilon\} \quad \text{and} \quad \tau_\varepsilon^n = \inf\{t : \left\|X_n^x(t)\right\|_H > \varepsilon\}.$$

Applying Itô's formula to $\Psi(X_n^x(t))$ and taking the expectations yield

$$E\psi\left(X_n^x(t \wedge \tau_\varepsilon^n)\right) - \psi(x) = E \int_0^{t \wedge \tau_\varepsilon} \mathscr{L}_n \Psi\left(X_n^x(s)\right) ds,$$

where

$$\mathscr{L}_n \Psi(x) = \left\langle \Psi'(x), Ax + R_n F(x) \right\rangle_H + \frac{1}{2} \operatorname{tr}\left(\Psi''(x)\left(R_n B(x)\right) Q\left(R_n B(x)\right)^*\right).$$

Let $\varepsilon < \delta$. Then for $x \in B_\varepsilon$, using condition (3), we get

$$\mathscr{L}_n \Psi(x) \le -\mathscr{L}\Psi(x) + \mathscr{L}_n \Psi(x)$$
$$= \left\langle \Psi'(x), (R_n - I) F(x) \right\rangle_H$$
$$+ \frac{1}{2} \operatorname{tr}\left\{\Psi''(x)\left[\left(R_n B(x)\right) Q\left(R_n B(x)\right)^* - B(x) Q\left(B(x)\right)^*\right]\right\}.$$

Hence,

$$E\Psi\left(X_n^x(t \wedge \tau_\varepsilon^n)\right) - \Psi(x)$$
$$\le E \int_0^{t \wedge \tau_\varepsilon^n} \left\{\left\langle \Psi'\left(X_n^x(s)\right), (R_n - I) F\left(X_n^x(s)\right)\right\rangle_H\right.$$
$$+ \frac{1}{2} \operatorname{tr}\left\{\Psi''\left(X_n^x(s)\right)\left[\left(R_n B\left(X_n^x(s)\right)\right) Q\left(R_n B\left(X_n^x(s)\right)\right)^*\right.\right.$$
$$\left.\left.\left. - B\left(X_n^x(s)\right) Q\left(B\left(X_n^x(s)\right)\right)^*\right]\right\}\right\} ds. \tag{6.35}$$

Using (6.20) and passing to the limit, as in the proof of Theorem 6.4, show that the RHS in (6.35) converges to zero. Using condition (2), we conclude that for $x \in \mathscr{D}(A) \cap B_\varepsilon$ and any n,

$$\Psi(x) \ge E\left(\Psi\left(X_n^x(t \wedge \tau_\varepsilon^n)\right)\right) \ge \lambda_\varepsilon P\left(\tau_\varepsilon^n < t\right). \tag{6.36}$$

By the a.s. (and hence weak) convergence of τ_ε^n to τ_ε, we have that

$$\Psi(x) \geq \lambda_\varepsilon \liminf_{n\to\infty} P\left(\tau_\varepsilon^n < t\right) \geq \lambda_\varepsilon P(\tau_\varepsilon < t).$$

To remove the restriction that $x \in \mathscr{D}(A)$, recall that

$$\sup_{0 \leq t \leq T} E\left\|X^x(t) - X^y(t)\right\|_H^2 \to 0 \quad \text{as } \|y - x\|_H \to 0.$$

We can select a sequence $y_n \to x$, $y_n \in \mathscr{D}(A)$, such that $X^{y_n}(t) \to X^x(t)$ a.s. for all t. Now using the assumptions on Ψ and the Lebesgue DCT, we obtain (6.36) for all $x \in H$. Inequality (6.36), together with conditions (2) and (1), implies that for $x \in B_\varepsilon$,

$$P\left(\sup_{t \geq 0}\|X_t^x\|_H > \varepsilon\right) \leq \frac{\Psi(x)}{\lambda_\varepsilon} \to 0, \quad \|x\|_H \to 0,$$

giving (6.34). $\qquad\square$

The following results are now obvious from Theorems 6.5 and 6.6.

Theorem 6.8 *Assume that A generates a pseudo-contraction semigroup of operators $\{S(t), t \geq 0\}$ on H. If the solution $X_0^x(t)$ of the linear equation (6.22) is exponentially stable in the m.s.s., then the zero solution of (6.22) is stable in probability.*

Theorem 6.9 *Assume that A generates a pseudo-contraction semigroup of operators $\{S(t), t \geq 0\}$ on H. If the solution $X_0^x(t)$ of the linear equation (6.22) is exponentially stable in the m.s.s. and condition (6.33) holds for a sufficiently small neighborhood of $x = 0$, then the zero solution of (6.10) is stable in probability.*

We note that the exponential stability gives degenerate invariant measures. To obtain nondegenerate invariant measures, we use a more general concept introduced in Chap. 7.

6.3 Stability in the Variational Method

We consider a Gelfand triplet of real separable Hilbert spaces

$$V \hookrightarrow H \hookrightarrow V^*.$$

The space V^* is the continuous dual of V, V is dense in H, and all embeddings are continuous. With $\langle \cdot, \cdot \rangle$ denoting the duality between V and V^*, we assume that for $h \in H$,

$$\langle v, h \rangle = \langle v, h \rangle_H.$$

Let K be a real separable Hilbert space, and Q a nonnegative definite trace-class operator on K. We consider $\{W_t, t \geq 0\}$, a K-valued Q-Wiener process defined on a filtered probability space $(\Omega, \mathscr{F}, \{\mathscr{F}\}_{t\geq0}, P)$.

Let $M^2([0, T], V)$ denote the space of all V-valued measurable processes satisfying

(1) $u(t, \cdot)$ is $\mathscr{F}_t$-measurable,

(2) $E \int_0^T \|u(t, \omega)\|_V^2 \, dt < \infty$.

Throughout this section we consider the following equation:

$$\begin{cases} dX(t) = A(X(t)) \, dt + B(X(t)) \, dW_t, \\ X(0) = x \in H, \end{cases} \tag{6.37}$$

where A and B are in general nonlinear mappings, $A : V \to V^*$, $B : V \to \mathscr{L}(K, H)$, and

$$\|A(v)\|_{V^*} \leq a_1 \|v\|_V \quad \text{and} \quad \|B(v)\|_{\mathscr{L}(K,H)} \leq b_1 \|v\|_V, \quad v \in V, \tag{6.38}$$

for some positive constants a_1, b_1.

We recall the coercivity and weak monotonicity conditions from Chap. 4, which we impose on the coefficients of (6.37)

(C) (Coercivity) There exist $\alpha > 0$, γ, $\lambda \in \mathbb{R}$ such that for all $v \in V$,

$$2\langle v, A(v)\rangle + \text{tr}\big(B(v)QB^*(v)\big) \leq \lambda \|v\|_H^2 - \alpha \|v\|_V^2 + \gamma. \tag{6.39}$$

(WM) (Weak Monotonicity) There exists $\lambda \in \mathbb{R}$ such that for all $u, v \in V$,

$$2\langle u - v, A(u) - A(v)\rangle + \text{tr}\big((B(u) - B(v))Q(B(u) - B(v))^*\big) \\ \leq \lambda \|u - v\|_H^2. \tag{6.40}$$

Since conditions (6.38), (6.39), and (6.40) are stronger than the assumptions in Theorem 4.7 (also in Theorem 4.4) of Chap. 4, we conclude that there exists a unique strong solution $\{X^x(t), t \geq 0\}$ of (6.37) such that

$$X^x(\cdot) \in L^2\big(\Omega, C([0, T], H)\big) \cap M^2\big([0, T], V\big).$$

Furthermore, the solution $X^x(t)$ is Markovian, and the corresponding semigroup has the Feller property.

The major tool we will use will be the Itô formula due to Pardoux [62]. It was introduced in Part I, Sect. 4.2, Theorem 4.3, for the function $\Psi(u) = \|u\|_H^2$.

Theorem 6.10 (Itô Formula) *Suppose that* $\Psi : H \to \mathbb{R}$ *satisfies the following conditions:*

(1) Ψ *is twice Fréchet differentiable, and* Ψ, Ψ', Ψ'' *are locally bounded.*

(2) Ψ *and* Ψ' *are continuous on* H.

(3) *For all trace-class operators* T *on* H, $\text{tr}(T\Psi''(\cdot)) : H \to \mathbb{R}$ *is continuous.*

(4) *If $v \in V$, then $\Psi'(v) \in V$, and for any $v' \in V^*$, the function $\langle \Psi'(\cdot), v' \rangle : V \to \mathbb{R}$ is continuous.*

(5) $\|\Psi'(v)\|_V \leq c_0(1 + \|v\|_V)$ *for some constant $c_0 > 0$ and any $v \in V$.*

Let $X^x(t)$ *be a solution of (6.37) in* $L^2(\Omega, C([0, T], H)) \cap M^2([0, T], V)$. *Then*

$$\Psi\big(X^x(t)\big) = \Psi(x) + \int_0^t \mathscr{L}\Psi\big(X^x(s)\big) \, ds + \int_0^t \big\langle \Psi'\big(X^x(s)\big), B\big(X^x(s)\big) dW_s \big\rangle_H,$$

where

$$\mathscr{L}\Psi(u) = \big\langle \Psi'(u), A(u) \big\rangle + \frac{1}{2} \operatorname{tr}\big(\Psi''(u) B(u) Q B^*(u)\big).$$

We extend the notion of exponential stability in the m.s.s. to the variational case.

Definition 6.3 We say that the strong solution of the variational equation (6.37) in the space $L^2(\Omega, C([0, T], H)) \cap M^2([0, T], V)$ is *exponentially stable in the m.s.s.* if it satisfies condition (6.16) in Definition 6.1.

The following is the analogue of Theorem 6.4 in the variational context. The proof for a strong solution is a simplified version of the proof of Theorem 6.4 and is left to the reader as an exercise.

Theorem 6.11 *The strong solution of the variational equation (6.37) in the space* $L^2(\Omega, C([0, T], H)) \cap M^2([0, T], V)$ *is exponentially stable in the m.s.s. if there exists a function Ψ satisfying conditions (1)–(5) of Theorem 6.10, and the following two conditions hold:*

(1) $c_1 \|x\|_H^2 \leq \Psi(x) \leq c_2 \|x\|_H^2$, $c_1, c_2 > 0$, $x \in H$.
(2) $\mathscr{L}\Psi(v) \leq -c_3 \Psi(v)$, $c_3 > 0$, $v \in V$, *with $\mathscr{L}$ defined in Theorem 6.10.*

Exercise 6.2 Prove Theorem 6.11.

We now consider the linear problem analogous to (6.37). Let $A_0 \in \mathscr{L}(V, V^*)$ and $B_0 \in \mathscr{L}(V, \mathscr{L}(K, H))$. In order to construct a Lyapunov function directly from the solution, we assume a more restrictive coercivity condition

(C') (Coercivity) There exist $\alpha > 0$, $\lambda \in \mathbb{R}$ such that for all $v \in V$,

$$2\langle v, A_0 v \rangle + \operatorname{tr}\big((B_0 v) Q (B_0 v)^*\big) \leq \lambda \|v\|_H^2 - \alpha \|v\|_V^2. \tag{6.41}$$

We denote by $\mathscr{L}_0$ the operator $\mathscr{L}$ with A and B replaced by A_0 and B_0. Consider the following linear problem:

$$\begin{cases} dX(t) = A_0 X(s) \, ds + B_0(X(s)) \, dW_s, \\ X(0) = x \in H. \end{cases} \tag{6.42}$$

Theorem 6.12 *Under the coercivity condition (6.41), the solution of the linear equation (6.42) is exponentially stable in the m.s.s. if and only if there exists a* *function Ψ satisfying conditions (1)–(5) of Theorem 6.10 and conditions (1) and (2) of Theorem 6.11.*

Remark 6.2 A function Ψ satisfying conditions in Theorem 6.12 is called a *Lyapunov function.*

Proof It remains to prove the necessity. By the Itô formula applied to $\|x\|_H^2$, taking expectations, and using condition (6.41), we have

$$E\|X^x(t)\|_H^2 = \|x\|_H^2 + 2E\int_0^t \langle A_0 X^x(s), X^x(s)\rangle ds$$

$$+ E\int_0^t \mathrm{tr}\big(B_0 X^x(s) Q (B_0 X^x(s))^*\big) ds$$

$$\leq \|x\|_H^2 + |\lambda|\int_0^t E\|X^x(s)\|_H^2 ds - \alpha \int_0^t E\|X^x(s)\|_V^2 ds$$

$$\leq \|x\|_H^2 \big(1 + |\lambda|c\big)\int_0^t e^{-\beta s}\, ds - \alpha \int_0^t E\|X^x(s)\|_V^2 ds$$

by exponential stability in the m.s.s. Let $t \to \infty$. Then

$$\int_0^\infty E\|X^x(s)\|_V^2 ds \leq \frac{1}{\alpha}\big(1 + |\lambda|c/\beta\big)\|x\|_H^2.$$

Define

$$T(x, y) = \int_0^\infty E\langle X^x(s), X^y(s)\rangle_V\, ds.$$

Then, by the preceding inequality and the Schwarz inequality, it is easy to see that T is a continuous bilinear form on $H \times H$. Since the embedding $V \hookrightarrow H$ is continuous, T is also a continuous bilinear form on $V \times V$. This fact can be used to show that conditions (1)–(5) of Theorem 6.10 are satisfied by the function $\Psi(x) = T(x, x)$. Clearly, $\Psi(x) \leq c_2\|x\|_H^2$. To prove the lower bound on $\Psi(x)$, we observe that

$$\mathscr{L}_0\|v\|_H^2 = 2\langle v, A_0 v\rangle + \mathrm{tr}\big((B_0 v)Q(B_0 v)^*\big),$$

so that, for some constants m, c_0',

$$|\mathscr{L}_0\|v\|_H^2| \leq c_0\|v\|_V^2 + m\,\mathrm{tr}(Q)\|v\|_V^2 \leq c_0'\|v\|_V^2.$$

Again, by Itô's formula, after taking the expectations, we obtain that

$$E\|X^x(t)\|_H^2 - \|x\|_H^2 = \int_0^t E\mathscr{L}_0\|X^x(s)\|_H^2 ds$$

$$\geq -c_0' \int_0^t E\|X^x(s)\|_V^2 ds.$$

As $t \to \infty$, using exponential stability in the m.s.s., we can see that

$$\Psi(x) \geq c_1 \|x\|_H^2,$$

where $c_1 = 1/c_0'$.

To prove the last condition, observe that, similarly as in (6.25), the uniqueness of the solution and the Markov property (3.59) yield

$$E\Psi\left(X^x(t)\right) = \int_0^\infty E\left\|X^x(s+t)\right\|_V^2 ds$$

$$= \int_t^\infty E\left\|X^x(s)\right\|_V^2 ds$$

$$\leq \int_0^\infty E\left\|X^x(s)\right\|_V^2 ds - k \int_0^t E\left\|X^x(s)\right\|_H^2 ds,$$

since $k\|x\|_H^2 \leq \|x\|_V^2$ for some constant k. Hence, by taking the derivatives of both sides at $t = 0$, we get

$$\mathscr{L}_0\Psi(x) \leq -k\|x\|_H^2 \leq -\frac{k}{c_2}\Psi(x). \qquad \square$$

Remark 6.3 Note that in case where $t \to E\|X^x(t)\|_V^2$ is continuous at zero, in the last step of the proof of Theorem 6.12, we obtain that $\mathscr{L}_0\Psi(v) = -\|v\|_V^2$ for $v \in V$.

Let us now state analogues of Theorem 6.6 for the solutions in variational case.

Theorem 6.13 *Let $\{X_0(t)\}_{t \geq 0}$ be the solution of the linear equation (6.42) with the coefficients satisfying condition (6.41). Assume that the function $t \to E\|X_0(t)\|_V^2$ is continuous and that the solution $X_0(t)$ is exponentially stable in the m.s.s. If for a sufficiently small constant c,*

$$2\|v\|_V \|A(v) - A_0v\|_{V*} + \tau\left(B(v)QB^*(v) - B_0vQ(B_0v)^*\right) \leq c\|v\|_V^2, \qquad (6.43)$$

then the strong solution of (6.37) is exponentially stable in the m.s.s.

For the zero solution of (6.37) to be stable in probability, it is enough to assume (6.43) for $v \in (V, \|\cdot\|_V)$ in a sufficiently small neighborhood of zero.

Theorem 6.14 *Let $\{X_0(t)\}_{t \geq 0}$ be the solution of the linear equation (6.42) with the coefficients satisfying condition (6.41). Assume that the solution $X_0(t)$ is exponentially stable in the m.s.s. Let for $v \in V$, $A(v) - A_0v \in H$. If for a sufficiently small constant c,*

$$2\|v\|_H \|A(v) - A_0v\|_H + \tau\left(B(v)QB^*(v) - B_0vQ(B_0v)^*\right) \leq c\|v\|_H^2, \qquad (6.44)$$

then the strong solution of (6.37) is exponentially stable in the m.s.s.

For the zero solution of (6.37) to be stable in probability, it is enough to assume (6.44) for $v \in (H, \|\cdot\|_H)$ in a sufficiently small neighborhood of zero.

Exercise 6.3 Verify that Theorem 6.7 holds for 6.37 (replacing (6.10)), under additional assumptions (1)–(5) of Theorem 6.10.

Exercise 6.4 Prove Theorems 6.13 and 6.14.

Remark 6.4 Using an analogue of Theorem 6.7 with the function Ψ satisfying conditions (1)–(5) of Theorem 6.10, we can also prove conclusions in Theorems 6.8 and 6.9 for (6.37) and its linear counterpart (6.42) under conditions (6.43) and (6.44).

Appendix: Stochastic Analogue of the Datko Theorem

Theorem 6.15 *Let A generate a pseudo-contraction C_0 semigroup $\{S(t), t \geq 0\}$ on a real separable Hilbert space H, and $B : H \to \mathscr{L}(K, H)$. A mild solution $\{X^x(t), t \geq 0\}$ of the stochastic differential equation (6.22) is exponentially stable in the m.s.s. if and only if there exists a nonnegative definite operator $R \in \mathscr{L}(H)$ such that*

$$\mathscr{L}_0\langle Rx, y\rangle_H = -\langle x, y\rangle_H \quad \text{for all } x, y \in H,$$

where $\mathscr{L}_0$ is defined in (6.24).

Proof The necessity part was already proved in Sect. 6.2, Theorem 6.5, with

$$\langle Rx, y\rangle_H = \int_0^\infty E\langle X^x(t), X^y(t)\rangle_H \, dt,$$

which, under stability assumption, is well defined by the Schwarz inequality. To prove the sufficiency, assume that R as postulated exists; then

$$2\langle Rx, Ay\rangle_H = -\langle (I + \Delta(R))x, y\rangle_H, \tag{6.45}$$

where $\Delta(R) = \mathrm{tr}(R(B_0x)Q(B_0x)^*)\,I$. The operator $I + \Delta(R)$ is invertible, so that we get

$$2\langle R(I + \Delta(R))^{-1}x, y\rangle_H = \langle x, y\rangle_H.$$

By Corollary 6.1,

$$\|S(t)\|_{\mathscr{L}(H)} \leq Me^{-\lambda t}, \quad \lambda > 0.$$

We consider the solutions $\{X_n^x(t), t \geq 0\}$ obtained by using the Yosida approximations $A_n = AR_n$ of A. Let us apply Itô's formula to $\langle RX_n^x(t), X_n^x(t)\rangle_H$ and take the expectations of both sides to arrive at

$$E\langle RX_n^x(t), X_n^x(t)\rangle_H = \langle Rx, x\rangle_H + 2E\int_0^t \langle RX_n^x(s), A_n X_n^x(s)\rangle_H ds$$

$$+ E\int_0^t \langle \Delta(R)X_n^x(s), X_n^x(s)\rangle_H ds.$$

From (6.45) with $y = R_n X_n^x$ it follows that

$$2\langle RX_n^x(s), AR_n X_n^x(s)\rangle_H = -\langle \Delta(R)X_n^x(s), R_n X_n^x(s)\rangle_H - \langle X_n^x(s), R_n X_n^x(s)\rangle_H.$$

Hence,

$$E\langle RX_n^x(t), X_n^x(t)\rangle_H = \langle Rx, x\rangle_H - E\int_0^t \langle X_n^x(s), R_n X_n^x(s)\rangle_H ds$$

$$+ E\int_0^t \langle \Delta(R)X_n^x(s), X_n^x(s) - R_n X_n^x(s)\rangle_H ds.$$

We let $n \to \infty$ and use the fact that $\sup_n \sup_{t \le T} E\|X_n^x(t)\|_H^2 < \infty$ to obtain

$$E\langle RX^x(t), X^x(t)\rangle_H = \langle Rx, x\rangle_H - \int_0^t E\|X^x(s)\|_H^2 ds.$$

Let $\Xi(t) = E\langle RX^x(t), X^x(t)\rangle_H$. Then

$$\Xi(t) \le \|R\|_{\mathscr{L}(H)} E\|X^x(t)\|_H^2$$

and

$$\Xi'(t) = -E\|X^x(t)\|_H^2 \le \frac{-1}{\|R\|_{\mathscr{L}(H)}}\Xi(t),$$

so that

$$\Xi(t) \le \langle Rx, x\rangle_H e^{\frac{-1}{\|R\|_{\mathscr{L}(H)}}t},$$

since $\Xi(0) = \langle Rx, x\rangle_H$. Hence,

$$E\|X^x(t)\|_H^2 \le 2\|S(t)x\|_H^2 + 2E\left\|\int_0^t S(t-s)BX^x(s)dW_s\right\|_H^2$$

$$\le 2M^2 e^{-2\lambda t}\|x\|_H^2 + 2\operatorname{tr}(Q)M^2\|B\|_{\mathscr{L}(H)}^2 \int_0^t e^{-2\lambda(t-s)}E\|X^x(s)\|_H^2 ds.$$

We complete the proof by using

$$\langle Rx, x\rangle_H \le \|R\|_{\mathscr{L}(H)}\|x\|_H^2 \quad \text{and} \quad E\|X^x(s)\|_H^2 = -\Xi'(s). \qquad \square$$

As shown in Sect. 6.1, Example 6.1, $x \to \langle Rx, x\rangle$, however, is not a Lyapunov function, so that we cannot study stability of nonlinear equations using Theorem 6.15.

Chapter 7
Ultimate Boundedness and Invariant Measure

We introduce in this chapter the concept of ultimate boundedness in the mean square sense (m.s.s.) and relate it to the problem of the existence and uniqueness of invariant measure. We consider semilinear stochastic differential equations in a Hilbert space and their mild solutions under the usual linear growth and Lipschitz conditions on the coefficients. We also study stochastic differential equations in the variational case, assuming that the coefficients satisfy the coercivity condition, and study their strong solutions which are exponentially ultimately bounded in the m.s.s.

7.1 Exponential Ultimate Boundedness in the m.s.s.

Definition 7.1 We say that the mild solution of (6.10) is exponentially ultimately bounded in the mean square sense (m.s.s.) if there exist positive constants c, β, M such that

$$E\left\|X^x(t)\right\|_H^2 \le ce^{-\beta t}\|x\|_H^2 + M \quad \text{for all } x \in H. \tag{7.1}$$

Here is an analogue of Theorem 6.4.

Theorem 7.1 *The mild solution $\{X^x(t),\, t \ge 0\}$ of (6.10) is exponentially ultimately bounded in the m.s.s. if there exists a function $\Psi \in C_{2p}^2(H)$ satisfying the following conditions:*

(1) $c_1\|x\|_H^2 - k_1 \le \Psi(x) \le c_2\|x\|_H^2 - k_2$,
(2) $\mathscr{L}\Psi(x) \le -c_3\Psi(x) + k_3$,

for $x \in H$, where c_1, c_2, c_3 are positive constants, and $k_1, k_2, k_3 \in \mathbb{R}$.

Proof Similarly as in the proof of Theorem 6.4, using Itô's formula for the solutions of the approximating equations (6.17) and utilizing condition (2), we arrive at

$$E\Psi\left(X^x(t)\right) - E\Psi\left(X^x(0)\right) \le \int_0^t \left(-c_3\,E\Psi\left(X^x(s)\right) + k_3\right)ds.$$

L. Gawarecki, V. Mandrekar, *Stochastic Differential Equations in Infinite Dimensions*, Probability and Its Applications, DOI 10.1007/978-3-642-16194-0_7, © Springer-Verlag Berlin Heidelberg 2011

Hence, $\Phi(t) = E\Psi(X^x(t))$ satisfies

$$\Phi'(t) \le -c_3 \Phi(t) + k_3.$$

By Gronwall lemma,

$$\Phi(t) \le \frac{k_3}{c_3} + \left(\Phi(0) - \frac{k_3}{c_3}\right)e^{-c_3 t}.$$

Using condition (1), we have, for all $x \in H$,

$$c_1 E\|X^x(t)\|_H^2 - k_1 \le E\Psi(X^x(t)) \le \frac{k_3}{c_3} + \left(c_2\|x\|_H^2 - k_2 - \frac{k_3}{c_3}\right)e^{-c_3 t},$$

and (7.1) follows.   □

Theorem 7.2 *Assume that A generates a pseudo-contraction semigroup of operators $\{S(t), t \ge 0\}$ on H. If the mild solution $\{X_0^x(t), t \ge 0\}$ of the linear equation (6.22) is exponentially ultimately bounded in the m.s.s., then there exists a function $\Psi_0 \in C_{2p}^2(H)$ satisfying conditions (1) and (2) of Theorem 7.1, with the operator $\mathscr{L}_0$ replacing $\mathscr{L}$ in condition (2).*

Proof Since the mild solution $X_0^x(t)$ is exponentially ultimately bounded in the m.s.s., we have

$$E\|X_0^x(t)\|_H^2 \le ce^{-\beta t}\|x\|_H^2 + M \quad \text{for all } x \in H.$$

Let

$$\Psi_0(x) = \int_0^T E\|X_0^x(s)\|_H^2 \, ds + \alpha\|x\|_H^2,$$

where T and α are constants to be determined later.

First, let us show that $\Psi_0 \in C_{2p}^2(H)$. It suffices to show that

$$\varphi_0(x) = \int_0^T E\|X_0^x(s)\|_H^2 \, ds \in C_{2p}^2(H).$$

Now,

$$\varphi_0(x) \le \frac{c}{\beta}(1 - e^{-\beta T})\|x\|_H^2 + MT \le \frac{c}{\beta}\|x\|_H^2 + MT.$$

If $\|x\|_H^2 = 1$, then $\varphi_0(x) \le c/\beta + MT$.

Since $X_0^x(t)$ is linear in x, we have that, for any positive constant k, $X_0^{kx}(t) = kX_0^x(t)$. Hence, $\varphi_0(kx) = k^2\varphi_0(x)$, and for any $x \in H$,

$$\varphi_0(x) = \|x\|_H^2 \varphi\left(\frac{x}{\|x\|_H}\right) \le \left(\frac{c}{\beta} + MT\right)\|x\|_H^2.$$

Let $c' = c/\beta + MT$. Then $\phi_0(x) \leq c' \|x\|_H^2$ for all $x \in H$. For $x, y \in H$, define

$$\tau(x, y) = \int_0^T E\langle X_0^x(s), X_0^y(s)\rangle_H\, ds.$$

Then $\tau(x, y)$ is a nonnegative definite bounded bilinear form on $H \times H$ since $\varphi_0(x) \leq c' \|x\|_H^2$. Hence, $\tau(x, y) = \langle Cx, y\rangle_H$, where C is a nonnegative definite bounded linear operator on H with $\|C\|_{\mathscr{L}(H)} \leq c'$. Therefore, $\varphi_0 = \langle Cx, x\rangle_H \in C_{2p}^2(H)$, and $\Psi_0 \in C_{2p}^2(H)$. Clearly Ψ_0 satisfies condition (1) of Theorem 7.1. To prove (2), observe that by the continuity of the function $t \to E\|X_0^x(t)\|_H^2$ and because

$$E\varphi_0\big(X_0^x(r)\big) = \int_0^T E\big\|X_0^x(r+s)\big\|_H^2\, ds = \int_r^{T+r} E\big\|X_0^x(s)\big\|_H^2\, ds,$$

we have

$$\begin{aligned}
\mathscr{L}_0\varphi_0(x) &= \frac{d}{dr}\big(E\varphi_0\big(X_0^x(r)\big)\big)\Big|_{r=0} \\
&= \lim_{r\to 0} \frac{E\varphi_0(X_0^x(r)) - E\varphi_0(x)}{r} \\
&= \lim_{r\to 0}\left(-\frac{1}{r}\int_0^r E\big\|X_0^x(s)\big\|_H^2\, ds + \frac{1}{r}\int_T^{r+T} E\big\|X_0^x(s)\big\|_H^2\, ds\right) \\
&= -\|x\|_H^2 + E\big\|X_0^x(T)\big\|_H^2 \\
&\leq -\|x\|_H^2 + ce^{-\beta T}\|x\|_H^2 + M \\
&\leq \big(-1 + ce^{-\beta T}\big)\|x\|_H^2 + M.
\end{aligned}$$

Therefore, since by (6.32), $\mathscr{L}_0\|x\|_H^2 \leq (2\lambda + d^2\operatorname{tr}(Q))\|x\|_H^2$, we have

$$\begin{aligned}
\mathscr{L}_0\Psi_0(x) &= \mathscr{L}_0\varphi_0(x) + \mathscr{L}_0\|x\|_H^2 \\
&\leq \big(-1 + ce^{-\beta T}\big)\|x\|_H^2 + \alpha\big(2\lambda + d^2\operatorname{tr}(Q)\big)\|x\|_H^2 + M. \quad (7.2)
\end{aligned}$$

If $T > \ln(c/\beta)$, then one can choose α small enough such that $\Psi_0(x)$ satisfies condition (2) with $\mathscr{L}$ replaced by $\mathscr{L}_0$. $\qquad\square$

The following theorem is a counterpart of Remark 6.1 in the framework of exponential ultimate boundedness.

Theorem 7.3 *If the mild solution of* (6.10) *is exponentially ultimately bounded in the m.s.s. and, for some $T > 0$,*

$$\varphi(x) = \int_0^T E\big\|X^x(t)\big\|_H^2\, dt \in C_{2p}^2(H),$$

then there exists a (Lyapunov) function $\Psi \in C^2_{2p}(H)$ *satisfying conditions* (1) *and* (2) *of Theorem* 7.1.

Theorem 7.4 *Suppose that the mild solution* $X^x_0(t)$ *of the linear equation* (6.22) *satisfies condition* (7.1). *Then the mild solution* $X^x(t)$ *of* (6.10) *is exponentially ultimately bounded in the m.s.s. if*

$$2\|x\|_H \big\| F(x) \big\|_H + \tau \big(B(x)QB^*(x) - B_0 x Q(B_0 x)^* \big) < \tilde{\omega} \|x\|^2_H + M_1, \qquad (7.3)$$

where $\tilde{\omega} < \max_{s > \ln(c/\beta)}(1 - ce^{-\beta s})/(c/\beta + Ms)$.

Proof Let $\Psi_0(x)$ be the Lyapunov function as defined in Theorem 7.2, with $T > \ln(c/\beta)$, such that the maximum in the definition of $\tilde{\omega}$ is achieved. It remains to show that

$$\mathscr{L}\Psi_0(x) \leq -c_3 \Psi_0(x) + k_3.$$

Since $\Psi_0(x) = \langle Cx, x \rangle_H + \alpha \|x\|^2_H$ for some $C \in \mathscr{L}(H)$ with $\|C\|_{\mathscr{L}(H)} \leq c/\beta + MT$ and α sufficiently small, we have

$$\mathscr{L}\Psi_0(x) - \mathscr{L}_0\Psi_0(x)$$
$$\leq \big(\|C\|_{\mathscr{L}(H)} + \alpha \big)\big(2\|x\|_H \big\| F(x) \big\|_H + \tau \big(B(x)QB^*(x) - B_0 x Q(B_0 x)^* \big) \big)$$
$$\leq (c/\beta + MT + \alpha)\big(\tilde{\omega} \|x\|^2_H + M_1 \big).$$

Using (7.2), we have

$$\mathscr{L}\Psi_0(x) \leq \big(-1 + ce^{-\beta T} \big)\|x\|^2_H + \alpha \big(2\lambda + d^2 \,\mathrm{tr}(Q) \big)\|x\|^2_H + M$$
$$+ (c/\beta + MT + \alpha)\big(\tilde{\omega} \|x\|^2_H + M_1 \big)$$
$$\leq \big(-1 + ce^{-\beta T} + \tilde{\omega}(c/\beta + MT) \big)\|x\|^2_H$$
$$+ \alpha \big(2\lambda + d^2 \,\mathrm{tr}(Q) + \tilde{\omega} \big)\|x\|^2_H + M + (c/\beta + MT + \alpha).$$

Using the bound for $\tilde{\omega}$, we have $-1 + ce^{-\beta T} + \tilde{\omega}(c/\beta + MT) < 0$, so that we can choose α small enough to obtain condition (2) of Theorem 7.1. $\qquad\square$

Corollary 7.1 *Suppose that the mild solution* $X^x_0(t)$ *of the linear equation* (6.22) *is exponentially ultimately bounded in the m.s.s. If, as* $\|x\|_H \to \infty$,

$$\big\| F(x) \big\|_H = o\big(\|x\|_H \big) \quad and \quad \tau \big(B(x)QB^*(x) - B_0 x Q(B_0 x)^* \big) = o\big(\|x\|_H \big),$$

then the mild solution $X^x(t)$ *of* (6.10) *is exponentially ultimately bounded in the m.s.s.*

Proof We fix $\tilde{\omega} < \max_{s > \ln(c/\beta)}(1 - ce^{-\beta t}/(c/\beta + Ms)$, and using the assumptions, we choose a constant K such that for $\|x\|_H \geq K$, condition (7.3) holds. But for

$\|x\|_H \le K$, by appealing to the growth conditions on F and B,

$$2\|x\|_H \|F(x)\|_H + \tau \left(B(x) Q B^*(x) - B_0 x Q (B_0 x)^* \right)$$

$$\le \|x\|_H^2 + \|F(x)\|_H^2 + \tau \left(B(x) Q B^*(x) - B_0 x Q (B_0 x)^* \right)$$

$$\le \|x\|_H^2 + \ell \left(1 + \|x\|_H^2 \right) + \left(\|B_0 x\|_{\mathscr{L}(H)}^2 \right) \mathrm{tr}(Q)$$

$$\le \|x\|_H^2 + \ell \left(1 + \|x\|_H^2 \right) + d^2 \|x\|_H^2 \, \mathrm{tr}(Q)$$

$$\le K^2 + \ell \left(1 + K^2 \right) + M'.$$

Hence, condition (7.3) holds with the constant $M_1 = K^2 + \ell(1 + K^2) + M'$, and the result follows from Theorem 7.4. ∎

Example 7.1 (Dissipative Systems) Consider SSDE (6.10) and, in addition to assumptions (1)–(3) in Sect. 6.2, impose the following *dissipativity condition*:

(D) (Dissipativity) There exists a constant $\omega > 0$ such that for all $x, y \in H$ and $n = 1, 2, \ldots,$

$$2\langle A_n(x - y), x - y \rangle_H + 2\langle F(x) - F(y), x - y \rangle_H + \|B(x) - B(y)\|_{\mathscr{L}_2(K_Q, H)}$$

$$\le -\omega \|x - y\|_H^2, \tag{7.4}$$

where $A_n x = A R_n x$, $x \in H$, are the Yosida approximations of A defined in (1.22).

Then the mild solution to (6.10) is ultimately exponentially bounded in the m.s.s. (Exercise 7.1).

Exercise 7.1 (a) Show that condition (D) implies that for any $\varepsilon > 0$, there exists a constant $C_\varepsilon > 0$ such that for any $x \in H$ and $n = 1, 2, \ldots,$

$$2\langle A_n x, x \rangle_H + 2\langle F(x), x \rangle_H + \|B(x)\|_{\mathscr{L}_2(K_Q, H)} \le -(\omega - \varepsilon)\|x\|_H^2 + C_\varepsilon$$

with A_n, the Yosida approximations of A. Use this fact to prove that the strong solutions $X_n^x(t)$ of the approximating SDEs (6.12) are ultimately exponentially bounded in the m.s.s. Conclude that the mild solution $X^x(t)$ of (6.10) is ultimately exponentially bounded in the m.s.s.

   (b) Prove that if zero is a solution of (6.10), then the mild solution $X^x(t)$ of (6.10) is exponentially stable in the m.s.s.

7.2 Exponential Ultimate Boundedness in Variational Method

We study in this section strong solutions to (6.37) whose coefficients satisfy linear growth, coercivity, and monotonicity assumptions (6.38)–(6.40).

Definition 7.2 We extend Definition 7.1 of exponential ultimate boundedness in the m.s.s. to the strong solution $\{X^x(t), t \geq 0\}$ of (6.37) and say that $X^x(t)$ is exponentially ultimately bounded in the m.s.s. if it satisfies condition (7.1).

Let us begin by noting that the proof of Theorem 7.1 can be carried out in this case if we assume that the function Ψ satisfies conditions (1)–(5) of Theorem 6.10 and that the operator $\mathscr{L}$ is defined by

$$\mathscr{L}\Psi(u) = \langle \Psi'(u), A(u) \rangle + \mathrm{tr}\big(\Psi''(u)B(u)QB^*(u)\big). \tag{7.5}$$

Hence, we have the following theorem.

Theorem 7.5 *The strong solution* $\{X^x(t), t \geq 0\}$ *of* (6.37) *is exponentially ultimately bounded in the m.s.s. if there exists a function* $\Psi : H \to \mathbb{R}$ *satisfying conditions* (1)–(5) *of Theorem* 6.10 *and, in addition, such that*

(1) $c_1 \|x\|_H^2 - k_1 \leq \Psi(x) \leq c_2 \|x\|_H^2 + k_2$ *for some positive constants* c_1, c_2, k_1, k_2 *and for all* $x \in H$,
(2) $\mathscr{L}\Psi(x) \leq -c_3 \Psi(x) + k_3$ *for some positive constants* c_3, k_3 *and for all* $x \in V$.

In the linear case, we have both, sufficiency and necessity, and the Lyapunov function has an explicit form under the general coercivity condition (C).

Theorem 7.6 *A solution* $\{X_0^x(t), t \geq 0\}$ *of the linear equation* (6.42) *whose coefficients satisfy coercivity condition* (6.39) *is exponentially ultimately bounded in the m.s.s. if and only if there exists a function* $\Psi_0 : H \to \mathbb{R}$ *satisfying conditions* (1)–(5) *of Theorem* 6.10 *and, in addition, such that*

(1) $c_1 \|x\|_H^2 - k_1 \leq \Psi_0(x) \leq c_2 \|x\|_H^2 + k_2$ *for some positive constants* c_1, c_2, k_1, k_2 *and for all* $x \in H$,
(2) $\mathscr{L}_0\Psi_0(x) \leq -c_3 \Psi_0(x) + k_3$ *for some positive constants* c_3, k_3 *and for all* $x \in V$.

This function can be written in the explicit form

$$\Psi_0(x) = \int_0^T \int_0^t E\|X_0^x(s)\|_V^2 \, ds \, dt \tag{7.6}$$

with $T > \alpha_0(c|\lambda|/(\alpha\beta) + 1/\alpha)$, *where* α_0 *is such that* $\|v\|_H^2 \leq \alpha_0 \|v\|_V^2$, $v \in V$.

Proof Assume that the solution $\{X_0^x(t), t \geq 0\}$ of the linear equation (6.42) is exponentially ultimately bounded in the m.s.s., so that

$$E\|X_0^x(t)\|_H^2 \leq ce^{-\beta t}\|x\|_H^2 + M \quad \text{for all } x \in H.$$

Applying Itô's formula to the function $\|x\|_H^2$, taking the expectations, and using the coercivity condition (6.39), we obtain

$$E\|X_0^x(t)\|_H^2 - \|x\|_H^2 = \int_0^t E\mathscr{L}_0\|X_0^x(s)\|_H^2\,ds$$

$$\leq \lambda \int_0^t E\|X_0^x(s)\|_H^2\,ds - \alpha \int_0^t E\|X_0^x(s)\|_V^2\,ds + \gamma t. \quad (7.7)$$

Hence,

$$\int_0^t E\|X_0^x(s)\|_V^2 \leq \frac{1}{\alpha}\left(\lambda \int_0^t E\|X_0^x(s)\|_H^2\,ds + \|x\|_H^2 + \gamma t\right).$$

Applying condition (7.1), we have

$$\int_0^t E\|X_0^x(s)\|_V^2 \leq \frac{1}{\alpha}\left(\frac{c|\lambda|}{\beta}(1 - e^{-\beta t})\|x\|_H^2 + \|x\|_H^2 + (|\lambda|M + \gamma)t\right)$$

$$\leq \left(\frac{c|\lambda|}{\alpha\beta} + \frac{1}{\alpha}\right)\|x\|_H^2 + \frac{|\lambda|M + \gamma}{\alpha}t.$$

Therefore, with Ψ_0 defined in (7.6),

$$\Psi_0(x) = \int_0^T\int_0^t E\|X_0^x(s)\|_V^2\,ds\,dt \leq \left(\frac{1}{\alpha} + \frac{c|\lambda|}{\alpha\beta}\right)T\|x\|_H^2 + \frac{|\lambda|M + \gamma}{2\alpha}T^2. \quad (7.8)$$

Now

$$\left|\mathscr{L}_0\|v\|_H^2\right| \leq 2a_1\|v\|_V^2 + b_1^2\,\mathrm{tr}(Q)\|v\|_V^2 \leq c'\|v\|_V^2$$

for some positive constant c'. Therefore, we conclude that

$$\mathscr{L}_0\|v\|_H^2 \geq -c'\|v\|_V^2.$$

From (7.7) we get

$$E\|X_0^x(t)\|_H^2 - \|x\|_H^2 \geq -c'\int_0^t E\|X_0^x(s)\|_V^2\,ds.$$

Using (7.1), we have

$$c'\int_0^t E\|X_0^x(t)\|_V^2\,ds \geq (1 - e^{-\beta t})\|x\|_H^2 - M.$$

Hence,

$$\Psi_0(x) \geq \frac{1}{c'}\int_0^T \|x\|_H^2(1 - e^{-\beta t})\,dt - MT \geq \frac{1}{c'}\left(T - \frac{c}{\beta}\right)\|x\|_H^2 - \frac{MT}{c'}.$$

Choose $T > c/\beta$ to obtain condition (1).

To prove that condition (2) holds, consider

$$E\Psi_0\big(X_0^x(r)\big) = \int_0^T \int_0^t E\big\|X_0^{X_0^x(r)}(s)\big\|_V^2 \, ds \, dt.$$

By the Markov property of the solution and the uniqueness of the solution,

$$E\Psi_0\big(X_0^x(r)\big) = \int_0^T \int_0^t E\big\|X_0^x(s+r)\big\|_V^2 \, ds \, dt = \int_0^T \int_r^{t+r} E\big\|X_0^x(s)\big\|_V^2 \, ds \, dt.$$

We now need the following technical lemma that will be proved later.

Lemma 7.1 *If $f \in L^1([0, T])$, $T > 0$, is a nonnegative real-valued function, then*

$$\lim_{\Delta t \to 0} \int_0^T \frac{\int_t^{t+\Delta t} f(s)\,ds}{\Delta t}\,dt = \int_0^T \lim_{\Delta t \to 0} \frac{\int_t^{t+\Delta t} f(s)\,ds}{\Delta t}\,dt = \int_0^T f(t)\,dt.$$

Assuming momentarily that Ψ_0 satisfies conditions (1)–(5) of Theorem 6.10, we have

$$\mathscr{L}_0\Psi_0(x) = \frac{\mathrm{d}}{\mathrm{d}r}\big(E\Psi_0\big(X_0^x(r)\big)\big)\Big|_{r=0}$$

$$= \lim_{r \to 0} \int_0^T \frac{\int_t^{t+r} E\|X_0^x(s)\|_V^2\,ds}{r}\,dt - \lim_{r \to 0}\frac{T}{r}\int_0^r E\|X_0^x(s)\|_V^2\,ds$$

$$\le \int_0^T E\|X_0^x(t)\|_V^2\,dt - \lim_{r \to 0}\frac{T}{\alpha_0}\frac{1}{r}\int_0^r E\|X_0^x(s)\|_H^2\,ds$$

for α_0 such that $\|v\|_H^2 \le \alpha_0\|v\|_V^2$. This gives

$$\mathscr{L}_0\Psi_0(x) \le \left(\frac{c|\lambda|}{\alpha\beta} + \frac{1}{\alpha} - \frac{T}{\alpha_0}\right)\|x\|_H^2 + \frac{|\lambda|M + \gamma}{\alpha}T. \tag{7.9}$$

With $T > \alpha_0(\frac{c|\lambda|}{\alpha\beta} + \frac{1}{\alpha})$, condition (2) holds.

It remains to prove that Ψ_0 satisfies conditions (1)–(5) of Theorem 6.10. We use linearity of (6.42) to obtain, for any positive constant k,

$$X_0^{kx}(t) = kX_0^x(t).$$

Then $\Psi_0(kx) = k^2\Psi_0(x)$, and by (7.8), for $\|x\|_H = 1$,

$$\Psi_0(x) \le \left(\frac{1}{\alpha} + \frac{c|\lambda|}{\alpha\beta}\right)T + \frac{|\lambda|M + \gamma}{2\alpha}T^2.$$

Hence, for $x \in H$,

$$\Psi_0(x) \le \|x\|_H^2 \Psi_0\left(\frac{x}{\|x\|_H}\right) \le \left[\left(\frac{1}{\alpha} + \frac{c|\lambda|}{\alpha\beta}\right)T + \frac{|\lambda|M + \gamma}{2\alpha}T^2\right]\|x\|_H^2. \tag{7.10}$$

which implies that $\Psi_0(x) \le c''\|x\|_H^2$ for all $x \in H$. For $x, y \in H$, denote

$$\tau(x, y) = \int_0^T \int_0^t E\langle X_0^x(s), X_0^y(s)\rangle_H \, ds \, dt \le \Psi_0^{\frac{1}{2}}(x)\Psi_0^{\frac{1}{2}}(y) \le c''\|x\|_H\|y\|_H.$$

Then τ is a continuous bilinear form on $H \times H$, and there exists $C \in \mathscr{L}(H)$, with $\|C\|_{\mathscr{L}(H)} \le c''$, such that

$$\tau(x, y) = \langle Cx, y\rangle_H. \qquad (7.11)$$

Using the continuity of the embedding $V \hookrightarrow H$, we conclude that $\tau(x, y)$ is a continuous bilinear form on $V \times V$, and hence,

$$\tau(x, y) = \langle \tilde{C}x, y\rangle_V \quad \text{for } x, y \in V, \qquad (7.12)$$

with $\tilde{C} \in \mathscr{L}(V)$. Now it is easy to verify that Ψ_0 satisfies conditions (1)–(5) of Theorem 6.10. $\qquad \square$

Proof of Lemma 7.1 We are going to use the Fubini theorem to change the order of integrals,

$$\int_0^T \frac{\int_t^{t+\Delta t} f(s)\,ds}{\Delta t} \, dt = \frac{1}{\Delta t}\int_0^T \left(\int_0^{t+\Delta t} f(s)\,ds\right) dt$$

$$= \frac{1}{\Delta t}\left[\int_0^{\Delta t}\left(\int_0^s f(s)\,dt\right)ds + \int_{\Delta t}^T\left(\int_{s-\Delta t}^T f(s)\,dt\right)ds\right.$$

$$\left. + \int_T^{T+\Delta t}\left(\int_{s-\Delta t}^T f(s)\,dt\right)ds\right]$$

$$= \frac{1}{\Delta t}\left[\int_0^{\Delta t} sf(s)\,ds + \int_{\Delta t}^T f(s)\Delta t\,ds + \int_T^{T+\Delta t} f(s)(T + \Delta t - s)\,ds\right]$$

$$\le \frac{1}{\Delta t}\left[\Delta t\int_0^{\Delta t} f(s)\,ds + \Delta t\int_{\Delta t}^T f(s)\,ds + \Delta t\int_T^{T+\Delta t} f(s)\,ds\right]$$

$$= \int_0^{\Delta t} f(s)\,ds + \int_{\Delta t}^T f(s)\,ds + \int_T^{T+\Delta t} f(s)\,ds.$$

The first and third terms converge to zero as $\Delta t \to 0$, so that

$$\lim_{\Delta t \to 0}\int_0^T \frac{\int_t^{t+\Delta t} f(s)\,ds}{\Delta t}\,dt \le \int_0^T f(t)\,dt.$$

The opposite inequality follows directly from Fatou's lemma. $\qquad \square$

By repeating the proof of Theorem 7.6, we obtain a partial converse of Theorem 7.5.

Theorem 7.7 *Let the strong solution* $\{X^x(t),\, t \geq 0\}$ *of* (6.37) *be exponentially ultimately bounded in the m.s.s. Let*

$$\Psi(x) = \int_0^T \int_0^t E\|X^x(s)\|_V^2 \, ds \, dt \qquad (7.13)$$

with $T > \alpha_0(c|\lambda|/(\alpha\beta) + 1/\alpha)$, *where* α_0 *is such that* $\|v\|_H^2 \leq \alpha_0\|v\|_V^2$, $v \in V$. *Suppose that* $\Psi(x)$ *satisfies conditions* (1)–(5) *of Theorem* 6.10. *Then* $\Psi(x)$ *satisfies conditions* (1) *and* (2) *of Theorem* 7.5.

To study exponential ultimate boundedness, i.e., condition (7.1), for the strong solution of (6.37), we use linear approximation and the function Ψ_0 of the corresponding linear equation (6.42) as the Lyapunov function. We will prove the following result.

Theorem 7.8 *Suppose that the coefficients of the linear equation* (6.42) *satisfy the coercivity condition* (6.39) *and its solution* $\{X_0^x(t),\, t \geq 0\}$ *is exponentially ultimately bounded in the m.s.s. Let* $\{X^x(t),\, t \geq 0\}$ *be the solution of the nonlinear equation* (6.37). *Furthermore, we suppose that*

$$A(v) - A_0 v \in H \quad \text{for all } v \in V$$

and that, for $v \in V$,

$$2\|v\|_H \|A(v) - A_0 v\|_H + \tau\left(B(v)QB^*(v) - B_0 v Q(B_0 v)^*\right) \leq \tilde{\omega}\|v\|_H^2 + k,$$

where $\tilde{\omega}$ *and* k *are constants, and*

$$\tilde{\omega} < \frac{c}{\alpha_0\beta\left[\left(\frac{1}{\alpha} + \frac{c|\lambda|}{\alpha\beta}\right) + \left(\frac{1}{\alpha} + \frac{c|\lambda|}{\alpha\beta} + \frac{c}{\beta}\right) + \frac{|\lambda|M}{2\alpha}\left(\frac{1}{\alpha} + \frac{c|\lambda|}{\alpha\beta} + \frac{c}{\beta}\right)^2\right]}.$$

Then $X^x(t)$ *is exponentially ultimately bounded in the m.s.s.*

Proof Let

$$\Psi_0(x) = \int_0^{T_0} \int_0^t E\|X_0^x(t)\|_V^2 \, ds \, dt$$

with $T_0 = \alpha_0(c|\lambda|/(\alpha\beta) + 1/\alpha) + c/\beta$. Then $\Psi_0(s)$ satisfies conditions (1)–(5) of Theorem 6.10, and for all $x \in H$,

$$c_1\|x\|_H^2 - k_1 \leq \Psi_0(x) \leq c_2\|x\|_H^2 + k_2.$$

It remains to prove that, for all $x \in V$,

$$\mathscr{L}\Psi_0(x) \leq -c_3\Psi_0(s) + k_3.$$

Then we can conclude the result by Theorem 7.5. Now, for $x \in V$,

$$\mathscr{L}\Psi_0(x) - \mathscr{L}_0\Psi_0(x)$$

$$= \langle \Psi_0'(x), A(x) - A_0 x \rangle + \frac{1}{2} \operatorname{tr}\big(\Psi_0''(x)\big(B(x)QB^*(x) - B_0 x Q(B_0 x)^*\big)\big)$$

$$= \langle \Psi_0'(x), A(x) - A_0 x \rangle_H + \frac{1}{2} \operatorname{tr}\big(\Psi_0''(x)\big(B(x)QB^*(x) - B_0 x Q(B_0 x)^*\big)\big).$$

But $\Psi_0'(x) = 2Cx$ and $\Psi_0''(x) = 2C$ for $x \in V$, where C is defined in (7.11). By inequality (7.10),

$$\|C\|_{\mathscr{L}(H)} \le \left(\frac{1}{\alpha} + \frac{c|\lambda|}{\alpha\beta}\right) T_0 + \frac{|\lambda|M + \gamma}{2\alpha} T_0^2.$$

Hence,

$$\mathscr{L}\Psi_0(x) - \mathscr{L}_0\Psi_0(x) \le 2\langle Cx, A(x) - A_0 x \rangle_H + \tau\big(C\big(B(x)QB^*(x) - B_0 x Q(B_0 x)^*\big)\big),$$

and we have

$$\mathscr{L}\Psi_0(x) \le \mathscr{L}_0\Psi_0(x) + \|C\|_{\mathscr{L}(H)}\Big[2\|x\|_H\|A(x) - A_0 x\|_H$$
$$+ \tau\big(B(x)QB^*(x) - B_0 x Q(B_0 x)^*\big)\Big].$$

When $T = T_0$, from (7.9) we have

$$\mathscr{L}_0\Psi_0(x) \le -\frac{c}{\alpha_0\beta}\|x\|_H^2 + \frac{|\lambda|M + \gamma}{\alpha} T_0,$$

giving

$$\mathscr{L}\Psi_0(x) \le -\frac{c}{\alpha_0\beta}\|x\|_H^2 + \frac{|\lambda|M + \gamma}{\alpha} T_0 + \|C\|_{\mathscr{L}(H)}\big(\tilde{\omega}\|x\|_H^2 + k\big)$$

$$\le \left(-\frac{c}{\alpha_0\beta} + \tilde{\omega}\|C\|_{\mathscr{L}(H)}\right)\|x\|_H^2 + \frac{|\lambda|M + \gamma}{\alpha} T_0 + k\|C\|_{\mathscr{L}(H)}.$$

Now, $-c/(\alpha_0\beta) + \tilde{\omega}\|C\|_{\mathscr{L}(H)} < 0$ if $\tilde{\omega}$ satisfies our original assumption, and we arrive at $\mathscr{L}\Psi_0(x) \le -c_3\Psi_0(x) + k_3$ with $c_3 > 0$. $\qquad\square$

Remark 7.1 Note that the function $\Psi_0(x)$ in Theorem 7.8 is the Lyapunov function for the nonlinear equation.

Corollary 7.2 *Suppose that the coefficients of the linear equation (6.42) satisfy the coercivity condition (6.39), and its solution $\{X_0^x(t), t \ge 0\}$ is exponentially ultimately bounded in the m.s.s. Let $\{X^x(t), t \ge 0\}$ be a solution of the nonlinear equation (6.37). Furthermore, suppose that*

$$A(v) - A_0 v \in H \quad \text{for all } v \in V$$

and that, for $v \in V$,

$$2\|v\|_H \|A(v) - A_0v\|_H + \tau\big(B(v)QB^*(v) - B_0vQ(B_0v)^*\big) \le k\big(1 + \|v\|_H^2\big) \quad (7.14)$$

for some $k > 0$. If for $v \in V$, as $\|v\|_H \to \infty$,

$$\|A(v) - A_0v\|_H = o\big(\|v\|_H\big)$$

and $\qquad\qquad\qquad\qquad\qquad\qquad\qquad\qquad\qquad\qquad$ (7.15)

$$\tau\big(B(v)QB^*(v) - B_0vQ(B_0v)^*\big) = o\big(\|v\|_H^2\big),$$

then $X^x(t)$ is exponentially ultimately bounded in the m.s.s.

Proof Under assumption (7.15), for a constant $\tilde{\omega}$ satisfying the condition of Theorem 7.8, there exists an $R > 0$ such that, for all $v \in V$ with $\|v\|_H > R$,

$$2\|v\|_H \|A(v) - A_0v\|_H + \tau\big(B(v)QB^*(v) - B_0vQ(B_0v)^*\big) \le \tilde{\omega}\|v\|_H^2.$$

For $v \in V$ and $\|v\|_H < R$, by (7.14),

$$2\|v\|_H \|A(v) - A_0v\|_H + \tau\big(B(v)QB^*(v) - B_0vQ(B_0v)^*\big)$$
$$\le k\big(1 + \|v\|_H^2\big) \le k\big(1 + R^2\big).$$

Hence, we have

$$2\|v\|_H \|A(v) - A_0v\|_H + \tau\big(B(v)QB^*(v) - B_0vQ(B_0v)^*\big)$$
$$\le \tilde{\omega}\|v\|_H^2 + (k+1)R^2.$$

An appeal to Theorem 7.8 completes the proof. $\qquad\qquad\qquad\qquad\square$

Theorem 7.9 *Suppose that the coefficients of the linear equation (6.42) satisfy the coercivity condition (6.39) and its solution $\{X_0^x(t),\, t \ge 0\}$ is exponentially ultimately stable in the m.s.s. with the function $t \to E\|X_0^x(t)\|_V^2$ being continuous for all $x \in V$. Let $\{X^x(t),\, t \ge 0\}$ be a solution of the nonlinear equation (6.37). If for $v \in V$,*

$$2\|v\|_V \|A(v) - A_0v\|_{V^*} + \tau\big(B(v)QB^*(v) - B_0vQ(B_0v)^*\big) \le \tilde{\omega}_0\|v\|_V^2 + k_0$$

for some constants $\tilde{\omega}_0$, k_0 such that

$$\tilde{\omega} < \frac{c}{(\alpha_0 + 1)\beta\big[\big(\frac{1}{\alpha} + \frac{c|\lambda|}{\alpha\beta}\big) + \big(\frac{1}{\alpha} + \frac{c|\lambda|}{\alpha\beta} + \frac{c}{\beta}\big) + \frac{|\lambda|M}{2\alpha}\big(\frac{1}{\alpha} + \frac{|\lambda|}{\alpha\beta} + \frac{c}{\beta}\big)^2\big]},$$

then $X^x(t)$ is exponentially ultimately bounded in the m.s.s.

Proof Let, as before,

$$\Psi_0(x) = \int_0^{T_0} \int_0^t E\|X_0^x(t)\|_V^2 \, ds\, dt$$

with $T_0 = \alpha_0(c|\lambda|/(\alpha\beta) + 1/\alpha) + \frac{c}{\beta}$. For $x \in V$,

$$\mathcal{L}\Psi_0(x) - \mathcal{L}_0\Psi_0(x)$$

$$= \langle \Psi_0'(x), A(x) - A_0x \rangle + \frac{1}{2} \operatorname{tr}\big(\Psi_0''(x)\big(B(x)QB^*(x) - B_0xQ(B_0x)^*\big)\big)$$

with $\Psi_0'(x) = 2\tilde{C}x$ and $\Psi_0''(x) = 2C$, where the operators C and $\tilde{C}$ are defined in (7.11) and (7.12). By inequality (7.10) and the continuity of the embedding $V \hookrightarrow H$,

$$\|C\|_{\mathscr{L}(H)} \leq \left(\frac{1}{\alpha} + \frac{c|\lambda|}{\alpha\beta}\right)T_0 + \frac{|\lambda|M + \gamma}{2\alpha}T_0^2,$$

$$\|\tilde{C}\|_{\mathscr{L}(V)} \leq \alpha_0\|C\|_{\mathscr{L}(V)}.$$

Hence,

$$\mathcal{L}\Psi_0(x) - \mathcal{L}_0\Psi_0(x) \leq 2\langle \tilde{C}x, Ax - A_0x \rangle_H + \operatorname{tr}\big(C\big(B(x)QB^*(x) - B_0xQ(B_0x)^*\big)\big),$$

and we have

$$\mathcal{L}\Psi_0(x) \leq \mathcal{L}_0\Psi_0(x) + 2\|\tilde{C}\|_{\mathscr{L}(V)}\|x\|_V\|Ax - A_0x\|_{V^*}$$

$$+ \operatorname{tr}\big(CB(x)QB^*(x) - B_0xQ(B_0x)^*\big)$$

$$\leq \mathcal{L}_0\Psi_0(x) + \big(\|C\|_{\mathscr{L}(H)} + \|\tilde{C}\|_{\mathscr{L}(V)}\big)\big(2\|x\|_V\|Ax - A_0x\|_{V^*}$$

$$+ \tau\big(B(x)QB^*(x) - B_0xQ(B_0x)^*\big)\big).$$

Since $s \to E\|X_0^x(s)\|_V^2$ is a continuous function, we obtain from earlier relations for $\mathcal{L}_0\Psi_0(x)$ that

$$\mathcal{L}_0\Psi_0(x) \leq -\frac{c}{\beta}\|x\|_V^2 + \frac{|\lambda|M + \gamma}{\alpha}T_0.$$

Hence,

$$\mathcal{L}\Psi_0(x) \leq -\frac{c}{\beta}\|x\|_V^2 + \frac{|\lambda|M}{\alpha}T_0 + \big(\|C\|_{\mathscr{L}(H)} + \|\tilde{C}\|_{\mathscr{L}(V)}\big)\big(\tilde{\omega}_0\|x\|_V^2 + k_0\big)$$

$$\leq \left(-\frac{c}{\beta} + \tilde{\omega}_0\big(\|C\|_{\mathscr{L}(H)} + \|\tilde{C}\|_{\mathscr{L}(V)}\big)\right)\|x\|_V^2$$

$$+ k_0\left(\|C\|_{\mathscr{L}(H)} + \|\tilde{C}\|_{\mathscr{L}(V)} + \frac{|\lambda|M + \gamma}{\alpha}T_0\right).$$

Since, with the condition on $\tilde{\omega}_0$, $-c/\beta + \tilde{\omega}_0(\|C\|_{\mathscr{L}(H)} + \|\tilde{C}\|_{\mathscr{L}(V)}) < 0$, we see that conditions analogous to those of Theorem 7.1 are satisfied by Ψ_0, giving the result. $\qquad\square$

Corollary 7.3 *Suppose that the coefficients of the linear equation* (6.42) *satisfy the coercivity condition* (6.39) *and its solution* $\{X_0^x(t),\ t \geq 0\}$ *is exponentially ultimately bounded in the m.s.s. with the function* $t \to E\|X_0^x(t)\|_V^2$ *being continuous for all* $x \in V$. *Let* $\{X^x(t),\ t \geq 0\}$ *be a solution of the nonlinear equation* (6.37). *If for* $v \in V$, *as* $\|v\|_V \to \infty$,

$$\|A(v) - A_0 v\|_{V^*} = o(\|v\|_V)$$

$$\quad and \tag{7.16}$$

$$\tau\big(B(v)QB^*(v) - B_0 v Q(B_0 v)^*\big) = o(\|v\|_V^2),$$

then $X^x(t)$ *is exponentially ultimately bounded in the m.s.s.*

Proof We shall use Theorem 7.9. Under assumption (7.16), for a constant $\tilde{\omega}_0$ satisfying the condition of Theorem 7.9, there exists an $R > 0$ such that, for all $v \in V$ with $\|v\|_V > R$,

$$2\|v\|_V\|A(v) - A_0 v\|_{V^*} + \tau\big(B(v)QB^*(v) - B_0 v Q(B_0 v)^*\big) \leq \tilde{\omega}_0\|v\|_V^2.$$

Using that $\|A(v)\|_{V^*}$, $\|A_0(v)\|_{V^*} \leq a_1\|v\|_V$ and $\|B(v)\|_{\mathscr{L}(K,H)}$, $\|B_0 v\|_{\mathscr{L}(K,H)} \leq b_1\|v\|_V$, we have, for $v \in V$ such that $\|v\|_H < R$,

$$2\|v\|_V\|A(v) - A_0 v\|_{V^*} + \tau\big(B(v)QB^*(v) - B_0 v Q(B_0 v)^*\big)$$

$$\leq 4a_1\|v\|_V^2 + \big(\|B(v)\|_{\mathscr{L}(K,H)}^2 + \|B_0 v\|_{\mathscr{L}(K,H)}^2\big)\operatorname{tr}(Q)$$

$$\leq \big(4a_1 + 2b_1^2\operatorname{tr}(Q)\big)\|v\|_V^2$$

$$\leq \big(4a_1 + 2b_1^2\operatorname{tr}(Q)\big)R^2.$$

Hence, for $v \in V$,

$$2\|v\|_V\|A(v) - A_0 v\|_{V^*} + \tau\big(B(v)QB^*(v) - B_0 v Q(B_0 v)^*\big)$$

$$\leq \tilde{\omega}_0\|v\|_V^2 + \big(4a_1 + 2b_1^2\operatorname{tr}(Q)\big)R^2.$$

An appeal to Theorem 7.9 completes the proof. $\square$

7.3 Abstract Cauchy Problem, Stability and Exponential Ultimate Boundedness

We present an analogue of a result of Zakai and Miyahara for the infinite-dimensional case.

Definition 7.3 A linear operator $A : V \to V^*$ is called *coercive* if it satisfies the following *coercivity condition*: for some $\alpha > 0$, $\gamma, \lambda \in \mathbb{R}$, and all $v \in V$,

$$2\langle v, Av \rangle \leq \lambda\|v\|_H^2 - \alpha\|v\|_V^2 + \gamma. \tag{7.17}$$

Proposition 7.1 *Consider a stochastic evolution equation,*

$$\begin{cases} dX(t) = A_0 X(t)\, dt + F(X(t))\, dt + B(X(t))\, dW_t, \\ X(0) = x \in H, \end{cases} \tag{7.18}$$

with the coefficients A_0 and F satisfying the following conditions:

(1) $A_0 : V \to V^*$ *is coercive.*
(2) $F : V \to H$, $B : V \to \mathscr{L}(K, H)$, *and there exists a constant $K > 0$ such that for all $v \in V$,*

$$\big\| F(v) \big\|_H^2 + \big\| B(v) \big\|_{\mathscr{L}(K,H)}^2 \leq K\big(1 + \|v\|_H^2\big).$$

(3) *There exists a constant $L > 0$ such that for all $v, v' \in V$,*

$$\big\| F(v) - F(v') \big\|_H^2 + \mathrm{tr}\big((B(v) - B(v'))Q(B^*(v) - B^*(v'))\big) \leq L\|v - v'\|_H^2.$$

(4) *For $v \in V$, as $\|v\|_H \to \infty$,*

$$\big\| F(v) \big\|_H = o\big(\|v\|_H\big), \quad \big\| B(v) \big\|_{\mathscr{L}(K,H)} = o\big(\|v\|_H\big).$$

If the classical solution $\{u^x(t), t \geq 0\}$ of the abstract Cauchy problem

$$\begin{cases} \dfrac{du(t)}{dt} = A_0 u(t), \\ u(0) = x \in H, \end{cases} \tag{7.19}$$

is exponentially stable (or even exponentially ultimately bounded), then the solution of (7.18) is exponentially ultimately bounded in the m.s.s.

Proof Let $A(v) = A_0 v + F(v)$ for $v \in V$. Since $F(v) \in H$,

$$2\langle v, A(v)\rangle + \mathrm{tr}\big(B(v)QB^*(v)\big)$$

$$= 2\langle v, A_0 v\rangle + 2\langle v, F(v)\rangle + \mathrm{tr}\big(B(v)QB^*(v)\big)$$

$$\leq \lambda\|v\|_H^2 - \alpha\|v\|_V^2 + 2\|v\|_H\big\| F(v) \big\|_H + \big\| B(v) \big\|_{\mathscr{L}(K,H)}^2 \mathrm{tr}(Q)$$

$$\leq \lambda'\|v\|_H^2 - \alpha\|v\|_H^2 + \gamma$$

for some constants λ' and γ. Hence, the evolution equation (7.18) satisfies the coercivity condition (6.39). Under assumption (2)

$$\big\| F(v) \big\|_H^2 + \mathrm{tr}\big(B(v)QB^*(v)\big) \leq \big\| F(v) \big\|_H^2 + \mathrm{tr}(Q)\big\| B(v) \big\|_{\mathscr{L}(K,H)}^2$$

$$\leq \big(1 + \mathrm{tr}(Q)\big)K\big(1 + \|v\|_H^2\big),$$

so that condition (7.14) holds, and since

$$\|F(v)\|_H = o(\|v\|_H) \quad \text{and} \quad \tau(B(v)QB^*(v)) = o(\|v\|_H^2) \quad \text{as } \|v\|_H \to \infty,$$

Corollary 7.2 gives the result.   □

Example 7.2 (Stochastic Heat Equation) Let S^1 be the unit circle realized as the interval $[-\pi, \pi]$ with identified points $-\pi$ and π. Denote by $W^{1,2}(S^1)$ the Sobolev space on S^1 and by $W(t, \xi)$ the Brownian sheet on $[0, \infty) \times S^1$, see Exercise 7.2. Let $\kappa > 0$ be a constant, and f and b be real-valued functions. Consider the following SPDE:

$$\begin{cases} \dfrac{\partial X(t)}{\partial t}(\xi) = \dfrac{\partial^2 X(t)}{\partial \xi^2}(\xi) - \kappa f(X(t)(\xi)) + b(X(t)(\xi))\dfrac{\partial^2 W}{\partial t \partial \xi}, \\ X(0)(\cdot) = x(\cdot) \in L^2(S^1). \end{cases} \tag{7.20}$$

Let $H = L^2(S^1)$ and $V = W^{1,2}(S^1)$. Consider

$$A_0(x) = \left(\frac{d^2}{d\xi^2} - \kappa\right)x$$

and mappings F, B defined for $\xi \in S^1$ and $x, y \in V$ by

$$F(x)(\xi) = f(x(\xi)), \quad (B(x)y)(\xi) = \langle b(x(\cdot)), y(\cdot)\rangle_{L^2(S^1)}.$$

Let

$$\|x\|_H = \left(\int_{S^1} x^2(\xi)\, d\xi\right)^{1/2} \quad \text{for } x \in H,$$

$$\|x\|_V = \left(\int_{S^1} \left(x^2(\xi) + \left(\frac{dx(\xi)}{d\xi}\right)^2\right) d\xi\right)^{1/2} \quad \text{for } x \in V.$$

Then we obtain the equation

$$dX(t) = A_0 X(t)\, dt + F(X(t))\, dt + B(X(t))\, d\tilde{W}_t,$$

where $\tilde{W}_t$ is a cylindrical Wiener process defined in Exercise 7.2. We have

$$2\langle x, A_0(x)\rangle = -2\|x\|_V^2 + (-2\kappa + 2)\|x\|_H^2$$

$$\leq -2\|x\|_H^2 + (-2\kappa + 2)\|x\|_H^2 = -2\kappa\|x\|_H^2.$$

By Theorem 6.3(a), with $\Lambda(x) = \|x\|_H^2$, the solution of (7.19) is exponentially stable. If we assume that f and b are Lipschitz continuous and bounded, then conditions (1)–(3) of Proposition 7.1 are satisfied. Using representation (2.35) of the stochastic integral with respect to a cylindrical Wiener process, we can conclude that the solution of the stochastic heat equation (7.20) is exponentially ultimately bounded in the m.s.s.

Exercise 7.2 Let S^1 be the unit circle realized as the interval $[-\pi, \pi]$ with identified points $-\pi$ and π. Denote by $\{f_j(\xi)\}$ an ONB in $L^2(S^1)$ and consider

$$W(t, \zeta) = \sum_{j=1}^{\infty} w_j(t) \int_{-\pi}^{\zeta} f_j(\xi) \, d\xi, \quad t \geq 0, -\pi \leq \zeta \leq \pi, \qquad (7.21)$$

where w_j are independent Brownian motions defined on $\{\Omega, \mathscr{F}, \{\mathscr{F}_t\}_{t \geq 0}\}, P\}$. Show that the series (7.21) converges P-a.s. and that

$$\text{Cov}\big(W(t_1, \zeta_1) W(t_2, \zeta_2)\big) = (t_1 \wedge t_2)(\zeta_1 \wedge \zeta_2).$$

Conclude that the Gaussian random field $W(\cdot, \cdot)$ has a continuous version. This continuous version is called the *Brownian sheet* on S^1.

Now, let $\Phi(t)$ be an adapted process with values in $L^2(S^1)$ (identified with $\mathscr{L}(L^2(S^1), \mathbb{R})$) and satisfying

$$E \int_0^\infty \|\Phi(t)\|^2_{L^2(S^1)} \, dt < \infty.$$

Consider a standard cylindrical Brownian motion $\tilde{W}_t$ in $L^2(S^1)$ defined by

$$\tilde{W}_t(k) = \sum_{j=1}^{\infty} w_j(t) \langle k, f_j \rangle_{L^2(S^1)}.$$

Show that the cylindrical stochastic integral process

$$\int_0^t \Phi(s) \, d\tilde{W}_s \qquad (7.22)$$

is well defined in $L^2(\Omega, \mathbb{R})$.

On the other hand, for an elementary processes of the form

$$\Phi(t, \xi) = 1_{[0,t]}(s) 1_{[-\pi, \zeta]}(\xi), \qquad (7.23)$$

define

$$\Phi \cdot W = \int_0^\infty \int_{S^1} \Phi(s, \xi) W(ds, d\xi). \qquad (7.24)$$

Clearly $\Phi \cdot W = W(t, \zeta)$. Extend the integral $\Phi \cdot W$ to general processes. Since

$$\Phi \cdot W = \int_0^\infty \Phi(s) \, d\tilde{W}_s$$

for elementary processes (7.23), conclude that the integrals are equal for general processes as well.

Example 7.3 Consider the following SPDE driven by a real-valued Brownian motion:

$$
\begin{cases}
d_t u(t, x) = \left(\alpha^2 \dfrac{\partial^2 u(t, x)}{\partial x^2} + \beta \dfrac{\partial u(t, x)}{\partial x} + \gamma u(t, x) + g(x) \right) dt \\[2mm]
\qquad\qquad + \left(\sigma_1 \dfrac{\partial u(t, x)}{\partial x} + \sigma_2 u(t, x) \right) dW_t, \\[2mm]
u(0, x) = \varphi(x) \in L^2((-\infty, \infty)) \cap L^1((-\infty, +\infty)),
\end{cases}
\tag{7.25}
$$

where we use the symbol d_t to signify that the differential is with respect to t. Let $H = L^2((-\infty, \infty))$ and $V = W_0^{1,2}((-\infty, \infty))$ with the usual norms

$$
\|v\|_H = \left(\int_{-\infty}^{+\infty} v^2 \, dx \right)^{1/2}, \quad v \in H,
$$

$$
\|v\|_V = \left(\int_{-\infty}^{+\infty} \left(v^2 + \left(\frac{dv}{dx} \right)^2 \right) dx \right)^{1/2}, \quad v \in V.
$$

Define the operators $A : V \to V^*$ and $B : V \to \mathscr{L}(H)$ by

$$
A(v) = \alpha^2 \frac{d^2 v}{dx^2} + \beta \frac{dv}{dx} + \gamma v + g, \quad v \in V,
$$

$$
B(v) = \sigma_1 \frac{dv}{dx} + \sigma_2 v, \quad v \in V.
$$

Suppose that $g \in L^2((-\infty, \infty)) \cap L^1((-\infty, \infty))$. Then, using integration by parts, we obtain for $v \in V$,

$$
2\langle v, A(v) \rangle + \mathrm{tr}\big(Bv(Bv)^* \big)
$$

$$
= 2\left\langle v, \alpha^2 \frac{d^2 v}{dx^2} + \beta^2 \frac{dv}{\partial x} + \gamma v + g \right\rangle + \left\| \sigma_1 \frac{dv}{dx} + \sigma_2 v \right\|_H^2
$$

$$
= (-2\alpha^2 + \sigma_1^2) \|v\|_V^2 + (2\gamma + \sigma_2^2 + 2\alpha^2 - \sigma_1^2) \|v\|_H^2 + 2\langle v, g \rangle_H
$$

$$
\leq (-2\alpha^2 + \sigma_1^2) \|v\|_V^2 + (2\gamma + \sigma_2^2 + 2\alpha^2 - \sigma_1^2 + \varepsilon) \|v\|_H^2 + \frac{1}{\varepsilon} \|g\|_H^2
$$

for any $\varepsilon > 0$. Similarly, for $u, v \in V$,

$$
2\langle u - v, A(u) - A(v) \rangle + \mathrm{tr}\big(B(u - v)(B(u - v))^* \big)
$$

$$
\leq (-2\alpha^2 + \sigma_1^2) \|u - v\|_V^2 + (2\gamma + \sigma_2^2 + 2\alpha^2 - \sigma_1^2) \|u - v\|_H^2.
$$

If $-2\alpha^2 + \sigma_1^2 < 0$, then the coercivity and weak monotonicity conditions, (6.39) and (6.40), hold, and we know from Theorem 4.7 that there exists a unique strong solution $u^\varphi(t)$ to (7.25) in $L^2(\Omega, C([0, T], H)) \cap M^2([0, T], V)$. Taking the Fourier transform yields

$$
\begin{aligned}
d_t \hat{u}^\varphi(t, \lambda) &= \left(-\alpha^2 \lambda^2 \hat{u}^\varphi(t, \lambda) + i\lambda\beta \hat{u}^\varphi(t, \lambda) + \gamma \hat{u}^\varphi(t, \lambda) + \hat{g}(\lambda)\right) dt \\
&\quad + \left(i\sigma_1 \lambda \hat{u}^\varphi(t, \lambda) + \sigma_2 \hat{u}^\varphi(t, \lambda)\right) dW_t \\
&= \left((-\alpha^2 \lambda^2 + i\lambda\beta + \gamma)\hat{u}^\varphi(t, \lambda) + \hat{g}(\lambda)\right) dt \\
&\quad + (i\sigma_1 \lambda + \sigma_2)\hat{u}^\varphi(t, \lambda) dW_t.
\end{aligned}
$$

For fixed λ,

$$
\begin{aligned}
a &= -\alpha^2 \lambda^2 + i\lambda\beta + \gamma, \\
b &= \hat{g}(\lambda), \\
c &= i\sigma_1 \lambda + \sigma_2.
\end{aligned}
$$

By simple calculation (see Exercise 7.3),

$$
\begin{aligned}
E\left|\hat{u}^\varphi(t, \lambda)\right|^2 &= E\left|\hat{\varphi}(\lambda)\right|^2 + 2\mathrm{Re}\left(\frac{b\bar{b} + \bar{b}\hat{\varphi}(\lambda)(a + \bar{a} + c\bar{c})}{(a + \bar{a} + c\bar{c})(\bar{a} + c\bar{c})} e^{(a+\bar{a}+c\bar{c})}\right) \\
&\quad - 2\mathrm{Re}\left(\frac{\bar{b}(a\hat{\varphi}(\lambda) + b)}{a(\bar{a} + c\bar{c})} e^{at}\right) + 2\mathrm{Re}\left(\frac{\beta\bar{b}}{a(a + \bar{a} + c\bar{c})}\right). \quad (7.26)
\end{aligned}
$$

By Plancherel's theorem,

$$
E\left\|u^\varphi(t)\right\|_H^2 = \int_{-\infty}^{+\infty} E\left|\hat{u}^\varphi(t, \lambda)\right|^2 d\lambda
$$

and

$$
\begin{aligned}
E\left\|u^\varphi(t)\right\|_V^2 &= E\left\|u^\varphi(t)\right\|_H^2 + E\left\|\frac{d}{dx} u^\varphi(t, x)\right\|_H^2 \\
&= \int_{-\infty}^{+\infty} \left(1 + \lambda^2\right) E\left|\hat{u}^\varphi(t, \lambda)\right|^2 d\lambda.
\end{aligned}
$$

For a suitable $T > 0$,

$$
\begin{aligned}
\Psi(\varphi) &= \int_0^T \int_0^t E\left\|u^\varphi(s)\right\|_V^2 ds \, dt \\
&= \int_{-\infty}^{+\infty} \left(1 + \lambda^2\right) \int_0^T \int_0^t E\left\|\hat{u}(s, \lambda)\right\|^2 ds \, dt \, d\lambda.
\end{aligned}
$$

Thus it is difficult to compute a Lyapunov function explicitly. In view of Remark 7.1, it is enough to compute a Lyapunov function of the linear SPDE

$$d_t u(t, x) = \left(\alpha^2 \frac{\partial^2 u(t, x)}{\partial x^2} + \beta \frac{\partial u(t, x)}{\partial x} + \gamma u(t, x) \right) du$$
$$+ \left(\sigma_1 \frac{\partial u(t, x)}{\partial x} + \sigma_2 u(t, x) \right) dW_t.$$

Define the operators $A_0 : V \to V^*$ and $B_0 : V \to \mathscr{L}(H)$ by

$$A_0(v) = \alpha^2 \frac{d^2 v}{dx^2} + \beta \frac{dv}{dx} + \gamma v, \quad v \in V,$$
$$B_0(v) = B(v), \quad v \in V$$

(since B is already linear). Taking the Fourier transform and solving explicitly, we obtain that the solution is the geometric Brownian motion

$$\hat{u}_0^{\varphi}(t, \lambda) = \hat{\varphi}(\lambda) e^{at - \frac{1}{2}c^2 t + cW_t},$$
$$E \left| \hat{u}_0^{\varphi}(t, \lambda) \right|^2 = \left| \hat{\varphi}(\lambda) \right|^2 e^{(a + \bar{a} + c\bar{c})t}.$$

The function $t \to E \| u_0^{\varphi}(t) \|_V^2$ is continuous for all $\varphi \in V$,

$$\left\| A(v) - A_0(v) \right\|_{V^*} = \| g \|_{V^*} = o \big(\| v \|_V \big) \quad \text{as } \| v \|_V \to \infty,$$

and

$$\tau \big(B(v) Q B^*(v) - (B_0 v) Q (B_0 v)^* \big) = 0.$$

Thus, if $\{ u_0(t), t \geq 0 \}$ is exponentially ultimately bounded in the m.s.s., then the Lyapunov function $\Psi_0(\varphi)$ of the linear system is the Lyapunov function of the nonlinear system, and

$$\Psi_0(\varphi) = \int_{-\infty}^{+\infty} (1 + \lambda^2) \left(\int_0^T \int_0^t E \left| \hat{u}_0(s, \lambda) \right|^2 ds\, dt \right) d\lambda$$
$$= \int_{-\infty}^{+\infty} \left\{ (1 + \lambda^2) |\hat{\varphi}(\lambda)|^2 \left(\frac{\exp\{(-2\alpha^2 + \sigma_1^2)\lambda^2 + 2\gamma + \sigma_2^2)T\}}{((-2\alpha^2 + \sigma_1^2)\lambda^2 + 2\gamma + \sigma_2^2)^2} \right) \right.$$
$$\left. - \frac{T}{(-2\alpha^2 + \sigma_1^2)\lambda^2 + 2\gamma + \sigma_2^2} - \frac{1}{(-2\alpha^2 + \sigma_1^2)\lambda^2 + 2\gamma + \sigma_2^2} \right\} d\lambda.$$

Using Theorem 7.8, we can conclude that the solution of the nonlinear system is exponentially ultimately bounded in the m.s.s.

Exercise 7.3 Complete the computations in (7.26).

Example 7.4 Consider an equation of the form

$$\begin{cases} dX_t = AX(t)\,dt + F(X(t))\,dt + B(X(t))\,dW_t, \\ X(0) = x \in H, \end{cases}$$

where F and B satisfy the conditions of Proposition 7.1. This example is motivated by the work of Funaki. If $-A$ is coercive, a typical case being $A = \Delta$, we conclude that the solution of the deterministic linear equation is exponentially stable since the Laplacian has negative eigenvalues. Thus, the solution of the deterministic equation is exponentially bounded, and hence, by Proposition 7.1, the solution of the nonlinear equation above is exponentially ultimately bounded in the m.s.s.

Example 7.5 Let $\mathcal{O} \subseteq \mathbb{R}^n$ be a bounded open domain with smooth boundary. Assume that $H = L^2(\mathcal{O})$ and $V = W_0^{1,2}(\mathcal{O})$, the Sobolev space. Suppose that $\{W_q(t, x); t \geq 0, x \in \mathcal{O}\}$ is an H-valued Wiener process with associated covariance operator Q, given by a continuous symmetric nonnegative definite kernel $q(x, y) \in L^2(\mathcal{O} \times \mathcal{O})$, $q(x, x) \in L^2(\mathcal{O})$,

$$(Qf)(x) = \int_{\mathcal{O}} q(x, y) f(y)\,dy.$$

By Mercer's theorem [41], there exists an orthonormal basis $\{e_j\}_{j=1}^{\infty} \subset L^2(\mathcal{O})$ consisting of eigenfunctions of Q such that

$$q(x, y) = \sum_{j=1}^{\infty} \lambda_j e_j(x) e_j(y)$$

with $\mathrm{tr}(Q) = \int_{\mathcal{O}} q(x, x)\,dx = \sum_{j=1}^{\infty} \lambda_j < \infty$.

Let $-A$ be a linear strongly elliptic differential operator of second order on $\mathcal{O}$, and $B(u) : L^2(\mathcal{O}) \to L^2(\mathcal{O})$ with $B(u)f(\cdot) = u(\cdot)f(\cdot)$. By Garding's inequality, $-A$ is coercive (see [63], Theorem 7.2.2). Then the infinite-dimensional problem is as follows:

$$d_t u(t, x) = Au(t, x)\,dt + u(t, x)\,d_t W_q(t, x),$$

and we choose $\Lambda(v) = \|v\|_H^2$ for $v \in W_0^{1,2}(\mathcal{O})$. We shall check conditions under which Λ is a Lyapunov function. With $\mathcal{L}$ defined in (6.15), using the spectral representation of $q(x, y)$, we have

$$\mathcal{L}\left(\|v\|_H^2\right) = 2\langle v, Av \rangle + \mathrm{tr}\left(B(v)QB^*(v)\right)$$

$$= 2\langle v, Av \rangle + \int_{\mathcal{O}} q(x, x)v^2(x)\,dx.$$

Let

$$\lambda_0 = \sup\left\{ \frac{\mathscr{L}(\|v\|_H^2)}{\|v\|_H^2}, \ v \in W_0^{1,2}(\mathcal{O}), \ \|v\|_H^2 \neq 0 \right\}$$

$$= \sup\left\{ \frac{2\langle v, Av \rangle + \langle Qv, v \rangle_H}{\|v\|_H^2}, \ v \in W_0^{1,2}(\mathcal{O}), \ \|v\|_H^2 \neq 0 \right\}.$$

If $\lambda_0 < 0$, then, by Theorem 6.4, the solution is exponentially stable in the m.s.s. Consider the nonlinear equation in $\mathcal{O}$,

$$\begin{cases} d_t u(t, x) = \tilde{A}(x, u(t, x)) \, dt + \tilde{B}(u(t, x)) d_t W_q(t, x), \\ u(0, x) = \varphi(x), \quad u(t, x)|_{\partial \mathcal{O}} = 0. \end{cases} \tag{7.27}$$

Assume that

$$\tilde{A}(x, v) = Av + \alpha_1(x, v), \quad \tilde{B}(x, v) = B(v) + \alpha_2(x, v),$$

where $\alpha_i(x, v)$ satisfy the Lipschitz-type condition

$$\sup_{x \in \mathcal{O}} \left| \alpha_i(x, v_1) - \alpha_i(x, v_2) \right| < c \|v_1 - v_2\|_H,$$

so that the nonlinear equation (7.27) has a unique strong solution. Under the assumption

$$\alpha_i(x, 0) = 0,$$

zero is a solution of (7.27), and if

$$\sup_{x \in \mathcal{O}} \left| \alpha_i(x, v) \right| = o(\|v\|_H), \quad \|v\|_H \to 0,$$

then, by Theorem 6.14, the strong solution of the nonlinear equation (7.27) is exponentially stable in the m.s.s.

On the other hand, let us consider the operator A as above and F and B satisfying the conditions of Proposition 7.1. Then, under the condition

$$\sup\left\{ \frac{2\langle v, Av \rangle}{\|v\|_H^2}, \ u \in W_0^{1,2}(\mathcal{O}), \ \|v\|_H^2 \neq 0 \right\} < 0,$$

the solution of the abstract Cauchy problem (7.19), with A_0 replaced by A, is exponentially stable, and we conclude that the solution of the equation

$$\begin{cases} dX(t) = AX(t) \, dt + F(X(t)) \, dt + B(X(t)) \, dW_t, \\ X(0) = x \in H, \end{cases}$$

is ultimately exponentially bounded in the m.s.s.

Consider now the SSDE (3.1) and assume that A is the infinitesimal generator of a pseudo-contraction C_0-semigroup $\{S(t), t \geq 0\}$ on H (see Chap. 3) with the coefficients $F : H \to H$ and $B : H \to \mathscr{L}(K, H)$, independent of t and ω. We assume that F and B are in general nonlinear mappings satisfying the linear growth condition (A3) and the Lipschitz condition (A4) (see Sect. 3.3). In addition, the initial condition is assumed deterministic, so that (3.1) takes the form

$$\begin{cases} dX(t) = (AX(t) + F(X(t)))\,dt + B(X(t))\,dW_t, \\ X(0) = x \in H. \end{cases} \tag{7.28}$$

By Theorem 3.5, there exists a unique continuous mild solution.

Using Corollary 7.1, we now have the following analogue of Proposition 7.1.

Proposition 7.2 *Suppose that the classical solution $\{u^x(t), t \geq 0\}$ of the abstract Cauchy problem (7.19) is exponentially stable (or even exponentially ultimately bounded) and, as $\|h\|_H \to \infty$,*

$$\left\| F(h) \right\|_H = o\big(\|h\|_H\big),$$

$$\left\| B(h) \right\|_{\mathscr{L}(K,H)} = o\big(\|h\|_H\big),$$

then the mild solution of (7.28) is exponentially ultimately bounded in the m.s.s.

7.4 Ultimate Boundedness and Invariant Measure

We are interested in the behavior of the law of a solution to an SDE as $t \to \infty$. Let us begin with a filtered probability space $(\Omega, \mathscr{F}, \{\mathscr{F}_t\}_{t \geq 0}, P)$ and an H-valued time-homogeneous Markov process $X^{\xi_0}(t)$, $X^{\xi_0}(0) = \xi_0$, where ξ_0 is $\mathscr{F}_0$-measurable random variable with distribution μ^{ξ_0}. Assume that its associated semigroup P_t is Feller. We can define for $A \in \mathscr{B}(H)$, the *Markov transition probabilities*

$$P(t, x, A) = P_t 1_A(x), \quad x \in H.$$

Since a regular conditional distribution of $X^{\xi_0}(t)$ exists (Theorem 3, Vol. I, Sect. 1.3 in [25]), we have that

$$P(t, x, A) = P\big(X^{\xi_0}(t) \in A | \xi_0 = x\big) = \int_H P\big(X^{\xi_0} \in A | \xi_0 = x\big)\,\mu^{\xi_0}(dx), \quad x \in H.$$

Then, for a bounded measurable function f on H $(f \in B_b(H))$,

$$(P_t f)(x) = \int_H f(y) P(t, x, dy). \tag{7.29}$$

The Markov property (3.52) takes the form

$$E\left(f\left(X^{\xi_0}(t+s)\right)\big|\mathscr{F}^{X^{\xi_0}_t}_t\right) = (P_s f)\left(X^{\xi_0}_t\right) = \int_H f(y) P\left(s, X^{\xi_0}(t), dy\right),$$

so that the transition probability $P(t, x, A)$ is a *transition function for a time-homogeneous Markov process* $X^{\xi_0}(t)$.

We observe that the following *Chapman–Kolmogorov* equation holds for Markov transition probabilities

$$P(t+s, x, A) = \int_H P(t, y, A) P(s, x, dy), \qquad (7.30)$$

which follows from the semigroup property of P_t, (3.58) applied to $\varphi(x) = 1_A(x)$ and from the fact that $P(t, x, dy)$ is the conditional law of $X^{\xi_0}(t)$.

Exercise 7.4 Show (7.30).

Let us now define an invariant probability measure and state a general theorem on its existence.

Definition 7.4 We say that a probability measure μ on H is invariant for a time-homogeneous Markov process $X^x(t)$ with the related Feller semigroup $\{P_t, \ t \geq 0\}$ defined by (7.29) if for all $A \in \mathscr{B}(H)$,

$$\mu(A) = \int_H P(t, x, A) \mu(dx),$$

or equivalently, since H is a Polish space, if for all $f \in C_b(H)$,

$$\int_H (P_t f) \, d\mu = \int_H f(y) \, d\mu.$$

Let μ be a probability measure on H and define

$$\mu_n(A) = \frac{1}{t_n} \int_0^{t_n} \int_H P(t, x, A) \, dt \, \mu(dx) \qquad (7.31)$$

for a sequence $\{t_n\}_{n=1}^{\infty} \subset \mathbb{R}_+$, $t_n \to \infty$. In particular, for a real-valued bounded Borel-measurable function $f(x)$ on H, we have

$$\int_H f(x) \mu_n(dx) = \frac{1}{t_n} \int_0^{t_n} \int_H \int_H f(y) P(t, x, dy) \, \mu(dx) \, dt. \qquad (7.32)$$

Theorem 7.10 *If v is weak limit of a subsequence of $\{\mu_n\}$, then v is an invariant measure.*

Proof We can assume without loss of generality that $\mu_n \Rightarrow \nu$. Observe that, by the Fubini theorem and the Chapman–Kolmogorov equation,

$$
\begin{aligned}
\int_H (P_t f)(x)\nu(dx) &= \lim_{n\to\infty} \int_H (P_t f)(x)\,\mu_n(dx) \\
&= \lim_{n\to\infty} \frac{1}{t_n} \int_0^{t_n} \int_H \int_H (P_t f)(y) P(s,x,dy)\,\mu(dx)\,ds \\
&= \lim_{n\to\infty} \frac{1}{t_n} \int_0^{t_n} \int_H (P_{t+s} f)(x)\,\mu(dx)\,ds \\
&= \lim_{n\to\infty} \left[\frac{1}{t_n} \left\{ \int_0^{t_n} \int_H (P_s f)(x)\,\mu(dx)\,ds \right. \right. \\
&\qquad \left. \left. + \int_{t_n}^{t_n+t} \int_H (P_s f)(x)\,\mu(dx)\,ds - \int_0^t \int_H (P_s f)(x)\,\mu(dx)\,ds \right\} \right].
\end{aligned}
$$

Since $\|P_s f(x_0)\|_H \le \|f(x_0\|_H$, the last two integrals are bounded by a constant, and hence, using (7.32),

$$
\begin{aligned}
\int_H (P_t f)(x)\,\nu(dx) &= \lim_{n\to\infty} \frac{1}{t_n} \int_0^{t_n} \int_H (P_s f)(x_0)\,\mu(dx)\,ds \\
&= \lim_{n\to\infty} \frac{1}{t_n} \int_0^{t_n} \int_H \int_H f(y) P(s,x,dy)\,\mu(dx)\,ds \\
&= \lim_{n\to\infty} \int_H f(x)\,\mu_n(dx) = \int_H f(x)\,\nu(dx). \qquad \square
\end{aligned}
$$

Corollary 7.4 *If the sequence $\{\mu_n\}$ is relatively compact, then an invariant measure exists.*

Exercise 7.5 Show that if, as $t \to \infty$, the laws of $X^x(t)$ converge weakly to a probability measure μ, then μ is an invariant measure for the corresponding semigroup P_t.

We shall now consider applications of the general results on invariant measures to SPDEs. In case where $\{X^{\xi_0}(t), \ t \ge 0\}$ is a solution of an SDE with a random initial condition ξ_0, taking in (7.31) $\mu = \mu^{\xi_0}$, the distribution of ξ_0, gives

$$
P\big(X^{\xi_0}(t) \in A\big) = \int_H P(t,x,A)\,\mu^{\xi_0}(dx). \tag{7.33}
$$

Thus, properties of the solution can be used to obtain tightness of the measures μ_n.

Exercise 7.6 Prove (7.33).

Before we apply the result on ultimate boundedness to obtain the existence of an invariant measure, let us consider some examples.

Example 7.6 (Navier–Stokes Equation [76]) Let $\mathscr{D} \subseteq \mathbb{R}^2$ be a bounded domain with smooth boundary $\partial \mathscr{D}$. Consider the equation

$$\begin{cases} \dfrac{\partial v_i(t,x)}{\partial t} + \sum_{j=1}^{2} v_j \dfrac{\partial v_i(t,x)}{\partial x_j} = -\dfrac{1}{\rho}\dfrac{\partial P(t,x)}{\partial x_i} + v\sum_{j=1}^{2}\dfrac{\partial^2 v_i(t,x)}{\partial x_j^2} + \sigma_i \dot{W}_t^i(x), \\[2mm] \sum_{j=1}^{2}\dfrac{\partial v_j}{\partial x_j} = 0, \quad x \in \mathscr{D},\ v > 0, \\[2mm] i = 1,2. \end{cases} \tag{7.34}$$

Let $C_0^\infty = \{v \in C_0^\infty(\mathscr{D}) \times C_0^\infty(\mathscr{D});\ \nabla v = 0\}$, with ∇ denoting the gradient. Let $H = \overline{C_0^\infty}$ in $L^2(\mathscr{D}) \times L^2(\mathscr{D})$, and $V = \{v : W_0^{1,2}(\mathscr{D}) \times W_0^{1,2}(\mathscr{D}),\ \nabla v = 0\}$. Then $V \subseteq H \subseteq V^*$ is a Gelfand triplet, and the embedding $V \hookrightarrow H$ is compact.

It is known [76] that

$$L^2(\mathscr{D}) \times L^2(\mathscr{D}) = H \oplus H^\perp,$$

where $H^\perp$ is characterized by $H^\perp = \{v : v = \nabla(p) \text{ for some } p \in W^{1,2}(\mathscr{D})\}$.

Denote by Π the orthogonal projection of $L^2(\mathscr{D}) \times L^2(\mathscr{D})$ onto $H^\perp$, and for $v \in C_0^\infty$, define

$$A(v) = v\Pi\Delta v - \Pi\big[(v \cdot \nabla)v\big].$$

Then A can be extended as a continuous operator form V to V^*.

Equation (7.34) can be recast as an evolution equation in the form

$$\begin{cases} dX(t) = A(X(t))\,dt + \sigma\,dW_t, \\ X(0) = \xi, \end{cases}$$

where W_t is an H-valued Q-Wiener process, and $\xi \in V$ a.e. is an $\mathscr{F}_0$-measurable H-valued random variable. It is known (see [76]) that the above equation has a unique strong solution $\{u^\xi(t),\ t \geq 0\}$ in $C([0,T], H) \cap L^2([0,T], V)$, which is a homogeneous Markov and Feller process, satisfying for $T < \infty$,

$$E\big\|u^\xi(T)\big\|_H^2 + v\int_0^T \sum_{i=1}^{2} \left\|\frac{\partial u^\xi(t)}{\partial x_i}\right\|_H^2 dt \leq E\|\xi\|_H^2 + \frac{T}{2}\operatorname{tr}(Q).$$

Using the fact that $\|u^\xi(t)\|_V$ is equivalent to $(\sum_{i=1}^{2}\|\frac{\partial u(\xi)}{\partial x_i}\|_H^2)^{1/2}$, we have

$$\sup_T \frac{1}{T}\int_0^T E\big(\|u^\xi(t)\|_V^2\big)\,dt \leq \frac{c}{2v}\operatorname{tr}(Q)$$

with some constant c. By the Chebychev inequality,

$$\lim_{R\to\infty}\sup_T \frac{1}{T}\int_0^T P\big(\|u^\xi(t)\|_V > R\big)\,dt = 0.$$

Hence, for $\varepsilon > 0$, there exists an R_ε such that

$$\sup_T \frac{1}{T} \int_0^T P\big(\|u^\xi(t)\|_V > R_\varepsilon\big) \, dt < \varepsilon.$$

Thus, as $t_n \to \infty$,

$$\sup_n \frac{1}{t_n} \int_H \int_0^{t_n} P\big(t, x, \widetilde{B}_{R_\varepsilon}\big) \, dt \, \mu^\xi(dx) < \varepsilon,$$

where $\widetilde{B}_{R_\varepsilon}$ is the image of the set $\{v \in V; \ \|v\|_V > R_\varepsilon\}$ under the compact embedding $V \hookrightarrow H$, and μ^ξ is the distribution of ξ on H. Since $\widetilde{B}_{R_\varepsilon}$ is a complement of a compact set, we can use Prokhorov's theorem and Corollary 7.4 to conclude that an invariant measure exists. Note that its support is in V, by the weak convergence.

Example 7.7 (Linear equations with additive noise [79]) Consider the mild solution of the equation

$$\begin{cases} dX(t) = AX(t) \, dt + dW_t, \\ X(0) = x \in H, \end{cases}$$

where A is an infinitesimal generator of a strongly continuous semigroup $\{S(t), t \geq 0\}$ on H. Denote

$$Q_t = \int_0^t S(r) Q S^*(r) \, dr,$$

and assume that $\mathrm{tr}(Q_t) < \infty$. We know from Theorems 3.1 and 3.2 that

$$X(t) = S(t)x + \int_0^t S(t - s) \, dW_s \tag{7.35}$$

is the mild solution of the above equation. The stochastic convolution $\int_0^t S(t - s) \, dW_s$ is an H-valued Gaussian process with covariance

$$Q_t = \int_0^t S(u) Q S^*(u) \, du$$

for any t. The Gaussian process $X(t)$ is also Markov and Feller, and it is called an *Ornstein–Uhlenbeck process*. The probability measure μ on H is invariant if for $f \in C_b(H)$ and any $t \geq 0$,

$$\int_H f(x) \mu(dx) = \int_H E\big(f(X^x(t))\big) \mu(dx)$$

$$= \int_H Ef\left(S(t)x + \int_0^t S(t - s) \, dW_s\right) \mu(dx).$$

For $f(x) = e^{i\langle \lambda, x \rangle_H}$, $\lambda \in H$, we obtain

$$\hat{\mu}(\lambda) = \hat{\mu}\big(S^*(t)\lambda\big)e^{-\frac{1}{2}\langle Q_t\lambda, \lambda \rangle_H},$$

where $\hat{\mu}$ denotes the characteristic function of μ. It follows that

$$\big|\hat{\mu}(\lambda)\big| \leq e^{-\frac{1}{2}\langle Q_t\lambda, \lambda \rangle_H},$$

or

$$\langle Q_t\lambda, \lambda \rangle_H \leq -2\ln\big|\hat{\mu}(\lambda)\big| = 2\ln\left(\frac{1}{|\hat{\mu}(\lambda)|}\right).$$

Since $\hat{\mu}(\lambda)$ is the characteristic function of a measure μ on H, then by Sazonov's theorem [74], for $\varepsilon > 0$, there exists a trace-class operator S_0 on H such that $|\hat{\mu}(\lambda)| \geq 1/2$ whenever $\langle S_0\lambda, \lambda \rangle_H \leq 1$. Thus, we conclude that

$$\langle Q_t\lambda, \lambda \rangle_H \leq 2\ln 2$$

if $\langle S_0\lambda, \lambda \rangle_H \leq 1$. This yields

$$0 \leq Q_t \leq (2\ln 2)S_0.$$

Hence, $\sup_t \operatorname{tr}(Q_t) < \infty$.

On the other hand, if $\sup_t \operatorname{tr}(Q_t) < \infty$, let us denote by $\overline{P}$ the limit in trace norm of Q_t and observe that

$$S(t)\overline{P}S^*(t) = \int_0^\infty S(t+r)QS^*(t+r)\,dr = \int_t^\infty S(u)QS(u)\,du = \overline{P} - Q_t.$$

Thus,

$$\frac{1}{2}\big\langle S(t)\overline{P}S^*(t)\lambda, \lambda \big\rangle_H = \frac{1}{2}\big\langle \overline{P}\lambda, \lambda \big\rangle_H - \frac{1}{2}\langle Q_t\lambda, \lambda \rangle_H,$$

implying

$$e^{-\frac{1}{2}\langle \overline{P}\lambda, \lambda \rangle_H} = e^{-\frac{1}{2}\langle \overline{P}S^*(t)\lambda, S^*(t)\lambda \rangle_H}e^{-\frac{1}{2}\langle Q_t\lambda, \lambda \rangle_H}.$$

In conclusion, μ with the characteristic functional $e^{-\frac{1}{2}\langle \overline{P}\lambda, \lambda \rangle_H}$ is an invariant measure. We observe that the invariant measure exists for the Markov process $X(t)$ defined in (7.35) if and only if $\sup_t \operatorname{tr}(Q_t) < \infty$. Also, if $S(t)$ is an exponentially stable semigroup (i.e., $\|S(t)\|_{\mathscr{L}(H)} \leq Me^{-\mu t}$ for some positive constants M and μ) or if $S_t x \to 0$ for all $x \in H$ as $t \to \infty$, then the Gaussian measure with covariance $\overline{P}$ is the invariant (Maxwell) probability measure.

Let $\{X(t), t \geq 0\}$ be exponentially ultimately bounded in the m.s.s., then, clearly,

$$\limsup_{t\to\infty} E\big\|X(t)\big\|_H^2 \leq M < \infty \quad \text{for all } x \in H. \tag{7.36}$$

Definition 7.5 A stochastic process $X(t)$ satisfying condition (7.36) is called ultimately bounded in the m.s.s.

7.4.1 Variational Equations

We focus our attention now on the variational equation with a deterministic initial condition,

$$\begin{cases} dX(t) = A(X(t))\,dt + B(X(t))\,dW_t, \\ X(0) = x \in H, \end{cases} \tag{7.37}$$

which is driven by a Q-Wiener process W_t. The coefficients $A : V \to V^*$ and $B : V \to \mathcal{L}(K, H)$ are independent of t and ω, and they satisfy the linear growth, coercivity (C), and weak monotonicity (WM) conditions (6.38), (6.39), (6.40). By Theorem 4.8 and Remark 4.2 the solution is a homogeneous Markov process, and the associated semigroup is Feller.

We note that in Theorem 7.5, we give conditions for exponential ultimate boundedness in the m.s.s. in terms of the Lyapunov function. Assume that $\Psi : H \to \mathbb{R}$ satisfies the conditions of Theorem 6.10 (Itô's formula) and define

$$\mathcal{L}\psi(u) = \langle \psi'(u), A(u) \rangle + (1/2)\,\mathrm{tr}\big(\psi''(u)B(u)QB^*(u)\big). \tag{7.38}$$

Let $\{X^x(t), t \geq 0\}$ be the solution of (7.37). We apply Itô's formula to $\Psi(X^x(t))$, take the expectation, and use condition (2) of Theorem 7.5 to obtain

$$E\Psi\big(X^x(t)\big) - E\Psi\big(X^x(t')\big) = E \int_{t'}^{t} \mathcal{L}\Psi\big(X^x(s)\big)\,ds$$

$$\leq \int_{t'}^{t} \big(-c_3 E\Psi\big(X^t(s)\big) + k_3\big)\,ds.$$

Let $\Phi(t) = E\Psi(X^x(t))$, then $\Phi(t)$ is continuous, so that

$$\Phi'(t) \leq -c_3\Phi(t) + k_3.$$

Hence,

$$E\Psi\big(X_t^x\big) \leq \frac{k_3}{c_3} + \left(\Psi(x) - \frac{k_3}{c_3}\right)e^{-c_3 t}.$$

Assuming that $\Psi(x) \geq c_1\|x\|_H^2 - k_1$, we obtain

$$c_1 E\big\|X^x(t)\big\|_H^2 - k_1 \leq \frac{k_3}{c_3} + \left(c_2\|x\|_H^2 - \frac{k_3}{c_3}\right)e^{-c_3 t}.$$

Thus we have proved the following:

Proposition 7.3 Let $\Psi : H \to \mathbb{R}$ satisfy conditions (1)–(5) of Theorem 6.10 and assume that condition (2) of Theorem 7.5 holds and that $c_1\|x\|_H^2 - k_1 \leq \Psi(x)$ for $x \in H$ and some constants $c_1 > 0$ and $k_1 \in \mathbb{R}$. Then

$$\limsup_{t \to \infty} E\big\|X^x(t)\big\|_H^2 \leq \frac{1}{c_1}\left(k_1 + \frac{k_3}{c_3}\right).$$

In particular, $\{X^x(t), t \geq 0\}$ is ultimately bounded.

Let us now state the theorem connecting the ultimate boundedness with the existence of invariant measure.

Theorem 7.11 *Let* $\{X^x(t), t \geq 0\}$ *be a solution of (7.37). Assume that the embedding* $V \hookrightarrow H$ *is compact. If* $X^x(t)$ *is ultimately bounded in the m.s.s., then there exists an invariant measure* μ *for* $\{X^x(t), t \geq 0\}$.

Proof Applying Itô's formula to the function $\|x\|_H^2$ and using the coercivity condition, we have

$$E\|X^x(t)\|_H^2 - \|x\|_H^2 = \int_0^t E\mathscr{L}\|X^x(t)\|_H^2 \, ds$$

$$\leq \lambda \int_0^t E\|X^x(s)\|_H^2 ds - \alpha \int_0^t E\|X^x(s)\|_V^2 + \gamma t$$

with $\mathscr{L}$ defined in (7.38). Hence,

$$\int_0^t E\|X^x(s)\|_V^2 \, ds \leq \frac{1}{\alpha}\left(\lambda \int_0^t E\|X^x(s)\|_H^2 \, ds + \|x\|_H^2 + \gamma t\right).$$

Therefore,

$$\frac{1}{T}\int_0^T P\left(\|X^x(t)\|_V > R\right) dt \leq \frac{1}{T}\int_0^T \frac{E\|X^x(t)\|_V^2}{R^2} \, dt$$

$$\leq \frac{1}{\alpha R^2}\frac{1}{T}\left(|\lambda| \int_0^T E\|X^x(t)\|_H^2 \, dt + \|x\|_H^2 + \gamma T\right).$$

Now, by (7.36), $E\|X^x(t)\|_H^2 \leq M$ for $t \geq T_0$ and some $T_0 \geq 0$. But

$$\sup_{t \leq T_0} E\|X^x(t)\|_H^2 \leq M'$$

by Theorem 4.7, so that

$$\lim_{R \to \infty} \sup_T \frac{1}{T}\int_0^T P\left(\|X^x(t)\|_V > R\right) dt$$

$$\leq \lim_{R \to \infty} \sup_T \frac{|\lambda|}{\alpha R^2}\frac{1}{T}\left(\int_0^{T_0} E\|X^x(t)\|^2 \, dt + \int_{T_0}^T E\|X^x(t)\|_H^2 \, dt\right)$$

$$\leq \lim_{R \to \infty} \sup_T \frac{|\lambda|}{\alpha R^2}\left(\frac{T_0}{T}M' + \frac{T - T_0}{T}M\right)$$

$$\leq \lim_{R \to \infty} \frac{|\lambda|}{\alpha R^2}(M' + M), \quad 0 \leq T_0 \leq T.$$

Hence, given $\varepsilon > 0$, there exists an R_ε such that

$$\sup_T \frac{1}{T}\int_0^T P\left(\|X^x(t)\|_V > R_\varepsilon\right) dt < \varepsilon.$$

By the assumption that the embedding $V \hookrightarrow H$ is compact, the set $\{v \in V : \|v\|_V \le R_\varepsilon\}$ is compact in H, and the result is proven. $\qquad\square$

Remark 7.2 Note that a weaker condition on the second moment of $X^x(t)$, i.e.,

$$\sup_{T > T_0} \frac{1}{T} \int_0^T E\|X^x(t)\|_H^2 \, dt < M \quad \text{for some } T_0 \ge 0,$$

is sufficient to carry out the proof of Theorem 7.11.

In Examples 7.2–7.6, we consider equations whose coefficients satisfy the conditions imposed on the coefficients of (7.37) and the embedding $V \hookrightarrow H$ is compact, so that an invariant measure exists if the solution is ultimately bounded in the m.s.s.

Theorem 7.12 *Suppose that $V \hookrightarrow H$ is compact and the solution of $\{X^x(t), t \ge 0\}$ of (7.37) is ultimately bounded in the m.s.s. Then any invariant measure μ satisfies*

$$\int_V \|x\|_V^2 \, \mu(dx) < \infty.$$

Proof Let $f(x) = \|x\|_V^2$ and $f_n(x) = 1_{[0,n]}(f(x))$. Now $f_n(x) \in L^1(V, \mu)$. We use the ergodic theorem for a Markov process with an invariant measure (see [78], p. 388). This gives

$$\lim_{T \to \infty} \frac{1}{T} \int_0^T (P_t f_n)(x) \, dt = f_n^*(x) \quad \mu\text{-a.e.}$$

and $E_\mu f_n^* = E_\mu f_n$, where $E_\mu f_n = \int_V f_n(x) \mu(dx)$.

By the assumption of ultimate boundedness, we have, as in the proof of Theorem 7.11,

$$\limsup_{T \to \infty} \frac{1}{T} \int_0^T E\|X^x(t)\|_V^2 \, dt \le \frac{C|\lambda|}{\alpha}, \quad C < \infty.$$

Hence,

$$f_n^*(x) = \lim_{T \to \infty} \frac{1}{T} \int_0^T (P_t f_n)(x) \, dt$$

$$\le \limsup_{T \to \infty} \frac{1}{T} \int_0^T (P_t f(x)) \, dt$$

$$= \limsup_{T \to \infty} \frac{1}{T} \int_0^T E\|X^x(t)\|_V^2 dt \le \frac{C|\lambda|}{\alpha}.$$

But $f_n(x) \uparrow f(x)$, so that

$$E_\mu f = \lim_{n \to \infty} E_\mu f_n = \lim_{n \to \infty} E_\mu f_n^* \le \frac{C|\lambda|}{\alpha}. \qquad\square$$

Remark 7.3 (a) For parabolic Itô equations, one can easily derive the result using $\Psi(x) = \|x\|_H^2$ and Theorem 7.11.

(b) Note that if $\mu_n \Rightarrow \mu$ and the support of μ_n is in V with the embedding $V \hookrightarrow H$ being compact, then by the weak convergence the support of μ is in V by the same argument as in Example 7.6.

Let us now consider the problem of uniqueness of the invariant measure.

Theorem 7.13 *Suppose that for ε, δ, and $R > 0$, there exists a constant $T_0(\varepsilon, \delta, R) > 0$ such that for $T \geq T_0$,*

$$\frac{1}{T} \int_0^T P\left(\|X^x(t) - X^y(t)\|_V \geq \delta\right) dt < \varepsilon$$

for all $x, y \in V_R = \{v \in V : \|v\|_V \leq R\}$ with the embedding $V \hookrightarrow H$ being compact. If there exists an invariant measure μ for a solution of (7.37), $\{X^{x_0}(t), t \geq 0\}$, $X(0) = x_0$, with support in V, then it is unique.

Proof Suppose that μ, ν are invariant measures with support in V. We need to show that

$$\int_H f(x)\mu(dx) = \int_H f(x)\nu(dx)$$

for f uniformly continuous bounded on H, since such functions form a determining class.

For $G \in \mathscr{B}(H)$, define

$$\mu_T^x(G) = \frac{1}{T} \int_0^T P\left(X^x(t) \in G\right) dt, \quad x \in H, \ T > 0.$$

Then, using invariance of μ and ν, we have

$$\left| \int_H f(x)\,\mu(dx) - \int_H f(x)\,\nu(dx) \right|$$

$$= \left| \int_H \int_H f(x)\left[\mu_T^y(dx)\mu(dy) - \mu_T^z(dx)\nu(dz)\right] \right|$$

$$\leq \int_{H \times H} \left| \int_H f(x)\,\mu_T^y(dx) - \int_H f(x)\mu_T^z(dx) \right| \mu(dy)\nu(dz).$$

Let

$$F(y, z) = \left| \int_H f(x)\mu_T^y(dx) - \int_H f(x)\mu_T^z(dx) \right|.$$

Then, using the fact that μ, ν have the supports in V, we have

$$\left| \int_H f(x)\,\mu(dx) - \int_H f(x)\,\nu(dx) \right| \leq \int_{V \times V} |F(y, z)|\,\mu(dy)\nu(dz).$$

Let $V_R^c = V \setminus V_R$ and choose $R > 0$ such that

$$\mu\left(V_R^c\right) + v\left(V_R^c\right) < \varepsilon.$$

Then,

$$\left| \int_H f(x)\, \mu(dx) - \int_H f(x)\, v(dx) \right| \leq \int_{V_R \times V_R} \left| F(y,z) \right| \mu(dy) v(dz) + \left(4\varepsilon + 2\varepsilon^2\right) M,$$

where $M = \sup_{x \in H} |f(x)|$. But for $\delta > 0$,

$$\int_{V_R \times V_R} \left| F(y,z) \right| \mu(dy) v(dz)$$

$$\leq \int_{V_R \times V_R} \left\{ \frac{1}{T} \int_0^T E \left| f(X^y(t)) - f(X^z(t)) \right| \mu(dy) v(dz) \right\}$$

$$\leq 2M \sup_{y,z \in V_R} \frac{1}{T} \int_0^T P\left(\left\| X^y(t) - X^z(t) \right\|_V > \delta \right) + \sup_{\substack{y,z \in V_R \\ \|y-z\| < \delta}} \left| f(y) - f(z) \right|$$

$$\leq 2M\varepsilon + \varepsilon$$

for $T \geq T_0$, since f is uniformly continuous.

Using the last inequality and the bound for $\left| \int_H f(x)\, \mu(dx) - \int_H f(x)\, v(dx) \right|$, we obtain the result. $\qquad\square$

Let us now give a condition on the coefficients of the SDE (7.37) which guarantees the uniqueness of the invariant measure. We have proved in Theorem 7.11 (see Remark 7.3), that the condition

$$\sup_{T > T_0} \left\{ \frac{1}{T} \int_0^T E \left\| X^x(t) \right\|_H^2 dt \right\} \leq M \quad \text{for some } T_0 \geq 0$$

implies that there exists an invariant measure to the strong solution $\{X^x(t), t \geq 0\}$, whose support is in V.

Theorem 7.14 *Suppose that* $V \hookrightarrow H$ *is compact, the coefficients of* (7.37) *satisfy the coercivity condition* (6.39), *and that for* $u, v \in V$,

$$2\langle u - v, A(u) - A(v) \rangle + \left\| B(u) - B(v) \right\|_{\mathscr{L}_2(K_Q, H)}^2 \leq -c\|u - v\|_V^2,$$

where the norm $\| \cdot \|_{\mathscr{L}_2(K_Q, H)}$ *is the Hilbert–Schmidt norm defined in* (2.7). *Assume that the solution* $\{X^x(t), t \geq 0\}$ *of* (7.37) *is ultimately bounded in the m.s.s. Then there exists a unique invariant measure.*

Proof By Itô's formula, we have, for $t > 0$,

$$E \left\| X^x(t) \right\|_H^2 = \|x\|_H^2 + 2E \int_0^t \langle X^x(s), A(X^x(s)) \rangle ds + E \int_0^t \left\| B(X^x(s)) \right\|_{\mathscr{L}_2(K_Q, H)}^2 ds.$$

Using the coercivity condition (C), (6.39), we have

$$E\|X^x(t)\|_H^2 + \alpha E \int_0^t \|X^x(s)\|_V^2 \, ds \le \left(\|x\|_H^2 + \gamma t\right) + \lambda E \int_0^t \|X^x(s)\|_H^2 \, ds.$$

It follows, similarly as in the proof of Theorem 7.11, that

$$\sup_{T > T_0} \frac{1}{T} \int_0^T E\|X^x(s)\|_V^2 \, ds \le \frac{|\gamma| + \|x\|_H^2/T_0}{\alpha} + \frac{|\lambda|}{\alpha} \sup_{T > T_0} \int_0^T E\|X^x(s)\|_H^2 \, ds.$$

By the Chebychev inequality, we know that

$$\frac{1}{T} \int_0^T P\left(\|X^x(s)\|_V > R\right) \le \frac{1}{R^2}\left\{\frac{1}{T} \int_0^T E\|X^x(s)\|_V^2 \, ds\right\}.$$

Hence, using the arguments in Example 7.6, an invariant measure exists and is supported on V. To prove the uniqueness, let $X^{x_1}(t)$, $X^{x_2}(t)$ be two solutions with initial values x_1, x_2. We apply Itô's formula to $X(t) = X^{x_1}(t) - X^{x_2}(t)$ and obtain

$$E\|X(t)\|_H^2 \le \|x_1 - x_2\|_H^2 + 2E \int_0^t \left\langle X(s) - A\left(X^{x_1}(s)\right) - A\left(X^{x_2}(s)\right)\right\rangle ds$$

$$+ E \int_0^t \left\|B\left(X^{x_1}(s)\right) - B\left(X^{x_2}(s)\right)\right\|_{\mathscr{L}_2(K_Q,H)}^2 ds.$$

Using the assumption, we have

$$E\|X(t)\|_H^2 \le \|x_1 - x_2\|_H^2 - c \int_0^t E\|X(s)\|_V^2 \, ds,$$

which implies that

$$\int_0^t E\|X(s)\|_V^2 \le \frac{1}{c}\|x_1 - x_2\|_H^2.$$

It now suffices to refer to the Chebychev inequality and Theorem 7.13 to complete the proof.   □

7.4.2 Semilinear Equations Driven by a Q-Wiener Process

Let us consider now the existence of an invariant measure for a mild solution of a semilinear SDE with deterministic initial condition

$$\begin{cases} dX(t) = (AX(t) + F(X(t))) \, dt + B(X(t)) \, dW_t, \\ X(0) = x \in H, \end{cases} \tag{7.39}$$

where A is the infinitesimal generator of a pseudo-contraction C_0-semigroup $S(t)$ on H, and the coefficients $F : H \to H$ and $B : H \to \mathcal{L}(K, H)$, independent of t and ω, are in general nonlinear mappings satisfying the linear growth condition (A3) and the Lipschitz condition (A4) in Sect. 3.3. We know from Theorem 3.6 that the solution is a homogeneous Markov process and from Theorem 3.7 that it is continuous with respect to the initial condition, so that the associated semigroup is Feller.

We studied a special case in Example 7.7. Here we look at the existence under the assumption of exponential boundedness in the m.s.s. We will use the Lyapunov function approach developed earlier in Theorem 7.8 and Corollary 7.3. We first give the following proposition.

Proposition 7.4 *Suppose that the mild solution* $\{X^x(t)\}$ *of* (7.39) *is ultimately bounded in the m.s.s. Then any invariant measure* ν *of the Markov process* $\{X^x(t), t \geq 0\}$ *satisfies*

$$\int_H \|y\|_H^2 \nu(dy) \leq M,$$

where M is as in (7.36).

The proof is similar to the proof of Theorem 7.12 and is left to the reader as an exercise.

Exercise 7.7 Prove Proposition 7.4.

Theorem 7.15 *Suppose that the solution* $\{X^x(t), t \geq 0\}$ *of* (7.39) *is ultimately bounded in the m.s.s. If for all* $R > 0$, $\delta > 0$, *and* $\varepsilon > 0$, *there exists* $T_0 = T_0(R, \delta, \varepsilon) > 0$ *such that for all* $t \geq T_0$,

$$P\left(\left\|X^x(t) - X^y(t)\right\|_H > \delta\right) < \varepsilon \quad for \ x, y \in B_H(R) \tag{7.40}$$

with $B_H(R) = \{x \in H, \|x\| \leq R\}$, *then there exists at most one invariant measure for the Markov process* $X^x(t)$.

Proof Let $\mu_i, i = 1, 2$, be two invariant measures. Then, by Proposition 7.4, for each $\varepsilon > 0$, there exists $R > 0$ such that $\mu_i(H \setminus B_H(R)) < \varepsilon$. Let f be a bounded weakly continuous function on H. We claim that there exists a constant $T = T(\varepsilon, R, f) > 0$ such that

$$\left|P_t f(x) - P_t f(y)\right| \leq \varepsilon \quad for \ x, y \in B_H(R) \quad if \ t \geq T.$$

Let C be a weakly compact set in H. The weak topology on C is given by the metric

$$d(x, y) = \sum_{k=1}^{\infty} \frac{1}{2^k} \left|\langle e_k, x - y \rangle_H\right|, \quad x, y \in C, \tag{7.41}$$

where $\{e_k\}_{k=1}^{\infty}$ in an orthonormal basis in H.

By the ultimate boundedness, there exists $T_1 = T_1(\varepsilon, R) > 0$ such that for $T \geq T_1$,

$$P\big(X^x(t) \in B_H(R)\big) > 1 - \varepsilon/2 \quad \text{for } x \in B_H(R).$$

Now f is uniformly continuous w.r.t. the metric (7.41) on $B_H(R)$. Hence, there exists $\delta' > 0$ such that $x, y \in H_R$ with $d(x, y) < \delta'$ imply that $|f(x) - f(y)| \leq \delta$, and there exists $J > 0$ such that

$$\sum_{k=J+1}^{\infty} \frac{1}{2^k} |\langle e_k, x - y \rangle_H| \leq \delta'/2 \quad \text{for } x, y \in B_H(R).$$

Since $P(|\langle e_k, X^x(t) - X^y(t)\rangle| > \delta) \leq P(\|X^x(t) - X^y(t)\|_H > \delta)$, by the given assumption we can choose $T_2 \geq T_1$ such that for $t \geq T_2$,

$$P\left\{ \sum_{k=1}^{J} \big(\langle e_k, X^x(t)\rangle - \langle e_k, X^y(t)\rangle\big)^2 > \delta'/2 \right\} \geq 1 - \varepsilon/3 \qquad (7.42)$$

for $x, y \in B_H(R)$. Hence, for $t \geq T_2$,

$$P\{|f(X^x(t)) - f(X^y(t))| \leq \delta\}$$
$$\geq P\{X^x(t), X^y(t) \in B_H(R), d(X^x(t), X^y(t)) \leq \delta'\}$$
$$\geq P\left\{ X^x(t), X^y(t) \in B_H(R), \sum_{k=1}^{J} \frac{1}{2^k} |\langle e_k, X^x(t) - X^y(t)\rangle_H| \leq \delta'/2 \right\}$$
$$\geq P\{X^x(t), X^y(t) \in B_H(R), |\langle e_k, X^x(t) - X^y(t)\rangle_H| \leq \delta'/2, k = 1, \dots, J\}$$
$$\geq 1 - \varepsilon/3 - \varepsilon/3 - \varepsilon/3 = 1 - \varepsilon,$$

since the last probability above is no smaller than that in (7.42).

Now, with $M_0 = \sup |f(x)|$, given $\varepsilon > 0$, choose T so that for $t \geq T$,

$$P\big(|f(X^x(t)) - f(X^y(t))| \leq \varepsilon/2\big) \geq 1 - \frac{\varepsilon}{4M_0}.$$

Then

$$E|f(X^x(t)) - f(X^y(t))| \leq \frac{\varepsilon}{2} + 2M_0 \frac{\varepsilon}{4M_0} = \varepsilon.$$

Note that for invariant measures μ_1, μ_2,

$$\int_H f(x)\mu_i(dx) = \int_H (P_t f)(x)\mu_i(dx), \quad i = 1, 2.$$

For $t \geq T$, we have

$$\left| \int_H f(x)\mu_1(dx) - \int_H f(y)\mu_2(dy) \right|$$

$$= \left| \int_H \int_H [f(x) - f(y)] \mu_1(dx) \mu_2(dy) \right|$$

$$= \left| \int_H \int_H [(P_t f)(x) - (P_t f)(y)] \mu_1(dx) \mu_2(y) \right|$$

$$= \left| \left(\int_{B_H(R)} + \int_{H \setminus B_H(R)} \right) \left(\int_{B_H(R)} + \int_{H \setminus B_H(R)} \right) \right.$$

$$\left. \times \left[(P_t f)(x) - (P_t f)(y) \right] \mu_1(dx) \mu_2(dy) \right|$$

$$\leq \varepsilon + 2(2M_0)\varepsilon + 2M_0\varepsilon^2.$$

Since $\varepsilon > 0$ is arbitrary, we conclude that

$$\int_H f(x)\,\mu_1(dx) = \int_H f(x)\,\mu_2(dx). \qquad \square$$

In case we look at the solution to (7.39), whose coefficients satisfy the linear growth and Lipschitz conditions (A3) and (A4) of Sect. 3.1 in Chap. 3, we conclude that under assumption (7.40) and conditions for exponential ultimate boundedness, there exists at most one invariant measure.

Note that in the problem of existence of the invariant measure, the relative weak compactness of the sequence μ_n in Theorem 7.10 is crucial. In the variational case, we achieved this condition, under ultimate boundedness in the m.s.s., assuming that the embedding $V \hookrightarrow H$ is compact. For mild solutions, Ichikawa [33] and Da Prato and Zabczyk [11], give sufficient conditions. Da Prato and Zabczyk use a factorization technique introduced in [10]. We start with the result in [32].

Theorem 7.16 *Assume that A is a self-adjoint linear operator with eigenvectors $\{e_k\}_{k=1}^{\infty}$ forming an orthonormal basis in H and that the corresponding eigenvalues $-\lambda_k \downarrow -\infty$ as $k \to \infty$. Let the mild solution of (7.39) satisfy*

$$\frac{1}{T} \int_0^T E \|X^x(s)\|_H^2 \, ds \leq M(1 + \|x\|_H^2). \qquad (7.43)$$

Then there exists an invariant measure for the Markov semigroup generated by the solution of (7.39).

Proof The proof depends on the following lemma.

Lemma 7.2 *Under the conditions of Theorem 7.16, the set of measures*

$$\mu_t(\cdot) = \frac{1}{t} \int_0^t P(s, x, \cdot) \quad \text{for } t \geq 0$$

with $P(s, x, A) = P(X^x(s) \in A)$ is relatively weakly compact.

Proof Let $y_k(t) = \langle X^x(t), e_k \rangle_H$. Then, by a well-known result about the weak compactness ([25], Vol. I, Chap. VI, Sect. 2, Theorem 2), we need to show that the expression

$$\frac{1}{T} \int_0^T \left(\sum_{k=1}^\infty E y_k^2(t) \right) dt$$

is uniformly convergent in T.

Let $S(t)$ be the C_0-semigroup generated by A. Since $S(t)e_k = e^{-\lambda_k t} e_k$ for each k, $y_k(t)$ satisfies

$$y_k(t) = e^{-\lambda_k t} x_k^0 + \int_0^t e^{-\lambda_k(t-s)} \langle e_k, F(X^x(s)) \rangle_H \, ds$$

$$+ \int_0^t e^{-\lambda_k(t-s)} \langle e_k, B(X^x(s)) \, dW(s) \rangle_H,$$

$$E y_k^2(t) \le 3 e^{-2\lambda_k t} (x_k^0)^2 + 3E \left| \int_0^t e^{-\lambda_k(t-s)} \langle e_k, F(X^x(s)) \rangle_H \, ds \right|^2$$

$$+ 3E \left| \int_0^t e^{-\lambda_k(t-s)} \langle e_k, B(X^x(s)) \, dW_s \rangle \right|^2.$$

For N large enough, so that $\lambda_N > 0$, and any $m > 0$, using Exercise 7.8 and assumption (7.43), we have

$$\sum_{k=N}^{N+m} \frac{1}{T} \int_0^T E \left| \int_0^t e^{-\lambda_k(t-s)} \langle e_k, F(X^x(s)) \rangle_H \, ds \right|^2 dt$$

$$\le \frac{1}{2\varepsilon T} \int_0^T \int_0^t e^{2(-\lambda_k+\varepsilon)(t-s)} |\langle e_k, F(X^x(s)) \rangle_H|^2 \, ds \, dt$$

$$= \frac{1}{T} \int_0^T \int_r^T e^{2(-\lambda_k+\varepsilon)(t-s)} \, dt \, |\langle e_k, F(X^x(s)) \rangle_H|^2 \, ds$$

$$\le \frac{\int_0^T E \| F(X^x(s)) \|_H^2 \, ds}{4\varepsilon(\lambda_N - \varepsilon)T} \le \frac{c_1(1 + \|x\|_H^2)}{\varepsilon(\lambda_N - \varepsilon)}$$

for some constants $\varepsilon > 0$ and $c_1 > 0$.

Utilizing the Hölder inequality, we also have that

$$\sum_N^{N+m} \frac{1}{T} \int_0^T E \left| \int_0^t e^{-\lambda_k(t-s)} \langle e_k, B(X^x(s)) \, dW(s) \rangle \right|^2 dt$$

$$\le \frac{\mathrm{tr}(Q) \int_0^T E \| B(X^x(t)) \|_{\mathscr{L}(K,H)}^2 \, dt}{2\lambda_N T} \le \frac{c_2 \, \mathrm{tr}(Q)(1 + \|x\|_H^2)}{\lambda_N}$$

for some constant $c_2 > 0$. Thus,

$$\sum_{N}^{N+m} \frac{1}{T} \int_0^T E y_k^2(t)\,dt \le \frac{3\|x\|_H^2}{2\lambda_N}$$

$$+ 3(c_1 + c_2)\big(1 + \|x\|_H^2\big)\bigg[\frac{1}{\delta(\lambda_N - \delta)} + \frac{\mathrm{tr}(Q)}{\lambda_N}\bigg].$$

Thus the condition in [25] holds.   □

The proof of Theorem 7.16 is an immediate consequence of the lemma.   □

Exercise 7.8 Let $p > 1$, and let g be a nonnegative locally p-integrable function on $[0, \infty)$. Then for all $\varepsilon > 0$ and real d,

$$\bigg(\int_0^t e^{d(t-r)} g(r)\,dr\bigg)^p \le \bigg(\frac{1}{q\varepsilon}\bigg)^{p/q} \int_0^t e^{p(d+\varepsilon)(t-r)} g^p(r)\,dr,$$

where $1/p + 1/q = 1$.

7.4.3 Semilinear Equations Driven by a Cylindrical Wiener Process

We finally present a result in [12], which uses an innovative technique to prove the tightness of the laws $\mathcal{L}(X^x(t))$. We start with the problem

$$\begin{cases} dX(t) = (AX(t) + F(X(t)))\,dt + B(X(t))\,d\tilde{W}_t, \\ X(0) = x \in H, \end{cases} \tag{7.44}$$

where $\tilde{W}_t$ is a cylindrical Wiener process in a separable Hilbert space K. Assume that the coefficients and the solution satisfy the following hypotheses.

Hypothesis (DZ) Let conditions (DZ1)–(DZ4) of Sect. 3.10 hold, and, in addition, assume that:

(DZ5) $\{S(t), t > 0\}$ is a compact semigroup.
(DZ6) For all $x \in H$ and $\varepsilon > 0$, there exists $R > 0$ such that for every $T \ge 1$,

$$\frac{1}{T} \int_0^T P\big(\|X^x(t)\|_H > R\big)\,dt < \varepsilon,$$

where $\{X^x(t), t \ge 0\}$ is a mild solution of (7.44).

Remark 7.4 (a) Condition (DZ6) holds if $\{X^x(t),\ t \geq 0\}$ is ultimately bounded in the m.s.s.

(b) In the special case where W_t is a Q-Wiener process, we can replace B with $\tilde{B} = BQ^{1/2}$.

Theorem 7.17 *Under Hypothesis* (DZ), *there exists an invariant measure for the mild solution of* (7.44).

Proof We recall the factorization formula used in Lemma 3.3. Let $x \in H$, and

$$Y^x(t) = \int_0^t (t-s)^{-\alpha} S(t-s) B\big(X^x(s)\big)\, dW_s.$$

Then

$$X^x(1) = S(1)x + G_1 F\big(X^x(\cdot)\big)(1) + \frac{\sin \pi \alpha}{\pi} G_\alpha Y^x(\cdot)(1) \quad P\text{-a.s.}$$

By Lemma 3.12, the compactness of the semigroup $\{S(t),\ t \geq 0\}$ implies that the operators G_α defined by

$$G_\alpha f(t) = \int_0^t (t-s)^{\alpha-1} S(t-s) f(s)\, ds, \quad f \in L^p\big([0,T], H\big),$$

are compact from $L^p([0,T], H)$ into $C([0,T], H)$ for $p \geq 2$ and $1/p < \alpha \leq 1$. Consider $\gamma : H \times L^p([0,1], H) \times L^p([0,1], H) \to H$,

$$\gamma(y, f, g) = S(1)y + G_1 f(1) + G_\alpha g(1).$$

Then γ is a compact operator, and hence, for $r > 0$, the set

$$K(r) = \left\{ x \in H : x = S(1)y + G_1 f(1) + G_\alpha g(1), \right.$$

$$\left. \|y\|_H \leq r, \|f\|_{L^p} \leq r, \|g\|_{L^p} \leq \frac{r\pi}{\sin \pi \alpha} \right\}$$

is relatively compact in H.

We now need the following lemma.

Lemma 7.3 *Assume that* $p > 2$, $\alpha \in (1/p, 1/2)$, *and that Hypothesis* (DZ) *holds. Then there exists a constant* $c > 0$ *such that for* $r > 0$ *and all* $x \in H$ *with* $\|x\|_H \leq r$,

$$P\big(X^x(1) \in K(r)\big) \geq 1 - cr^{-p}\big(1 + \|x\|_H^p\big).$$

Proof By Lemma 3.13, using Hypothesis (DZ3), we calculate

$$E \int_0^1 \big\|Y^x(s)\big\|_H^p\, ds$$

$$= E \int_0^1 \left\| \int_0^s (s-u)^{-\alpha} S(s-u) B\big(X^x(u)\big) \, dW_u \right\|_H^p ds$$

$$\leq k E \int_0^1 \left(\int_0^s (s-u)^{-2\alpha} \big\| S(s-u) B\big(X^x(u)\big) \big\|_{\mathscr{L}_2(K,H)}^2 \, du \right)^{p/2} ds$$

$$\leq k 2^{p/2} E \int_0^1 \left(\int_0^s (s-u)^{-2\alpha} \mathscr{K}^2(s-u) \big(1 + \big\| X^x(u) \big\|_H^2 \big) \, du \right)^{p/2} ds.$$

By (3.103) and Exercise 3.7,

$$E \int_0^1 \big\| Y^x(s) \big\|_H^p \, ds \leq k 2^{p/2} \left(\int_0^1 t^{-2\alpha} \mathscr{K}^2(t) \, dt \right)^{p/2} E \int_0^1 \big(1 + \big\| X^x(u) \big\|_H^2 \big)^{p/2} du$$

$$\leq k_1 \big(1 + \|x\|_H^p\big) \quad \text{for some } k_1 > 0.$$

Also, using Hypothesis (DZ2), we get

$$E \int_0^1 \big\| F\big(X^x(u)\big) \big\|_H^p du \leq k_2 \big(1 + \|x\|_H^p\big), \quad x \in H.$$

By the Chebychev inequality,

$$P\left(\big\| Y^x(\cdot) \big\|_{L^p} \leq \frac{\pi r}{\sin \alpha \pi} \right)$$

$$\geq 1 - r^{-p} \frac{\sin^p \alpha \pi}{\pi^p} E \big\| Y^x(\cdot) \big\|_{L^p}^p \geq 1 - r^{-p} \pi^{-p} k_1 \big(1 + \|x\|_H^p\big)$$

$$P\big(\big\| F\big(X^x(\cdot)\big) \big\|_{L^p} \leq r \big)$$

$$\geq 1 - r^{-p} E\big(\big\| F\big(X^x(\cdot)\big) \big\|_{L^p}^p \big) \geq 1 - r^{-p} k_2 \big(1 + \|x\|_H^p\big),$$

giving

$$P\big(X^x(1) \in K(r)\big) \geq P\left(\left\{ \big\| Y^x(\cdot) \big\|_{L^p} \leq \frac{\pi r}{\sin \alpha \pi} \right\} \cap \big\{ \big\| F\big(X^x(\cdot)\big) \big\|_{L^p} \leq r \big\} \right)$$

$$\geq 1 - r^{-p} \big(\pi^{-p} k_1 + k_2 \big) \big(1 + \|x\|_H^p\big). \qquad \square$$

We continue the proof of Theorem 7.17.

For any $t > 1$ and $r > r_1 > 0$, by the Markov property (recall Proposition 3.4) and Lemma 7.3, we have

$$P\big(X^x(t) \in K(r)\big) = P\big(t, x, K(r)\big)$$

$$= \int_H P\big(1, y, K(r)\big) P(t-1, x, dy)$$

$$\geq \int_{\|y\|_H \leq r_1} P\big(1, y, K(r)\big) P(t-1, x, dy)$$

$$\geq \left(1 - c\big(r^{-p}\big(1 + r_1^p\big)\big)\right) \int_{\|y\|_H \leq r_1} P(t - 1, x, dy)$$

$$= \left(1 - c\big(r^{-p}\big(1 + r_1^p\big)\big)\right) P\big(\big\|X^x(t - 1)\big\|_H \leq r_1\big),$$

giving

$$\frac{1}{T} \int_0^T P\big(X^x(t) \in K(r)\big)\, dt \geq 1 - cr^{-p}\big(1 + r_1^p\big) \frac{1}{T} \int_0^T P\big(\big\|X^x(t)\big\|_H \leq r_1\big)\, dt.$$

If we choose r_1 according to condition (DZ6) and take $r > r_1$ sufficiently large, we obtain that $\frac{1}{T} \int_0^T P(t, x, \cdot)\, dt$ is relatively compact, ensuring the existence of an invariant measure. $\qquad\square$

7.5 Ultimate Boundedness and Weak Recurrence of the Solutions

In Sect. 4.3 we proved the existence and uniqueness for strong variational solutions, and in Sect. 4.4 we showed that they are strong Markov and Feller processes. We will now study weak (positive) recurrence of the strong solution of (7.45), which is (exponentially) ultimately bounded in the m.s.s.

The weak recurrence property to a bounded set was considered in [59] for the solutions of SDEs in the finite dimensions and in [33] for solutions of stochastic evolution equations in a Hilbert space. This section is based on the work of R. Liu [51].

Let us consider a strong solution of the variational equation

$$\begin{cases} dX(t) = A(X(t))\, dt + B(X(t))\, dW_t, \\ X(0) = x \in H. \end{cases} \qquad (7.45)$$

We start with the definition of weak recurrence.

Definition 7.6 A stochastic process $X(t)$ defined on H is *weakly recurrent* to a compact set if there exists a compact set $C \subset H$ such that

$$P^x\big(X(t) \in C \text{ for some } t \geq 0\big) = 1 \quad \text{for all } x \in H,$$

where P^x is the conditional probability under the condition $X(0) = x$. The set C is called a *recurrent region*. From now on *recurrent* means *recurrent to a compact set*.

Theorem 7.18 *Suppose that $V \hookrightarrow H$ is compact and the coefficients of (7.45) satisfy the coercivity and the weak monotonicity conditions (6.39) and (6.40). If its solution $\{X^x(t),\ t \geq 0\}$ is ultimately bounded in the m.s.s., then it is weakly recurrent.*

Proof We prove the theorem using a series of lemmas.

Lemma 7.4 *Let* $\{X(t), t \geq 0\}$ *be a strong Markov process in* H. *If there exists a positive Borel-measurable function* $\rho : H \to \mathbb{R}_+$, *a compact set* $C \subset H$, *and a constant* $\delta > 0$ *such that*

$$P^x\big(X(\rho(x)) \in C\big) \geq \delta \quad \text{for all } x \in H,$$

then $X^x(t)$ *is weakly recurrent with the recurrence region* C.

Proof For a fixed $x \in H$, let $\tau_1 = \rho(x)$, $\Omega_1 = \{\omega : X(\tau_1) \notin C\}$, $\tau_2 = \tau_1 + \rho(X(\tau_1))$, $\Omega_2 = \{\omega : X(\tau_2) \notin C\}$, $\tau_3 = \tau_2 + \rho(X(\tau_2))$, etc. Define $\Omega_\infty = \bigcap_{i=1}^\infty \Omega_i$. Since

$$\{\omega : X(t, \omega) \notin C \text{ for any } t \geq 0\} \subset \Omega_\infty,$$

it suffices to show that $P^x(\Omega_\infty) = 0$. Note that

$$P^x(\Omega_1) < 1 - \delta < 1.$$

Since $\rho : H \to \mathbb{R}_+$ is Borel measurable and τ_i is a stopping time for each i, we can use the strong Markov property to get

$$
\begin{aligned}
P^x(\Omega_1 \cap \Omega_2) &= E^x\big(E^x\big(1_{\Omega_1}(\omega)1_{\Omega_2}(\omega)|\mathscr{F}_{\tau_1}\big)\big) \\
&= E^x\big(1_{\Omega_1}(\omega)E^x\big(1_{\Omega_2}(\omega)|\mathscr{F}_{\tau_1}\big)\big) \\
&= E^x\big(1_{\Omega_1}(\omega)E^x\big(1_{\Omega_2}(\omega)|X(\tau_1)\big)\big) \\
&= E^x\big(1_{\Omega_1}(\omega)P^{X(\tau_1)}\big(\{\omega : X(\rho(\tau_1)) \notin C\}\big)\big).
\end{aligned}
$$

But, by the assumption,

$$P^{X(\tau_1)}\big(\{\omega : X\big(\rho\big(X\big(\tau_1(\omega)\big)\big)\big) \notin C\}\big) < 1 - \delta,$$

so that

$$P^x(\Omega_1 \cap \Omega_2) < (1 - \delta)^2.$$

By repeating the above argument, we obtain

$$P^x\left(\bigcap_{i=1}^n \Omega_i\right) < (1 - \delta)^n,$$

which converges to zero, and this completes the proof. $\qquad\square$

Lemma 7.5 *Let* $\{X(t), t \geq 0\}$ *be a continuous strong Markov process. If there exists a positive Borel-measurable function* γ *defined on* H, *a closed set* C, *and a constant* $\delta > 0$ *such that*

$$\int_{\gamma(x)}^{\gamma(x)+1} P^x\big(X(t) \in C\big)\, dt \geq \delta \quad \text{for all } x \in H, \tag{7.46}$$

then, there exists a Borel-measurable function $\rho : H \to \mathbb{R}_+$ *such that* $\gamma(x) \leq \rho(x) \leq \gamma(x) + 1$ *and*

$$P^x\big(X(\rho(x)) \in C\big) \geq \delta \quad \text{for all } x \in H. \tag{7.47}$$

Proof By the assumption (7.46), there exists $t_x \in [\gamma(x), \gamma(x) + 1)$ such that

$$P^x\big(X(t_x) \in C\big) \geq \delta.$$

Define

$$\rho(x) = \inf\big\{t \in [\gamma(x), \gamma(x) + 1) : P^x\big(\{\omega : X(t, \omega) \in C\}\big) \geq \delta\big\}.$$

Since the mapping $t \to X(t)$ is continuous and the characteristic function of a closed set is upper semicontinuous, we have that the function

$$t \to P^x\big(X(t) \in C\big)$$

is upper semicontinuous for each x. Hence,

$$P^x\big(X(\rho(x)) \in C\big) \geq \delta.$$

We need to show that the function $x \to \rho(x)$ is Borel measurable. Let us define $\mathscr{B}_t(H) = \mathscr{B}(H)$, for $t > 0$. Since $\{X(t), 0 \leq t \leq T\}$ is a Feller process, the map $\Theta : (t, x) \to P^x(\omega : X(t) \in C)$ from $([0, T] \times H, \mathscr{B}([0, T] \times H))$ to $(\mathbb{R}^1, \mathscr{B}(\mathbb{R}^1))$ is measurable (see [54], [27]). Hence, Θ is a progressively measurable process with respect to $\{\mathscr{B}_t(H)\}$. By Corollary 1.6.12 in [16], $x \to \rho(x)$ is Borel measurable. $\square$

Let us now introduce some notation. Let $B_r = \{v \in V : \|v\|_V \leq r\}$ be a sphere in V with the radius r, centered at 0, and let $\overline{B}_r$ be its closure in $(H, \|\cdot\|_H)$. For $A \subset H$, denote its interior in $(H, \|\cdot\|_H)$ by A^0. If $B_r^c = H \setminus B_r$, then $(\overline{B}_r)^c = (B_r^c)^0$.

Lemma 7.6 *Suppose that the coefficients of (7.45) satisfy the coercivity condition (6.39) and, in addition, that its solution $\{X^x(t), t \geq 0\}$ exists and is ultimately bounded in the m.s.s. Then there exists a positive Borel-measurable function ρ on H such that*

$$P^x\big(\{\omega : X(\rho(x), \omega) \in \overline{B}_r\}\big) \geq 1 - \frac{1}{\alpha r^2}\big(|\lambda| M_1 + M_1 + |\gamma|\big), \quad x \in H, \tag{7.48}$$

and

$$P^x\big(\{\omega : X(\rho(x), \omega) \in (B_r^c)^0\}\big) \leq \frac{1}{\alpha r^2}\big(|\lambda| M_1 + M_1 + |\gamma|\big), \quad x \in H, \tag{7.49}$$

where α, λ, γ are as in the coercivity condition, and $M_1 = M + 1$ with M as in the ultimate boundedness condition (7.36).

Proof Since $\limsup_{t\to\infty} E^x \|X(t)\|_H^2 \le M < M_1$ for all $x \in H$, there exist positive numbers $\{T_x, \, x \in H\}$ such that

$$E^x \|X(t)\|_H^2 \le M \quad \text{for } t \ge T_x.$$

Hence, we can define

$$\gamma(x) = \inf\{t : E^x \|X(s)\|_H^2 \le M_1 \text{ for all } s \ge t\}.$$

Since $t \to E^x \|X(t)\|_H^2$ is continuous, $E^x \|X(\gamma(x))\|_H^2 \le M_1$. The set

$$
\begin{aligned}
\{x : \gamma(x) \le t\} &= \{x : E^x \|X(s)\|_H^2 \le M_1 \text{ for all } s \ge t\} \\
&= \bigcap_{\substack{s \ge t \\ s \in Q}} \{x : E^x \|X(s)\|_H^2 \le M_1\}
\end{aligned}
$$

is in $\mathscr{B}(H)$, since the function $x \to E^x \|X(s)\|^2$ is Borel measurable. Using Itô's formula (4.37) for $\|x\|_H^2$, then taking the expectations on both sides, and applying the coercivity condition (6.39), we arrive at

$$
\begin{aligned}
&E^x \|X(\gamma(x) + 1)\|_H^2 - E^x \|X(\gamma(x))\|_H^2 \\
&= E^x \int_{\gamma(x)}^{\gamma(x)+1} \left(2\langle X(s), A(X(s))\rangle + \text{tr}\big(B(X(s))Q(B(X(s))^*)\big)\right) ds \\
&\le \lambda \int_{\gamma(x)}^{\gamma(x+1)} E^x \|X(s)\|_H^2 \, ds - \alpha \int_{\gamma(x)}^{\gamma(x)+1} E^x \|X(s)\|_V^2 \, ds + \gamma.
\end{aligned}
$$

It follows that

$$\int_{\gamma(x)}^{\gamma(x)+1} E\|X(s)\|_V^2 \, ds \le \frac{1}{\alpha}\left(|\lambda|M_1 + M_1 + |\gamma|\right).$$

Using Chebychev's inequality, we get

$$\int_{\gamma(x)}^{\gamma(x+1)} P^x\big(\{\omega : \|X(t,\omega)\|_V > r\}\big) dt \le \frac{1}{\alpha r^2}\left(|\lambda|M_1 + M_1 + |\gamma|\right).$$

Hence,

$$\int_{\gamma(x)}^{\gamma(x)+1} P^x\big(\{\omega : X(t,\omega) \in (B_r^c)^0\}\big) \le \frac{1}{\alpha r^2}\left(|\lambda|M_1 + M_1 + |\gamma|\right),$$

and consequently,

$$\int_{\gamma(x)}^{\gamma(x)+1} P^x\big(\{\omega : X(t,\omega) \in \overline{B}_r\}\big) dt \ge 1 - \frac{1}{\alpha r^2}\left(|\lambda|M_1 + M_1 + |\gamma|\right).$$

Using Lemma 7.5, we can claim the existence of a positive Borel-measurable function $\rho(x)$ defined on H such that $\gamma(x) \leq \rho(x) \leq \gamma(x) + 1$, and (7.48) and then (7.49) follow for $r > 0$ and for all $x \in H$. $\qquad\square$

We now conclude the proof of Theorem 7.18. Using (7.48), we can choose r large enough such that

$$P^x\left(\{\omega : X(\rho(x), \omega) \in \overline{B}_r\}\right) \geq \frac{1}{2} \quad \text{for } x \in H.$$

Since the mapping $V \hookrightarrow H$ is compact, the set $\overline{B}_r$ is compact in H, giving that $X(t)$ is weakly recurrent to $\overline{B}_r$ by Lemma 7.4. $\qquad\square$

Definition 7.7 An H-valued stochastic process $\{X(t), t \geq 0\}$ is called *weakly positive recurrent to a compact set* if there exists a compact set $C \subset H$ such that $X(t)$ is weakly recurrent to C and the first hitting time to C,

$$\tau = \inf\{t \geq 0 : X(t) \in C\},$$

has finite expectation for any $x = X(0) \in H$.

Theorem 7.19 *Suppose that $V \hookrightarrow H$ is compact and the coefficients of (7.45) satisfy the coercivity condition (6.39) and the monotonicity condition (6.40). If its solution $\{X^x(t), t \geq 0\}$ is exponentially ultimately bounded in the m.s.s., then it is weakly positively recurrent.*

Proof We know that

$$E^x\|X(t)\|_H^2 \leq ce^{-\beta t}\|x\|_H^2 + M \quad \text{for all } x \in H.$$

Let $M_1 = M + 1$, and $w(r) = \frac{1}{\beta}\ln(1 + cr^2)$, $r \in \mathbb{R}$. Then we have

$$E^x\|X(t)\|_H^2 \leq M_1 \quad \text{for } x \in H \text{ and } t \geq w(\|x\|_H),$$

and

$$\sum_{l=1}^{\infty} \frac{w((l+1)N)}{l^2} < \infty \quad \text{for any } N \geq 0. \tag{7.50}$$

Let $K = (1 + \Delta)\sqrt{|\lambda|M_1 + M_1 + |\gamma|}/\sqrt{\alpha}$, and let us define the sets

$$E_0 = \overline{B}_K,$$

$$E_l = \overline{B}_{(l+1)K} - \overline{B}_{lK} = \overline{B}_{(l+1)K} \cap \left(B_{lK}^c\right)^0 \quad \text{for } l \geq 1,$$

where B_r is a sphere in V with the radius r, centered at 0. We denote $w'(l) = w(lK\alpha_0) + 1$ with α_0 such that $\|x\|_H \leq \alpha_0\|x\|_V$ for all $x \in V$. As in the proof of

Lemma 7.6, there exists a Borel-measurable function $\rho(x)$ defined on H satisfying $w(\|x\|_H) \leq \rho(x) \leq w(\|x\|_H) + 1$, and

$$P^x\left(\left\{\omega : X(\rho(x), \omega) \in \left(B_{lK}^c\right)^0\right\}\right) \leq \frac{1}{\alpha(lK)^2}\left(|\lambda|M_1 + M_1 + |\gamma|\right)$$

$$\leq \frac{1}{l^2(1+\Delta)^2} \quad \text{for all } x \in H. \quad (7.51)$$

Let

$$\tau_1 = \rho(x), \quad x_1(\omega) = X(\tau_1, \omega), \quad \Omega_1 = \left\{\omega : x_1(\omega) \notin E_0\right\},$$

$$\tau_2 = \tau_1 + \rho(x_1(\omega)), \quad x_2(\omega) = X(\tau_2, \omega), \quad \Omega_2 = \left\{\omega : x_2(\omega) \notin E_0\right\}, \quad \ldots,$$

and so on. Let $\Omega_\infty = \bigcap_{i=1}^\infty \Omega_i$. As in the proof of Lemma 7.4,

$$P^x\left(\bigcap_{i=1}^\infty \Omega_i\right) = 0.$$

Hence, Ω differs from

$$\bigcup_{i=1}^\infty \Omega_i^c = \bigcup_{i=1}^\infty \left\{\omega : x_i(\omega) \in E_0\right\}$$

by at most a set of P^x-measure zero. Let

$$A_i = \Omega_i^c - \bigcup_{j=1}^{i-1}\left\{\omega : x_j(\omega) \in E_0\right\} = \left\{\omega : x_1(\omega) \notin E_0, \ldots, x_{i-1} \notin E_0, \ x_i \in E_0\right\}.$$

Then Ω differs from $\bigcup_{i=1}^\infty A_i$ by at most a set of P^x-measure zero. For $i \geq 2$, let us further partition

$$A_i = \bigcup_{j_1, j_2, \ldots, j_{n-1}} A_i^{j_1, \ldots, j_{i-1}},$$

where

$$A_i^{j_1, \ldots, j_{i-1}} = \left\{\omega : x_1(\omega) \in E_{j_1}, \ldots, x_{i-1}(\omega) \in E_{j_{i-1}}, x_i(\omega) \in E_0\right\}.$$

Let $\tau(\omega)$ be first hitting time to E_0. Then for $\omega \in A_1 = \Omega_1^c$,

$$\tau(\omega) \leq \rho(x) \leq w(\|x\|_H) + 1,$$

and for $\omega \in A_i^{j_1, \ldots, j_{i-1}}$,

$$\tau(\omega) \leq \tau_i(\omega) \leq \tau_{i-1}(\omega) + \rho(x_{i-1}(\omega)).$$

Moreover, for $\omega \in A_i^{j_1,\ldots,j_{i-1}}$,

$$x_{i-1}(\omega) \in E_{j_{i-1}} \subset \overline{B}_{(j_{i-1}+1)K}.$$

Hence,

$$\|x_{i-1}(\omega)\|_H \leq \alpha_0 \|x_{i-1}(\omega)\|_V \leq \alpha_0(j_{i-1}+1)K,$$

giving

$$\rho(x_{i-1}(\omega)) \leq w(\|x_{i-1}(\omega)\|_H) + 1 \leq w(\alpha_0(j_{i-1}+1)K) + 1 = w'(j_{i-1}+1)$$

and

$$\tau(\omega) \leq \tau_{i-1} + w'(j_{i-1}+1).$$

Using induction,

$$\tau(\omega) \leq w(\|x\|_H) + 1 + w'(j_1+1) + \cdots + w'(j_{i-1}+1).$$

By the strong Markov property,

$$P^x\left(A_i^{j_1,\ldots,j_{i-1}}\right) = P^x\left(\{\omega : x_1(\omega) \in E_{j_1}, \ldots, x_{i-1}(\omega) \in E_{j_{i-1}}, x_i(\omega) \in E_0\}\right)$$
$$\leq P^x\left(\{\omega : x_1(\omega) \in E_{j_1}, \ldots, x_{i-1}(\omega) \in E_{j_{i-1}}\}\right)$$
$$= P^x\left(\{\omega : x_1(\omega) \in E_{j_1}, \ldots, x_{i-2}(\omega) \in E_{j_{i-2}}\} \cap \{x_{i-1}(\omega) \in E_{j_{i-1}}\}\right)$$
$$\leq E^x\left\{1_{\{\omega : x_1(\omega) \in E_{j_1}, \ldots, x_{i-2} \in E_{j_{i-2}}\}} P^{x_{i-2}(\omega)}\left(\{\tilde{\omega} : X(\rho(x_{i-2}(\omega)), \tilde{\omega}) \in E_{j_{i-1}}\}\right)\right\}.$$

Since $E_{j_{i-1}} = \overline{B}_{(j_{i-1}+1)K} \cap (B^c_{j_{i-1}K})^0$, we get by (7.51)

$$P^{x_{i-2}(\omega)}\left(\tilde{\omega} : X(\rho(x_{i-2}(\omega)), \tilde{\omega}) \in E_{j_{i-1}}\right)$$
$$\leq P^{x_{i-2}}\left(\tilde{\omega} : X(\rho(x_{i-2}(\omega)), \tilde{\omega}) \in (B^c_{j_{i-1}K})^0\right)$$
$$\leq \frac{1}{j_{i-1}^2(1+\Delta)^2}.$$

Hence,

$$P^x\left(A_i^{j_1,\ldots,j_{i-1}}\right) \leq \frac{1}{j_{i-1}^2(1+\Delta)^2} P^x\left(\{\omega : x_1(\omega) \in E_{j_1}, \ldots, x_{i-2}(\omega) \in E_{j_{i-2}}\}\right).$$

By induction,

$$P^x\left(A_i^{j_1,\ldots,j_{i-1}}\right) \leq \frac{1}{(1+\Delta)^{2(i-1)}} \frac{1}{j_1^2 \cdots j_{i-1}^2},$$

which implies that $P^x(A_i) < 1$, for Δ large enough.

Now

$$E^x(\tau) \leq \sum_{i,j_1,\ldots,j_{i-1}\geq 1} P^x\big(A_i^{j_1,\ldots,j_{i-1}}\big)$$

$$\times \big[w'(\|x\|_H) + 1 + w'(j_1 + 1) + \cdots + w'(j_{i-1} + 1)\big]$$

$$\leq w(\|x\|_H) + 1 + \left(\sum_{i=2}^{\infty} \frac{1}{(1+\Delta)^{2(i-1)}}\right)$$

$$\left(\sum_{j_1,\ldots,j_{i-1}\geq 1} \frac{w'(\|x\|_H) + 1 + w'(j_1 + 1) + \cdots + w'(j_{i-1} + 1)}{j_1^2 \cdots j_{i-1}^2}\right)$$

$$= w(\|x\|_H) + 1 + \left(\sum_{i=2}^{\infty} \frac{1}{(1+\Delta)^{2(i-1)}}\big(w(\|x\|_H) + 1\big)\right)$$

$$\left\{\left(\sum_{j_1,\ldots,j_{i-1}\geq 1} \frac{1}{j_1^2 \cdots j_{i-1}^2}\right) + (i-1) \sum_{j_1,\ldots,j_{i-1}\geq 1} \frac{w'(j_1 + 1)}{j_1^2 \cdots j_{i-1}^2}\right\}$$

$$= \big(w(\|x\|_H) + 1\big)\left(1 + \sum_{i=2}^{\infty} \left(\frac{A}{(1+\Delta)^2}\right)^{i-1}\right)$$

$$+ \frac{B}{(1+\Delta)^2} \sum_{i=2}^{\infty} \left(\frac{A}{(1+\Delta)^2}\right)^{i-2} (i-1),$$

where $A = \sum_{l=1}^{\infty} \frac{1}{l^2}$, and $B = \sum_{l=1}^{\infty} \frac{1}{l^2}w'(l+1)$, with both series converging due to (7.50).

Consequently, $E^x(\tau)$ is finite for Δ large enough. The set E_0 is compact since the embedding $V \hookrightarrow H$ is compact. $\qquad\square$

We have given precise conditions using a Lyapunov function for exponential ultimate boundedness in the m.s.s. We can thus obtain sufficient conditions for weakly (positive) recurrence of the solutions in terms of a Lyapunov function.

We close with important examples of stochastic reaction–diffusion equations. Let $\mathscr{O} \subset \mathbb{R}^n$ be a bounded domain with smooth boundary $\partial\mathscr{O}$, and p be a positive integer. Let $V = W^{1,2}(\mathscr{O})$ and $H = W^{0,2}(\mathscr{O}) = L^2(\mathscr{O})$. We know that $V \hookrightarrow H$ is a compact embedding. Let

$$A_0(x) = \sum_{|\alpha|\leq 2p} a_\alpha(x)\frac{\partial^{\alpha_1}}{\partial x_1^{\alpha_1}} \cdots \frac{\partial^{\alpha_n}}{\partial x_n^{\alpha_n}},$$

where $\alpha = (\alpha_1,\ldots,\alpha_n)$ is a multiindex. If A_0 is strongly elliptic, then by Garding inequality ([63], Theorem 7.2.2) A_0 is coercive.

Example 7.8 (Reaction–diffusion equation) Consider a parabolic Itô equation

$$
\begin{cases}
dX(t, x) = A_0 X(t, x)\, dt + f(X(t, x))\, dt + B(t, x)\, dW_t, \\
X(0, x) = \varphi(x) \in H, \quad X|_{\partial \mathcal{O}} = 0,
\end{cases}
\tag{7.52}
$$

where A_0, f, and B satisfy the following conditions:

(1) $A_0 : V \to V^*$ is a strongly elliptic operator.
(2) $f : H \to H$ and $B : H \to \mathscr{L}(K, H)$ satisfy

$$
\left\| f(h) \right\|_H^2 + \left\| B(h) \right\|_{\mathscr{L}(K,H)}^2 \leq K \left(1 + \|h\|_H^2 \right), \quad h \in H.
$$

(3) $\| f(h_1) - f(h_2) \|_H^2 + \mathrm{tr}((B(h_1) - B(h_2)) Q (B(h_1) - B(h_2))^*) \leq \lambda \| h_1 - h_2 \|$,
     $h_1, h_2 \in H$.

If the solution to the equation

$$
du(t, x) = A_0 u(t, x)\, dt
$$

is exponentially ultimately bounded and, as $\|h\|_H \to \infty$,

$$
\left\| f(h) \right\|_H = o\left(\|h\|_H \right),
$$

$$
\left\| B(h) \right\|_{\mathscr{L}(K,H)} = o\left(\|h\|_H \right),
$$

then the strong variational solution of (7.52) is exponentially ultimately bounded in the m.s.s. by Proposition 7.1, and consequently it is weakly positive recurrent.

Example 7.9 (Reaction–diffusion equation) Consider the following one-dimensional parabolic Itô equation

$$
\begin{cases}
dX(t, x) = \left(\alpha^2 \dfrac{\partial^2 X}{\partial x^2} + \beta \dfrac{\partial X}{\partial x} + \gamma X + g(x) \right) dt + \left(\sigma_1 \dfrac{\partial X}{\partial x} + \sigma_2 X \right) dW_t, \\
u(0, x) = \varphi(x) \in L^2(\mathcal{O}) \cap L^1(\mathcal{O}), \quad X|_{\partial \mathcal{O}} = 0,
\end{cases}
\tag{7.53}
$$

where $\mathcal{O} = (0, 1)$, and W_t is a standard Brownian motion.

Similarly as in Example 7.3, if $-2\alpha^2 + \sigma_1^2 < 0$, then the coercivity and weak monotonicity conditions (6.39) and (6.40) hold, and Theorem 4.7 implies the existence of a unique strong solution.

With $\Lambda(v) = \|v\|_H^2$ and $\mathscr{L}$ defined by (6.15), we get

$$
\mathscr{L}\Lambda(v) \leq \left(-2\alpha^2 + \sigma_1^2 \right) \left\| \frac{dv}{dx} \right\|_H^2 + \left(2\gamma + \sigma_2^2 + \epsilon \right) \|v\|_H^2 + \frac{1}{\epsilon} \|g\|_H^2.
$$

Since $\| \frac{dv}{dx} \|_H^2 \geq \|v\|_H^2$ (see Exercise 7.9), we have

$$
\mathscr{L}\Lambda(v) \leq \left(-2\alpha^2 + \sigma_1^2 + 2\gamma + \sigma 22 + \epsilon \right) \|v\|_H^2 + \frac{1}{\epsilon} \|g\|_H^2.
$$

Hence, if $-2\alpha^2 + \sigma_1^2 + 2\gamma + \sigma_2^2 < 0$, then the strong variational solution of (7.53) is exponentially ultimately bounded by Theorem 7.5, and hence it is weakly positive recurrent.

Exercise 7.9 Let $f \in W^{0,2}((a, b))$. Prove the Poincaré inequality

$$\int_a^b f^2(x)\, dx \le (b-a)^2 \int_a^b \left(\frac{df(x)}{dx} \right)^2 dx.$$

References

1. S. Agmon. Lectures on Elliptic Boundary Value Problems, *Mathematical Studies No. 2*, Van Nostrand, Princeton (1965).
2. S. Albeverio and R. Hoegh-Krohn. Homogeneous random fields and statistical mechanics, *J. Funct. Anal.* **19**, 242–272 (1975).
3. S. Albeverio, Yu. G. Kondratiev, M. Röckner, and T. V. Tsikalenko. Glauber dynamics for quantum lattice systems, *Rev. Math. Phys.* **13** No. 1, 51–124 (2001).
4. P. Billingsley. Convergence of Probability Measures, Wiley, New York (1968).
5. P. Billingsley. Probability and Measure, Wiley, New York (1979).
6. P.L. Butzer and H. Berens. Semi-Groups of Operators and Approximation, Springer, New York (1967).
7. J. R. Cannon. The One-Dimensional Heat Equation, *Encyclopedia of Mathematics and Its Applications* **23**, Addison–Wesley, Reading (1984).
8. S. Cerrai. Second Order PDE's in Finite and Infinite Dimension, *Lecture Notes in Mathematics* **1762**, Springer, Berlin (2001).
9. A. Chojnowska-Michalik. Stochastic Differential Equations in Hilbert Space, *Banach Center Publications* **5**, PWN–Polish Scientific Publishers, Warsaw (1979).
10. G. Da Prato, S. Kwapien, and J. Zabczyk. Regularity of solutions of linear stochastic equations in Hilbert spaces, *Stochastics* **23**, 1–23 (1987).
11. G. Da Prato and J. Zabczyk. Stochastic Equations in Infinite Dimensions, *Encyclopedia of Mathematics and its Applications* **44**, Cambridge University Press, Cambridge (1992).
12. G. Da Prato and J. Zabczyk. Ergodicity for Infinite Dimensional Systems, *London Mathematical Society Lecture Note Series* **229**, Cambridge University Press, Cambridge (1996).
13. R. Datko. Extending a theorem of A. M. Liapunov to Hilbert space, *J. Math. Anal. Appl.* **32**, 610–616 (1970).
14. J. Diestel and J.J. Uhl. Vector Measures, *Mathematical Surveys* **15**, AMS, Providence (1977).
15. J. Dieudonné. Treatise on Analysis, Academic Press, New York (1969).
16. R. J. Elliott. Stochastic Calculus and Applications, Springer, New York (1982).
17. S. N. Ethier and T. G. Kurtz. Markov Processes: Characterization and Convergence. Wiley, New York (1986).
18. B. Gaveau. Intégrale stochastique radonifiante, *C.R. Acad. Sci. Paris Ser. A* **276**, 617–620 (1973).
19. L. Gawarecki. Extension of a stochastic integral with respect to cylindrical martingales, *Stat. Probab. Lett.* **34**, 103–111 (1997).
20. L. Gawarecki and V. Mandrekar. Stochastic differential equations with discontinuous drift in Hilbert space with applications to interacting particle systems, *J. Math. Sci.* **105**, No. 6, 2550–2554 (2001). Proceedings of the Seminar on Stability Problems for Stochastic Models, Part I (Naleczow, 1999).

21. L. Gawarecki and V. Mandrekar. Weak solutions to stochastic differential equations with discontinuous drift in Hilbert space. In: Stochastic Processes, Physics and Geometry; New Interplays, II (Leipzig, 1999), *CMS Conf. Proc.* **29**, Amer. Math. Soc., Providence, 199–205 (2000).

22. L. Gawarecki, V. Mandrekar, and P. Richard. Existence of weak solutions for stochastic differential equations and martingale solutions for stochastic semilinear equations, *Random Oper. Stoch. Equ.* **7**, No. 3, 215–240 (1999).

23. L. Gawarecki, V. Mandrekar, and B. Rajeev. Linear stochastic differential equations in the dual to a multi-Hilbertian space, *Theory Stoch. Process.* **14**, No. 2, 28–34 (2008).

24. L. Gawarecki, V. Mandrekar, and B. Rajeev. The monotonicity inequality for linear stochastic partial differential equations, *Infin. Dimens. Anal. Quantum Probab. Relat. Top.* **12**, No. 4, 1–17 (2009).

25. I.I. Gikhman and A.V. Skorokhod. The Theory of Stochastic Processes, Springer, Berlin (1974).

26. A.N. Godunov. On Peano's theorem in Banach spaces, *Funct. Anal. Appl.* **9**, 53–55 (1975).

27. K. Gowrisankaran. Measurability of functions in product spaces, *Proc. Am. Math. Soc.* **31**, No. 2, 485–488 (1972).

28. M. Hairer. Ergodic properties of a class of non-Markovian processes. In: Trends in Stochastic Analysis, *London Mathematical Society Lecture Note Series* **353**. Ed. J. Blath et al. 65–98 (2009).

29. E. Hille. Lectures on Ordinary Differential Equations, Addison–Wesley, Reading (1969)

30. F. Hirsch and G. Lacombe. Elements of Functional Analysis, *Graduate Texts in Mathematics* **192**, Springer, New York (1999).

31. H. Holden, B. Øksendal, J. Uboe, and T. Zhang. Stochastic Partial Differential Equations: A Modeling, White Noise Functional Approach, Birkhauser, Boston (1996).

32. A. Ichikawa. Stability of semilinear stochastic evolution equations, *J. Math. Anal. Appl.* **90**, 12–44 (1982).

33. A. Ichikawa, Semilinear stochastic evolution equations: boundedness, stability and invariant measures, *Stochastics* **12**, 1–39 (1984).

34. A. Ichikawa. Some inequalities for martingales and stochastic convolutions, *Stoch. Anal. Appl.* **4**, 329–339 (1986).

35. K. Itô. Foundations of Stochastic Differential Equations in Infinite Dimensional Spaces, *CBMS–NSF* **47** (1984).

36. G. Kallianpur, I. Mitoma, and R. L. Wolpert. Diffusion equations in dual of nuclear spaces, *Stoch. Stoch. Rep.* **29**, 285–329 (1990).

37. G. Kallianpur and J. Xiong. Stochastic differential equations in infinite dimensions: a brief survey and some new directions of research. In: Multivariate Analysis: Future Directions, *North-Holland Ser. Statist. Probab.* **5**, North-Holland, Amsterdam, 267–277 (1993).

38. I. Karatzas and S. E. Shreve. Brownian Motion and Stochastic Calculus, *Graduate Texts in Mathematics*, Springer, New York (1991).

39. R. Khasminskii. Stochastic Stability of Differential Equations, Sijthoff and Noordhoff, Alphen aan den Rijn (1980).

40. R. Khasminskii and V. Mandrekar. On stability of solutions of stochastic evolution equations. In: The Dynkin Festschrift, *Progr. Probab.* Ed. M. Freidlin, Birkhäuser, Boston, 185–197 (1994).

41. H. König. Eigenvalue Distribution of Compact Operators, Birkhäuser, Boston (1986).

42. N. V. Krylov and B. L. Rozovskii. Stochastic evolution equations, *J. Sov. Math.* **16**, 1233–1277 (1981).

43. K. Kuratowski and C. Ryll-Nardzewski, A general theorem on selectors, *Bull. Acad. Pol. Sci.* **13**, 349–403 (1965).

44. S. Lang. Analysis II, Addison–Wesley, Reading (1969).

45. M. Ledoux and M. Talagrand. Probability in Banach Spaces, Springer, Berlin (1991).

46. G. Leha and G. Ritter. On diffusion processes and their semigroups in Hilbert spaces with an application to interacting stochastic systems, *Ann. Probab.* **12**, No. 4, 1077–1112 (1984).

47. G. Leha and G. Ritter. On solutions to stochastic differential equations with discontinuous drift in Hilbert space, *Math. Ann.* **270**, 109–123 (1985).
48. J. L. Lions. Équations Différentielles Opérationelles et Problèmes aux Limites, Springer, Berlin (1961).
49. R. S. Liptzer and A. N. Shiryaev. Statistics of Stochastic Processes, Nauka, Moscow (1974).
50. K. Liu. Stability of Infinite Dimensional Stochastic Differential Equations with Applications, *Chapman & Hall/CRC Monographs and Surveys in Pure and Applied Mathematics* **135** (2006).
51. R. Liu. Ultimate boundedness and weak recurrence of stochastic evolution equations, *Stoch. Anal. Appl.* **17**, 815–833 (1999).
52. R. Liu and V. Mandrekar. Ultimate boundedness and invariant measures of stochastic evolution equations, *Stoch. Stoch. Rep.* **56**, No. 1–2, 75–101 (1996).
53. R. Liu and V. Mandrekar. Stochastic semilinear evolution equations: Lyapunov function, stability, and ultimate boundedness, *J. Math. Anal. Appl.* **212**, No. 2, 537–553 (1997).
54. G.W. Mackey. A theorem of Stone and von Neuman, *Duke Math. J.* **16**, No. 2, 313–326 (1949).
55. V. Mandrekar. On Lyapunov stability theorems for stochastic (deterministic) evolution equations. In: Stochastic Analysis and Applications in Physics, *NATO Adv. Sci. Inst. Ser. C Math. Phys. Sci.* **449**, Kluwer, Dordrecht, 219–237 (1994).
56. M. Metivier. Stochastic Partial Differential Equations in Infinite Dimensional Spaces, *Scuola Normale Superiore*, Quaderni, Pisa (1988).
57. M. Metivier and J. Pellaumail. Stochastic Integration, Academic Press, New York (1980).
58. M. Metivier and M. Viot. On weak solutions of stochastic partial differential equations. In: Stochastic Analysis, *LNM* **1322**. Ed. M. Metivier, S. Watanabe, Springer, Berlin, 139–150 (1988).
59. Y. Miyahara. Ultimate boundedness of the system governed by stochastic differential equations, *Nagoya Math. J.* **47**, 111–144 (1972).
60. E. Nelson. Probability Theory and Euclidian Field Theory, *Lecture Notes in Phys.* **25**, Springer, Berlin, 94–124 (1973).
61. B. Øksendal. Stochastic Differential Equations, Springer, New York (1998).
62. E. Pardoux. Stochastic partial differential equations and filtering of diffusion processes, *Stochastics* **3**, 127–167 (1979).
63. A. Pazy. Semigroups of Linear Operators and Applications to Partial Differential Equations, *Applied Mathematical Sciences*, **44**, Springer, New York (1983).
64. C. Prévôt and M. Röckner. A Concise Course on Stochastic Partial Differential Equations, *LNM* **1905**, Springer, Berlin (2007).
65. Yu. V. Prokhorov. Convergence of random processes and limit theorems in probability theory, *Theory Probab. Appl.* **1**, 157–214 (1956).
66. B. L. Rozovskii. Stochastic Evolution Systems: Linear Theory and Applications to Non-Linear Filtering, Kluwer, Boston (1983).
67. M. Röckner, B. Schmuland, and X. Zhang. Yamada–Watanabe theorem for stochastic evolution equations in infinite dimensions, *Condens. Matter Phys.* **11**, No. 2(54), 247–259 (2008).
68. R. S. Schatten. Norm Ideals of Continuous Operators, Springer, New York (1970).
69. A.V. Skorokhod. Personal communication.
70. D.W. Stroock and S.R.S. Varadhan. Multidimensional Diffusion Processes, Springer, New York (1979).
71. H. Tanabe. Equations of Evolution, Pitman, London (1979).
72. L. Tubaro. An estimate of Burkholder type for stochastic processes defined by the stochastic integral, *Stoch. Anal. Appl.* **62**, 187–192 (1984).
73. R. Wheeden, A. Zygmund. Measure and Integral, Marcel Dekker, New York (1977).
74. N. N. Vakhania, V. I. Tarieladze, and S. A. Chobanyan. Probability Distributions on Banach Spaces, *Mathematics and Its Applications (Soviet Series)* **14**, Reidel, Dordrecht (1987).
75. M. Viot. Solutions faibles d'équations aux dérivées partielles non linéaires, Thése, Université Pierre et Marie Curie, Paris (1976).

76. M. Vishik and A. Fursikov. Mathematical Problems of Statistical Hydrodynamics, Kluwer, London (1988).

77. M. Yor. Existence et unicité de diffusions à valeurs dans un espace de Hilbert, *Ann. Inst. H. Poincaré B* **10**, 55–88 (1974).

78. K. Yosida. Functional Analysis, Springer, New York (1980).

79. J. Zabczyk. Linear stochastic systems in Hilbert spaces; spectral properties and limit behaviour, *Banach Cent. Publ.* **41**, 591–609 (1985).

Index